Lecture Notes in Artificial Intelligence 9853

Subseries of Lecture Notes in Computer Science

More information about this series at http://www.springer.com/series/1244

Bettina Berendt · Björn Bringmann
Élisa Fromont · Gemma Garriga
Pauli Miettinen · Nikolaj Tatti
Volker Tresp (Eds.)

Machine Learning and Knowledge Discovery in Databases

European Conference, ECML PKDD 2016
Riva del Garda, Italy, September 19–23, 2016
Proceedings, Part III

 Springer

Editors

Bettina Berendt
Department of Computer Science
KU Leuven
Leuven
Belgium

Björn Bringmann
Deloitte GmbH
Munich
Germany

Élisa Fromont
Laboratoire Hubert Curien
Jean Monnet University
Saint-Etienne
France

Gemma Garriga
Allianz SE
Munich
Germany

Pauli Miettinen
Max Planck Institute for Informatics
Saarbrücken
Germany

Nikolaj Tatti
Aalto University School of Science
Espoo
Finland

Volker Tresp
Siemens AG and Ludwig Maximilians
  University of Munich
Munich
Germany

ISSN 0302-9743   ISSN 1611-3349   (electronic)
Lecture Notes in Artificial Intelligence
ISBN 978-3-319-46130-4   ISBN 978-3-319-46131-1   (eBook)
DOI 10.1007/978-3-319-46131-1

Library of Congress Control Number: 2016950748

LNCS Sublibrary: SL7 – Artificial Intelligence

Printed on acid-free paper

This Springer imprint is published by Springer Nature
The registered company is Springer International Publishing AG
The registered company address is: Gewerbestrasse 11, 6330 Cham, Switzerland

Foreword to the ECML PKDD 2016 Demo Track

It is our great pleasure to introduce the Demo Track of ECML PKDD 2016. This year's track continues its tradition of providing a forum for researchers and practitioners to demonstrate novel systems and research prototypes, using data mining and machine learning techniques in a variety of application domains. Besides the live demonstrations during the conference period, each selected demo is allocated a 4-page paper in the proceedings. The Demo Track of ECML PKDD 2016 solicited working systems based on state-of-the-art machine learning and data mining technology. Both innovative prototype implementations and mature systems were welcome, provided that they used machine learning techniques and knowledge discovery processes in a real setting. The evaluation criteria encompassed innovation, interestingness for the target users, and whether it would be of interest mainly for researchers, mainly for practitioners, or both. Each submission was evaluated by at least three reviewers. This year we received 29 submissions. Sixteen demos were presented at the Demo Session during the conference in Riva del Garda. The accepted demos cover a wide range of machine learning and data mining techniques, as well as a very diverse set of real-world application domains. The success of the Demo Track of ECML PKDD 2016 is due to the effort of several people. First and foremost, we thank the authors for their submissions and their engagement in turning data mining and machine learning methods to software that can be presented and tried by others. We would like to thank all members of our Program Committee for helping us in the difficult task of selecting the most interesting sub-missions. Finally, we would like to thank the ECML PKDD General Chairs and the Program Chairs for entrusting us with this track, and the whole Organizing Committee for the practical support and the logistics for the Demo Track. We hope that the readers will enjoy this set of short papers and the demonstrated systems, and that the Demo Session will inspire further ECML PKDD participants to turn their research ideas into working prototypes that can be used by other researchers and practitioners in machine learning and data mining.

September 2016

Élisa Fromont
Nikolaj Tatti

Foreword to the ECML PKDD 2016 Industry Track

We are pleased to present the proceedings of the Industrial, Governmental, and NGO Track of ECML PKDD 2016. This track aims at bringing together participants from academia, industry, government, and NGOs (non-governmental organizations) in a venue that promotes industrial experiences and real-world applications of machine learning and data science.

The program included two invited talks given by Michael May (Siemens) and Matthias Seeger (Amazon), and 10 high-quality papers featuring the technical talks. Given a total of 50 submissions, this year's industrial track was highly selective: only 10 papers could be accepted for publication and for presentation at the conference; corresponding to an acceptance rate of 20 %. Each of the 50 submissions was thoroughly reviewed, and accepted papers were chosen both for their originality and for the application they promoted.

The accepted papers focus on topics ranging from machine learning methods and data science processes to dedicated applications. Topics covered include time series mining and multi-target classification, visualization, software engineering, robotics, bioinformatics, steel production, grammar-based text analysis for purposes such as plagiarism or bible analysis, music recommendation, search task extraction, crowdsourcing and social networks, and discrimination discovery.

We thank all the authors who submitted the 50 papers for their work and effort to bring machine learning and data science to industry. We also thank all the PC members for their substantial efforts to guarantee the quality of these proceedings.

We hope that this program will inspire the growth of machine learning in all areas of industry.

September 2016

Björn Bringmann
Gemma Garriga
Volker Tresp

Foreword to the ECML PKDD 2016 Nectar Track

The goal of the ECML PKDD Nectar Track, started in 2012, is to offer conference attendees a compact overview of recent scientific advances at the frontier of machine learning and data mining with other disciplines, as already published in related conferences and journals. Submissions describing work that summarizes a line of work comprising older and more recent papers were particularly encouraged. Authors were invited to submit 4-page summaries of their published work.

We received 28 submissions, each of which was reviewed by two or three PC members. Thirteen submissions were selected for inclusion in the proceedings and presentation at the conference. The accepted papers range from interesting and important applications to data mining and machine learning methods and processes. Topics covered include methods and applications for time series mining and multi-target classification, visualization, software engineering, robotics, bioinformatics, steel production, grammar-based text analysis for purposes such as plagiarism or bible analysis, music recommendation, search task extraction, crowdsourcing and social networks, and discrimination discovery. The papers illustrate how the questions and methods of these areas pose new challenges for data mining and machine learning, thereby contributing also to the development of these core fields of ECML PKDD.

We thank all authors of submitted papers and all PC members for their excellent work. We are also very grateful to the ECML PKDD General Chairs and the Program Chairs for entrusting us with this track. We hope that the readers will enjoy this set of short papers and their pointers to work at the intersection of machine learning/data mining and their manifold application areas, and that the papers, presentations, and discussions will inspire further work in the different disciplines and at their joint boundaries.

September 2016

Bettina Berendt
Pauli Miettinen

Organization

ECML PKDD 2016 Organization

General Chairs

Andrea Passerini   University of Trento, Italy
Fosca Giannotti   National Research Council (ISTI-CNR), Italy

Program Chairs

Paolo Frasconi   University of Florence, Italy
Niels Landwehr   University of Potsdam, Germany
Giuseppe Manco   National Research Council (ICAR-CNR), Italy
Jilles Vreeken   Cluster of Excellence MMCI, Saarland University
    & Max Planck Institute for Informatics, Germany

Journal Track Chairs

Thomas Gärtner   University of Nottingham, UK
Mirco Nanni   National Research Council (ISTI-CNR), Italy
Andrea Passerini   University of Trento, Italy
Céline Robardet   National Institute of Applied Science in Lyon, France

Industrial Track Chairs

Björn Bringmann   Deloitte GmbH, Germany
Gemma Garriga   Allianz SE Munich, Germany
Volker Tresp   Siemens AG & Ludwig Maximilian University of Munich,
    Germany

Local Organization Chairs

Simone Marinai   University of Florence, Italy
Gianluca Corrado   University of Trento, Italy
Katya Tentori   University of Trento, Italy

Workshop and Tutorial Chairs

Matthijs van Leeuwen   Leiden University, Netherlands
Fabrizio Costa   University of Freiburg, Germany
Albrecht Zimmermann   University of Caen, France

Awards Committee Chairs

Toon Calders   Free University of Bruxelles, Belgium
Hendrik Blockeel   University of Leuven, Belgium

Nectar Track Chairs

Bettina Berendt   KU Leuven, Belgium
Pauli Miettinen   Max Planck Institute for Informatics, Germany

Demo Chairs

Nikolaj Tatti   Aalto University School of Science, Finland
Élisa Fromont   Jean Monnet University, France

Discovery Challenge Chairs

Elio Masciari   National Research Council (ICAR-CNR), Italy
Alessandro Moschitti   Qatar Computing Research Institute, HBKU,
                               Qatar & University of Trento, Italy

Sponsorship Chairs

Michelangelo Ceci   University of Bari, Italy
Chedy Raïssi   Inria, France

Publicity and Social Media Chairs

Olana Missura   University of Bonn, Germany
Nicola Barbieri   Yahoo!, UK
Gianluca Corrado   University of Trento, Italy

PhD Forum Chairs

Leman Akoglu   Stony Brook University, USA
Tijl De Bie   Ghent University, Belgium

Proceedings Chairs

Marco Lippi   University of Bologna, Italy
Stefano Ferilli   University of Bari, Italy

Web Chair

Daniele Baracchi   University of Florence, Italy

Demo Track Program Committee

Albert Bifet   Gustavo Carneiro   Mark Last
Antti Ukkonen   Jerzy Stefanowski   Mykola Pechenizkiy
Bernhard Pfahringer   Joao Papa   Nikolaj Tatti
Elisa Fromont   John Hipwell   Panagiotis Papapetrou
Ernestina Menasalvas   Luis Teixeira   Ricard Gavalda
Francois Jacquenet   Maguelonne Teisseire   Taneli Mielikainen
Grigorios Tsoumakas   Marcus Harz   Vincent Lemaire

Nectar Track Program Committee

Andreas Maletti   Luiza Antonie   Pierre Geurts
Chedy Raïssi   Luc De Raedt   Weike Pan
Dora Erdos   Michele Berlingerio   Balaraman Ravindran
Ernestina Menasalvas   Wagner Meira Jr.   Salvatore Ruggieri
Ricard Gavaldà   Rosa Meo
Kristian Kersting   Niina Haiminen

Industrial Track Program Committee

Kareem Aggour   Gianmarco De   Deguang Kong
Kamal Ali   Francisci Morales   Nicolas Kourtellis
Michele Berlingerio   Patrick Düssel   Hardy Kremer
Michael Berthold   Enrique Frias   Denis Krompass
Albert Bifet   Dinesh Garg   Mohit Kumar
Rui Cai   Aris Gkoulalas-Divanis   Mounia Lalmas
Berkant   Francesco Gullo   Ni Lao
   Barla Cambazoglu   Bing Hu   Kuang-chih Lee
Haifeng Chen   Georges Hébrail   Vincent Lemaire
Jiefeng Cheng   Hongxia Jin   Bangyong Liang
Michelangelo D'Agostino Anne Kao   Jiebo Luo
Kamalika Das   Alexandros Karatzoglou   Arun Maiya

Silviu Maniu	Barna Saha	Maguelonne Teisseire
Luis Matias	Krishna Sankar	Ingo Thon
Dimitrios Mavroeidis	Amy Shi-Nash	Antti Ukkonen
Sameep Mehta	Alkis Simitsis	Pinghui Wang
Veena Mendiratta	Papadimitriou Spiros	Xiang Wang
Xia Ning	Shivashankar	Ding Wei
Stelios Paparizos	Subramanian	Cheng Weiwei
Fabio Pinelli	Siqi Sun	YanChang Zhao

Sponsors

Gold Sponsors

Google	http://research.google.com
IBM	http://www.ibm.com

Silver Sponsors

Deloitte	http://www.deloitte.com
Siemens	http://www.siemens.com
Unicredit	http://www.unicreditgroup.eu
Zalando	http://www.zalando.com

Award Sponsors

Deloitte	http://www.deloitte.com
DMKD	http://link.springer.com/journal/10618
MLJ	http://link.springer.com/journal/10994

Badge Lanyard

Knime	http://www.knime.org

Additional Supporters

Springer	http://www.springer.com

Institutional Supporters

ICAR-CNR	http://www.icar.cnr.it
DISI-UNITN	http://www.disi.unitn.it
ISTI-CNR	http://www.isti.cnr.it
COGNET	http://www.cognet.5g-ppp.eu

Organizing Institutions

UNITN	http://www.unitn.it
UNIFI	http://www.unifi.it
ISTI-CNR	http://www.isti.cnr.it
ICAR-CNR	http://www.icar.cnr.it

Invited Talks Abstracts
(Industrial Track)

Towards Industrial Machine Intelligence

Michael May

Siemens Corporate Technology, Munich, Germany

Abstract. The next decade will see a deep transformation of industrial applications by big data analytics, machine learning and the internet of things. Industrial applications have a number of unique features, setting them apart from other domains. Central for many industrial applications in the internet of things is time series data generated by often hundreds or thousands of sensors at a high rate, e.g. by a turbine or a smart grid. In a first wave of applications this data is centrally collected and analyzed in Map-Reduce or streaming systems for condition monitoring, root cause analysis, or predictive maintenance. The next step is to shift from centralized analysis to distributed in-field or in situ analytics, e.g in smart cities or smart grids. The final step will be a distributed, partially autonomous decision making and learning in massively distributed environments.

In this talk I will give an overview on Siemens' journey through this transformation, highlight early successes, products and prototypes and point out future challenges on the way towards machine intelligence. I will also discuss architectural challenges for such systems from a Big Data point of view.

Bio. Michael May is Head of the Technology Field Business Analytics & Monitoring at Siemens Corporate Technology, Munich, and responsible for eleven research groups in Europe, US, and Asia. Michael is driving research at Siemens in data analytics, machine learning and big data architectures. In the last two years he was responsible for creating the Sinalytics platform for Big Data applications across Siemens' business.

Before joining Siemens in 2013, Michael was Head of the Knowledge Discovery Department at the Fraunhofer Institute for Intelligent Analysis and Information Systems in Bonn, Germany. In cooperation with industry he developed Big Data Analytics applications in sectors ranging from telecommunication, automotive, and retail to finance and advertising.

Between 2002 and 2009 Michael coordinated two Europe-wide Data Mining Research Networks (KDNet, KDubiq). He was local chair of ICML 2005, ILP 2005 and program chair of the ECML PKDD Industrial Track 2015. Michael did his PhD on machine discovery of causal relationships at the Graduate Programme for Cognitive Science at the University of Hamburg.

Machine Learning Challenges at Amazon

Matthias Seeger

Amazon, Berlin, Germany

Abstract. At Amazon, some of the world's largest and most diverse problems in e-commerce, logistics, digital content management, and cloud computing services are being addressed by machine learning on behalf of our customers. In this talk, I will give an overview of a number of key areas and associated machine learning challenges.

Bio. Matthias Seeger got his PhD from Edinburgh. He had academic appointments at UC Berkeley, MPI Tuebingen, Saarbruecken, and EPF Lausanne. Currently, he is a principal applied scientist at Amazon in Berlin. His interests are in Bayesian methods, large scale probabilistic learning, active decision making and forecasting.

Contents – Part III

Demo Track Contributions

A Tool for Subjective and Interactive Visual Data Exploration   3
    Bo Kang, Kai Puolamäki, Jefrey Lijffijt, and Tijl De Bie

GMMbuilder – User-Driven Discovery of Clustering Structure
for Bioarchaeology   8
    Markus Mauder, Yulia Bobkova, and Eirini Ntoutsi

Bipeline: A Web-Based Visualization Tool for Biclustering of Multivariate
Time Series   12
    Ricardo Cachucho, Kaihua Liu, Siegfried Nijssen, and Arno Knobbe

h(odor): Interactive Discovery of Hypotheses on the Structure-Odor
Relationship in Neuroscience   17
    Guillaume Bosc, Marc Plantevit, Jean-François Boulicaut,
    Moustafa Bensafi, and Mehdi Kaytoue

INSIGHT: Dynamic Traffic Management Using Heterogeneous Urban Data   22
    Nikolaos Panagiotou, Nikolas Zygouras, Ioannis Katakis,
    Dimitrios Gunopulos, Nikos Zacheilas, Ioannis Boutsis, Vana Kalogeraki,
    Stephen Lynch, Brendan O'Brien, Dermot Kinane, Jakub Mareček,
    Jia Yuan Yu, Rudi Verago, Elizabeth Daly, Nico Piatkowski,
    Thomas Liebig, Christian Bockermann, Katharina Morik,
    Francois Schnitzler, Matthias Weidlich, Avigdor Gal, Shie Mannor,
    Hendrik Stange, Werner Halft, and Gennady Andrienko

Coordinate Transformations for Characterization and Cluster Analysis
of Spatial Configurations in Football   27
    Gennady Andrienko, Natalia Andrienko, Guido Budziak,
    Tatiana von Landesberger, and Hendrik Weber

Leveraging Spatial Abstraction in Traffic Analysis and Forecasting
with Visual Analytics   32
    Natalia Andrienko, Gennady Andrienko, and Salvatore Rinzivillo

The SPMF Open-Source Data Mining Library Version 2   36
    Philippe Fournier-Viger, Jerry Chun-Wei Lin, Antonio Gomariz,
    Ted Gueniche, Azadeh Soltani, Zhihong Deng, and Hoang Thanh Lam

DANCer: Dynamic Attributed Network with Community Structure
Generator.   41
  Oualid Benyahia, Christine Largeron, Baptiste Jeudy,
  and Osmar R. Zaïane

Topy: Real-Time Story Tracking via Social Tags   45
  Gevorg Poghosyan, M. Atif Qureshi, and Georgiana Ifrim

Ranking Researchers Through Collaboration Pattern Analysis.   50
  Mario Cataldi, Luigi Di Caro, and Claudio Schifanella

Learning Language Models from Images with ReGLL.   55
  Leonor Becerra-Bonache, Hendrik Blockeel, Maria Galván,
  and François Jacquenet

Exploratory Analysis of Text Collections Through Visualization
and Hybrid Biclustering.   59
  Nicolas Médoc, Mohammad Ghoniem, and Mohamed Nadif

SITS-P2miner: Pattern-Based Mining of Satellite Image Time Series   63
  Tuan Nguyen, Nicolas Méger, Christophe Rigotti, Catherine Pothier,
  and Rémi Andreoli

Finding Incident-Related Social Media Messages for Emergency
Awareness   67
  Alexander Nieuwenhuijse, Jorn Bakker, and Mykola Pechenizkiy

TwitterCracy: Exploratory Monitoring of Twitter Streams for the 2016
U.S. Presidential Election Cycle   71
  M. Atif Qureshi, Arjumand Younus, and Derek Greene

Industrial Track Contributions

Using Social Media to Promote STEM Education: Matching College
Students with Role Models.   79
  Ling He, Lee Murphy, and Jiebo Luo

Concept Neurons – Handling Drift Issues for Real-Time Industrial
Data Mining.   96
  Luis Moreira-Matias, João Gama, and João Mendes-Moreira

PULSE: A Real Time System for Crowd Flow Prediction at Metropolitan
Subway Stations   112
  Ermal Toto, Elke A. Rundensteiner, Yanhua Li, Richard Jordan,
  Mariya Ishutkina, Kajal Claypool, Jun Luo, and Fan Zhang

Finding Dynamic Co-evolving Zones in Spatial-Temporal Time Series Data. . .   129
  Yun Cheng, Xiucheng Li, and Yan Li

ECG Monitoring in Wearable Devices by Sparse Models.   145
  Diego Carrera, Beatrice Rossi, Daniele Zambon, Pasqualina Fragneto,
  and Giacomo Boracchi

Do Street Fairs Boost Local Businesses? A Quasi-Experimental Analysis
Using Social Network Data   161
  Ke Zhang and Konstantinos Pelechrinis

Intelligent Urban Data Monitoring for Smart Cities   177
  Nikolaos Panagiotou, Nikolas Zygouras, Ioannis Katakis,
  Dimitrios Gunopulos, Nikos Zacheilas, Ioannis Boutsis,
  Vana Kalogeraki, Stephen Lynch, and Brendan O'Brien

Automatic Detection of Non-Biological Artifacts in ECGs Acquired
During Cardiac Computed Tomography   193
  Rustem Bekmukhametov, Sebastian Pölsterl, Thomas Allmendinger,
  Minh-Duc Doan, and Nassir Navab

Active Learning with Rationales for Identifying Operationally Significant
Anomalies in Aviation.   209
  Manali Sharma, Kamalika Das, Mustafa Bilgic, Bryan Matthews,
  David Nielsen, and Nikunj Oza

Engine Misfire Detection with Pervasive Mobile Audio   226
  Joshua Siegel, Sumeet Kumar, Isaac Ehrenberg, and Sanjay Sarma

Nectar Track Contributions

From Plagiarism Detection to Bible Analysis: The Potential of Machine
Learning for Grammar-Based Text Analysis.   245
  Michael Tschuggnall and Günther Specht

A KDD Process for Discrimination Discovery   249
  Salvatore Ruggieri and Franco Turini

Personality-Based User Modeling for Music Recommender Systems   254
  Bruce Ferwerda and Markus Schedl

Time and Again: Time Series Mining via Recurrence Quantification
Analysis.   258
  Stephan Spiegel and Norbert Marwan

Resource-Aware Steel Production Through Data Mining   263
  Hendrik Blom and Katharina Morik

Learning from Software Project Histories.   267
  Verena Honsel, Steffen Herbold, and Jens Grabowski

Practical Bayesian Inverse Reinforcement Learning for Robot Navigation . . .   271
  Billy Okal and Kai O. Arras

Machine Learning Challenges for Single Cell Data   275
  Sofie Van Gassen, Tom Dhaene, and Yvan Saeys

Multi-target Classification: Methodology and Practical Case Studies   280
  Mark Last

Query Log Mining for Inferring User Tasks and Needs   284
  Rishabh Mehrotra and Emine Yilmaz

Data Mining Meets HCI: Data and Visual Analytics of Frequent Patterns. . . .   289
  *Carson K. Leung, Christopher L. Carmichael, Yaroslav Hayduk,
  Fan Jiang, Vadim V. Kononov, and Adam G.M. Pazdor*

Machine Learning for Crowdsourced Spatial Data   294
  Musfira Jilani, Padraig Corcoran, and Michela Bertolotto

Local Exceptionality Detection on Social Interaction Networks   298
  Martin Atzmueller

Author Index   303

Demo Track Contributions

A Tool for Subjective and Interactive Visual Data Exploration

Bo Kang[1(✉)], Kai Puolamäki[2], Jefrey Lijffijt[1], and Tijl De Bie[1]

[1] Data Science Lab, Ghent University, Ghent, Belgium
{bo.kang,jefrey.lijffijt,tijl.de.bie}@ugent.be
[2] Finnish Institute of Occupational Health, Helsinki, Finland
kai.puolamaki@ttl.fi

Abstract. We present SIDE, a tool for Subjective and Interactive Visual Data Exploration, which lets users explore high dimensional data via subjectively informative 2D data visualizations. Many existing visual analytics tools are either restricted to specific problems and domains or they aim to find visualizations that align with user's belief about the data. In contrast, our generic tool computes data visualizations that are surprising given a user's current understanding of the data. The user's belief state is represented as a set of projection tiles. Hence, this user-awareness offers users an efficient way to interactively explore yet-unknown features of complex high dimensional datasets.

1 Introduction

Exploratory Data Mining is the process of using data mining methods to gain novel insights into data without having a specific goal in mind. To convey large amounts of complex information, it is a logical choice to present this information visually, as the information bandwidth of the eye is much larger than the other senses, and humans excel at spotting visual patterns [11]. Surprisingly, visual interactive data mining tools are still rare.

The few tools that exist are either designed for specific problems and domains (e.g., itemset and subgroup discovery [1,4,7], information retrieval [10], or analysis of networks [2]) and/or aim to present information that align with the user's beliefs (e.g., semi-supervised PCA [7]). However, users are typically interested in *finding structures in the data that contrast with their current knowledge* [5].

In this paper, we present a generic tool[1] that enables users to efficiently explore data via a sequence of 2D scatter plots, i.e., *projections*. It models the user's beliefs about data by iteratively incorporating their feedback, which in turn is utilized for calculating an updated data projection. SIDE operates iteratively, with three steps in each iteration (see Fig. 1). In step 1, it presents a user with a 'surprising' data projection. In step 2, the user provides feedback about the projection. Finally, in step 3, the *background model* is updated to reflect

---

[1] Our tool, SIDE, is freely accessible at http://www.interesting-patterns.net/forsied/ a-tool-for-subjective-and-interactive-visual-data-exploration.

© Springer International Publishing AG 2016
B. Berendt et al. (Eds.): ECML PKDD 2016, Part III, LNAI 9853, pp. 3–7, 2016.
DOI: 10.1007/978-3-319-46131-1_1

the user's current belief state. It then repeats from step 1, and shows a data projection that takes into account the updated background model.

2   Subjectively Interesting Projections

SIDE employs a generic method for interactive visual exploration of high dimensional data, with awareness of a user's belief sate about the data. Due to space constraints we limit ourselves to describe only the intuition and overview of the approach. For a full description, we refer the reader to our paper [9].

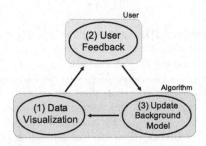

Fig. 1. This three-step cycle illustrates our tool's flow of action.

In order to present the user with subjectively informative data projections, there are two modeling problems [3]. First, we have to maintain a *background model* throughout the exploration process. This model accumulates the user's feedback, which represents the knowledge they learned from the data projections. Hence, this model represents a user's current belief about the data.

The second obstacle is quantification of the informativeness, for which we employ constrained randomization [6]. The idea is that we sample random data from the user's current belief state, where the beliefs are modeled as constraints to the randomization procedure. Then, we search for projections that contrast with the random data, and hence that contrast with the current beliefs. That is, we assume that a data projection that (maximally) deviates from the beliefs will reveal subjectively novel structures.

Then, an optimization problem arises to find a projection that makes the real data maximally different from the randomized data. Currently the tool employs the L1 distance, which can be optimized well using standard optimization toolboxes. We have not studied the choice of measure extensively yet.

3   User Interface

SIDE was designed according to three principles for visually controllable data mining [8], which essentially says that the model and the interactions should be transparent to users, and the analysis method should be fast enough such that the user does not lose their trail of thought. Figure 2 shows the user interface of our tool. The main component of this interface is the interactive scatter plot (Fig. 2a). The scatter plot visualizes the projected data (filled dots) and the randomized data (gray circles) using the same projection. By drawing circles (Fig. 2b), the user can highlight a *projection tile pattern* (i.e., a set of filled dots). Once a set of points is marked, the user can press either feedback button (Fig. 2c), indicating these points form a cluster. If the users believe the points are clustered only in the shown projection, they click '2D Constraint', while 'Cluster Constraint' indicates they are aware of the fact that these points will be clustered in other dimensions as well. To identify the defined clusters, data points associated with the same feedback (i.e., user's belief) are filled by the same color (Fig. 2d), and their statistics are shown in a table. The user can define multiple clusters in a single projection, and they can also undo (Fig. 2e) the feedback. Once a user finishes exploring the current projection, they can press 'Update Background Model' (Fig. 2f). Then, the background model is updated with the provided feedback and a new scatter plot is computed and presented to the user, etc.

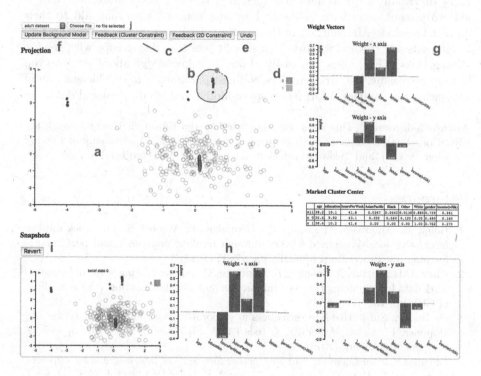

Fig. 2. Visual layout of interactive dimensionality tool, which contains interact area (a), projection meta information area (g), and snapshots area (h).

A few extra features are provided to assist the data exploration process: to gain an intuitive understanding of a projection, the weight vectors associated with the projection axes are plotted as bar charts (Fig. 2g). At the bottom of Fig. 2g, a table lists the mean vectors of each colored point set (i.e., cluster). The exploration history is maintained by taking snapshots of the background model when updated, together with the associated data projection (scatter plot) and bar charts (weight vectors). This history in reverse chronological order is illustrated in Fig. 2h. The tool also allows a user to click and revert (Fig. 2i) back to a certain snapshot, to restart from that time point. This allows the user to discover different aspects of a dataset more consistently. Finally, custom datasets can be selected for analysis from the drop-down menu (Fig. 2j). Currently our tool only works with CSV files and it also automatically sub-samples any data set so that the interactive experience is not compromised. By default, two datasets are preloaded so that users can get familiar with the tool.

4   Conclusions

We presented SIDE, an interactive exploratory data mining tool that allows users to visually explore data. By modeling a user's belief state, our tool is able to present users with views of data that contrast with and add to their current knowledge. In contrast to the existing visual analytics systems, our tool is automatically tailored towards each specific user and able to cope with generic mining tasks. Thus, users can easily obtain new knowledge about data on top of their increasingly accurate understandings, providing a more efficient way of navigating the complex information space hidden in high-dimensional data.

Acknowledgments. This work was supported by the European Union through the ERC Consolidator Grant FORSIED (project reference 615517), Academy of Finland (decision 288814), and Tekes (Revolution of Knowledge Work project).

References

1. Boley, M., Mampaey, M., Kang, B., Tokmakov, P., Wrobel, S.: One click mining: interactive local pattern discovery through implicit preference and performance learning. In: Proceedings of KDD, pp. 27–35 (2013)
2. Chau, D.H., Kittur, A., Hong, J.I., Faloutsos, C.: Apolo: making sense of large network data by combining rich user interaction and machine learning. In: Proceedings of CHI, pp. 167–176 (2011)
3. De Bie, T.: Subjective interestingness in exploratory data mining. In: Tucker, A., Höppner, F., Siebes, A., Swift, S. (eds.) IDA 2013. LNCS, vol. 8207, pp. 19–31. Springer, Heidelberg (2013)
4. Dzyuba, V., van Leeuwen, M.: Interactive discovery of interesting subgroup sets. In: Tucker, A., Höppner, F., Siebes, A., Swift, S. (eds.) IDA 2013. LNCS, vol. 8207, pp. 150–161. Springer, Heidelberg (2013)
5. Hand, D.J., Mannila, H., Smyth, P.: Principles of Data Mining. MIT Press, Cambridge (2001)
6. Lijffijt, J., Papapetrou, P., Puolamäki, K.: A statistical significance testing approach to mining the most informative set of patterns. DMKD **28**(1), 238–263 (2014)

7. Paurat, D., Gärtner, T.: InVis: a tool for interactive visual data analysis. In: Blockeel, H., Kersting, K., Nijssen, S., Železný, F. (eds.) ECML PKDD 2013, Part III. LNCS, vol. 8190, pp. 672–676. Springer, Heidelberg (2013)
8. Puolamäki, K., Papapetrou, P., Lijffijt, J.: Visually controllable data mining methods. In: Proceedings of ICDMW, pp. 409–417 (2010)
9. Puolamäki, K., Kang, B., Lijffijt, J., De Bie, T.: Interactive visual data exploration with subjective feedback. In: Proceedings of ECML-PKDD (2016, to appear)
10. Ruotsalo, T., Jacucci, G., Myllymäki, P., Kaski, S.: Interactive intent modeling: information discovery beyond search. CACM **58**(1), 86–92 (2015)
11. Ware, C.: Information Visualization: Perception for Design, 3rd edn. Morgan Kaufmann/Elsevier, San Francisco (2013)

GMMbuilder – User-Driven Discovery of Clustering Structure for Bioarchaeology

Markus Mauder[1]([⊠]), Yulia Bobkova[1], and Eirini Ntoutsi[2]

[1] Ludwig-Maximilians-University Munich, Munich, Germany
mauder@dbs.lmu.de, yulia.bobkova@campus.lmu.de
[2] Leibniz Universität Hannover, Hannover, Germany
ntoutsi@kbs.uni-hannover.de

Abstract. We present *GMMbuilder*, a tool that allows domain scientists to build Gaussian Mixture Models (GMM) that adhere to domain specific constraints like spatial coherence. Domain experts use this tool to generate different models, extract stable object communities across these models, and use these communities to interactively design a final clustering model that explains the data but also considers prior beliefs and expectations of the domain experts.

Keywords: Bioarchaeology · Isotopic mapping · Gaussian mixture models · Interactive clustering · Community detection · Demo

1 Introduction

Data mining has become an indispensable tool for social and the humanities. The *GMMbuilder* tool was developed in the context of the interdisciplinary research project FOR1670[1] that aims at building an isotopic fingerprint for bioarchaeological finds (human and animal remains) from excavation sites along the Inn-Eisack-Adige passage spanning Italy, Austria, and Germany. The data consists of spatial information on the location of the finds and the ratios of oxygen, strontium, and lead isotopes in the finds. Data mining methods were employed for the construction of a large scale isotopic map of the area to be used to differentiate local from non-local finds and to define the place of origin of the latter.

To be useful for origin prediction, the derived model must be based solely on isotopes, i.e. supplementary information like the spatial origin of the finds should not be used for model building. Domain knowledge however suggests that the derived models should also be spatially coherent. Intuitively, this means that finds coming from the same location should have similar isotope values. However, the task of building a model of plausible origins of the measured values is complicated by the noise introduced by the environment, range areas of animals, import of food, and further confounding factors. Additionally, displacement of live humans and animals and also animal trading in the past generates mixed

---

[1] www.for1670-transalpine.uni-muenchen.de.

© Springer International Publishing AG 2016
B. Berendt et al. (Eds.): ECML PKDD 2016, Part III, LNAI 9853, pp. 8–11, 2016.
DOI: 10.1007/978-3-319-46131-1_2

measurements and spatial outliers. Over the course of many months various clustering models were developed, discussed with the domain experts and refined based on their feedback. This approach was characterized by very slow turnaround. *GMMbuilder* was developed to allow model building and assessment in an integrated fashion and allow for immediate feedback by domain experts. The result is a model that fits the data well but it is also in accordance with domain knowledge (for example, spatial coherence of the models).

2 *GMMbuilder*

The model is built by identifying strong object communities in the data and incorporating the models of these communities into the final clustering model. To derive the strongly connected components in the data we rely on unsupervised learning. In particular, we generate multiple clusterings from the data and we find object formations that are stable across many clusterings. The intuition is that similar objects should be clustered together across the different clusterings. The domain expert has a very active role in the whole process: from the selection of the clusterings from which the communities will be extracted to the selection of the communities that will form the basis for the final clustering model. Figure 1 depicts the *GMMbuilder* architecture, consisting of several modules that will be presented hereafter. As it is shown in this figure the role of the domain expert is vital.

Clusterer module. The Clusterer module derives a clustering over a given dataset D. Domain knowledge suggests continuous values for the measurements, which can be best modeled as a mixture model of continuous distributions, like a Gaussian Mixture Model (GMM). Therefore, the Expectation-Maximization (EM)

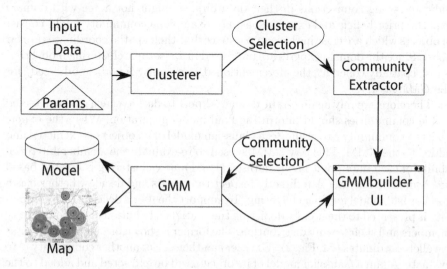

Fig. 1. An overview of *GMMbuilder*. Oval shapes depict user interaction.

algorithm [1] was applied to extract a robust indication of the data's structure in an unsupervised way. EM fits k multi-variate normal distributions over the given dataset, k is a user-defined input. The result is a soft-clustering; in our dataset though the assignment is typically fairly hard [2].

Input and Cluster Selection modules. The selection of the input data D for the GMM model is crucial as it affects the derived clustering model. Therefore, we rely on the domain experts to decide which of the generated models are acceptable. The decision is based on their expertise, however in order to facilitate their task, we provide a detailed clustering description, in terms of the spatial projection of the cluster members and the distribution of the isotope values in each cluster. The result of this step is a set of user-accepted clusterings $\mathbb{C}$.

Community Extractor module. By examining the different clusterings, we can identify objects that are frequently assigned to the same cluster. We call such object formations "stable" communities. More formally, a stable community c consists of a set of points $p \in D$ that are clustered into the same cluster across multiple clusterings $\mathbb{C}$:

$$c(C_1, C_2, \ldots, C_n) = \{p \mid p \in C_{1,i} \wedge p \in C_{2,j} \wedge \cdots \wedge p \in C_{n,m}\}$$

where $C_{i,j}$ is the set of points in the jth cluster in clustering C_i.

The idea is to use these strong components as building blocks for the final clustering, because their members have shown a strong adhesion to each other over a range of clusterings and therefore, they are more likely to represent a cluster in any final model-based clustering.

GMM module. The stable communities extracted from the previous step which indicate strong connections in their data objects might not agree with domain experts' prior beliefs and expectations. For example, a community might consist of objects which are close in the isotopic space, but their spatial coordinates are far apart. Since the domain experts are interested in an isotopic clustering model that is also spatially coherent, the aforementioned community is not a good "seed" for the *GMMbuilder*.

Therefore, we rely again on the domain expert to decide which of the detected stable communities should inform the final model generation. When the expert selects a community c to evaluate, a Gaussian model of its objects is extracted and added to the GMM. This new GMM is used to re-evaluate the membership probability of each data point in our dataset D and a new clustering is created based on c's model. The user can directly inspect the results and decide whether it is a good or bad model for final clustering. To support the user's decision, the community is presented to the user by depicting the spatial distribution of the community members and their feature distribution. The former is shown in a map, the latter as parallel coordinates (c.f. Fig. 2). The user can then select another community c' to evaluate. Again a Gaussian model of its objects will be extracted and added to the GMM. The old component c and the new component c' will be used to re-evaluate the membership probability of all points in D. This is an iterative process, the user can add or remove communities and directly inspect the effect on the final clustering. The output of this step is a set of user-accepted communities from which Gaussian models are extracted. All points in D will be assigned to these models, deriving the final clustering.

3   Demo Scenario

In our example scenario we generate different clusterings by varying the number of clusters for the EM algorithm. The users can inspect the individual clusterings, with the help of the clustering statistics and visualization window, and select those that should contribute to the final model. The stable communities, derived from the selected clusterings, will be presented to the user. The user can interactively choose which of these communities should be part of the final clustering model. User decisions are reflected in the final model so the user can directly inspect the effect of her decisions and proceed accordingly by removing or adding certain components. *GMMbuilder* is a web-based tool. A screenshot of the interactive model building step is shown in Fig. 2.

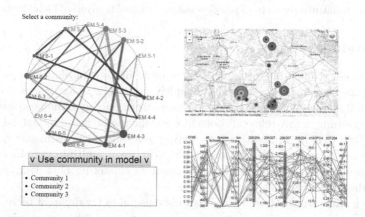

Fig. 2. *GMMbuilder*: Interactive GMM building - inspecting one of the communities found in all three clusterings (orange, green and blue). (Color figure online)

References

1. Dempster, A.P., Laird, N.M., Rubin, D.B.: Maximum likelihood from incomplete data via the EM algorithm. J. Roy. Stat. Soc. Ser. B (Methodological) **39**(1), 1–38 (1977)
2. Mauder, M., Ntoutsi, E., Kröger, P., Grupe, G.: Data mining for isotopic mapping of bioarchaeological finds in a central european alpine passage. In: 27th International Conference on Scientific and Statistical Database Management (2015)

Bipeline: A Web-Based Visualization Tool for Biclustering of Multivariate Time Series

Ricardo Cachucho[✉], Kaihua Liu, Siegfried Nijssen, and Arno Knobbe

LIACS, Leiden University, Leiden, The Netherlands
{r.cachucho,s.nijssen,a.j.knobbe}@liacs.leidenuniv.nl,
k.liu.5@umail.leidenuniv.nl

Abstract. Large amounts of multivariate time series data are being generated every day. Understanding this data and finding patterns in it is a contemporary task. To find prominent patterns present in multivariate time series, one can use biclustering, that is looking for patterns both in subsets of variables that show coherent behavior and in a number of time periods. For this, an experimental tool is needed.

Here, we present *Bipeline*, a web-based visualization tool that provides both experts and non-experts with a pipeline for experimenting with multivariate time series biclustering. With *Bipeline*, it is straightforward to save experiments and try different biclustering algorithms, enabling users to intuitively go from pre-processing to visual analysis of biclusters.

1 Introduction

The development of sensor networks has resulted in an explosion of time series data over the last years. These are large multivariate time series, where variables are collected synchronously over time. Thus, pattern mining of multivariate time series is becoming highly relevant, both in scientific research and industrial applications. Note that in the multivariate setting, not only patterns in one variable over time are relevant, but also relationships between multiple variables could provide useful insights. This task can be seen as clustering both time periods and variables, also know as biclustering [8–10].

Given a multivariate time series, it could be useful to try different biclustering algorithms. Also, one needs to optimize parameters across different steps, such as pre-processing, segmentation and biclustering itself. For each of these steps, there are many parameters to be optimized, leading to a large number of experiments. Furthermore, at each step, visual inspection is highly important for researchers to validate their findings. However, there is a lack of tools for this process.

We propose *Bipeline*, a web-based visualization tool that provides a pipeline for applying biclustering to multivariate time series. This tool is readily accessible to anyone via a web-based interface, allowing them to navigate through multiple experimental settings. Parameters can be interactively tuned, with web components such as checkboxes, sliders and drop-down menus. At each step of the biclustering process, feedback is provided be means of visualizations, with plots such as pre-processed time series, segmentation boundaries and biclusters. One or more biclusters can be plotted with a simple selection procedure.

© Springer International Publishing AG 2016
B. Berendt et al. (Eds.): ECML PKDD 2016, Part III, LNAI 9853, pp. 12–16, 2016.
DOI: 10.1007/978-3-319-46131-1_3

2   Related Work

Until now, biclustering software tools with a graphical user interface have been developed to deal with biological gene expression data. BicOverlapper [1] is a tool for visual inspection of gene expression biclusters, introducing a novel visualization algorithm *Overlapper* to represent biclusters. Similarly, BiCluster Viewer [2] is a visualization tool for efficient and interactive analysis of large gene expression datasets. BicAT [3] implements multiple biclustering algorithms, for visualization and analysis of biclusters for expression data. BiGGEsTS [4] provides an environment for biclustering time series gene expression data.

All tools mentioned above integrate techniques for pre-processing and biclustering analysis, specifically for gene expression data. Their main purpose is to support biologists with the analysis and exploration of the gene expression data. However, these tools do not support biclustering analysis for multivariate time series. Also, most of them do not provide a pipeline experiment environment. Bipeline provides such a pipeline, where intermediate results can be inspected and saved. Using a friendly and interactive plotting environment, both non-experts and experts can pre-process, segment and analyze biclusters for multivariate time series.

3   Tool Overview

Bipeline is a web-based application that provides a pipeline to pre-process, segment and bicluster multivariate time series. An online version is available [12], which is compatible with all modern web browsers and across different client platforms. Both the user interface in the web browser and the server are implemented using R Shiny package [5]. In Fig. 1, the system architecture illustrates the experimental pipeline and how each individual step relates to the other steps:

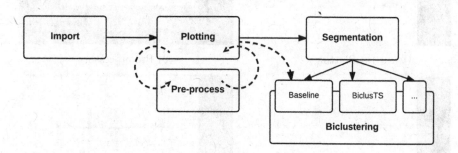

Fig. 1. A overview of *Bipeline* architecture.

Importing: Users can upload datasets and have a first view of the data table and descriptive statistics (*minimum, maximum, mean, . . .*). This first inspection, although useful, is not enough to assess the quality of the data.

Plotting: To gain further insight into the time series, it is crucial to have a visual inspection of the time series. The plotting panel includes multiple interactive plotting views, using a plotting R package dygraphs [6]. An example of these plots is illustrated in Fig. 2(a). These interactive plots allow zoom in and out functionality, which is a highly desirable functionality for visual inspection of large time series.

Pre-processing: This panel allows preliminary handling of data such as: excluding variables, normalization, conditional removal and replacement of data, and outlier removal. Users can alternate between *plotting* (Fig. 2(a)), and *pre-processing* (Fig. 2(b)) until satisfied, then export the pre-processed data by clicking the *Save* button.

Segmentation: This allows segmentation of the data, one of the steps necessary for the biclustering as suggested by [10]. By default, all variables share the same parameter settings: *window size*, *overlap* and *threshold* can be easily tuned. For greater flexibility, the user can dynamically create new tabs to set the parameters for individual variables. Additionally, a *minimum segment size* is customizable, and the tool will merge short segments to its most similar contiguous segment. Segmentation results can be visualized (Fig. 2(c)), saved and (re-)loaded, allowing the results to be used during the next step, biclustering.

Biclustering: In *Bipeline*, we implement a number of biclustering algorithms, group in three categories. The *baseline algorithms* allow users to try well-known

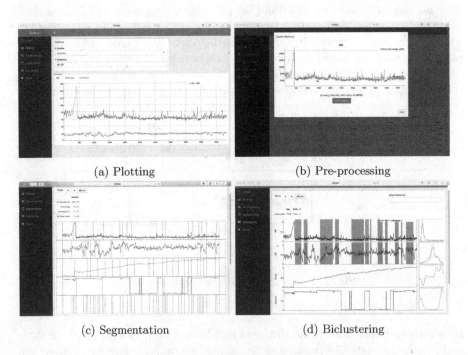

(a) Plotting   (b) Pre-processing

(c) Segmentation   (d) Biclustering

Fig. 2. *Bipeline* user interface. (Color figure online)

biclustering algorithms (e.g., Cheng & Church) [8,9], that have been implemented using R package biclust [7]. *Segmentation + Baseline* biclusters the time series using an average representation of each segment, instead of using individual rows. *Segmentation + BiclusTS* is a novel algorithm [10] introduced to recognize similarities between segments, using probability density-difference estimation [11]. All biclusters are plotted in colored blocks, as shown in Fig. 2(d). Users can select the biclusters they want to see, and the plot will respond with a real-time update.

Multiple features are shared by both *Segmentation* and *Biclustering*. Plots and parameter tables from different experiments are kept in history, allowing users to navigate back and forth to compare results and optimize parameters. During computationally expensive tasks, the front-end displays a progress bar, while the back-end server is busy carrying out the calculations. Furthermore, interactive web components can be saved into images with a single click.

4   Conclusion

We propose *Bipeline*, a web-based visualization tool, which provides a pipeline for applying biclustering to multivariate time series. Its main features include: visual inspection at multiple stages, interactive zoom in and out plotting, easy navigation, storage of results, and saving plots and experimental settings using a single click. *Bipeline*'s intuitive web-based design, makes it accessible both to experts and non-experts, and compatible across platforms.

References

1. Santamara, R., Thern, R., Quintales, L.: BicOverlapper: a tool for bicluster visualization. J. Bioinform. **24**(9), 1212–1213 (2008)
2. Heinrich, J., Seifert, R., Burch, M., Weiskopf, D.: BiCluster viewer: a visualization tool for analyzing gene expression data. In: Bebis, G., et al. (eds.) ISVC 2011, Part I. LNCS, vol. 6938, pp. 641–652. Springer, Heidelberg (2011)
3. Barkow, S., Bleuler, S., Prelic, A., Zimmermann, P., Zitzler, E.: BicAT: a biclustering analysis toolbox. J. Bioinform. **22**(10), 1282–1283 (2006)
4. Gonalves, J., Madeira, S., Oliveira, A.: BiGGEsTS: integrated environment for biclustering analysis of time series gene expression data. J. BMC, 1–11 (2009)
5. Chang, W., Cheng, J., Allaire, J., Xie, Y., McPherson, J.: Shiny: Web Application Framework for R. R package version 0.13.1 (2016)
6. Vanderkam, D., Allaire, J., Owen, J., Gromer, D., Shevtsov, P., Thieurmel, B.: Dygraphs: Interface to 'Dygraphs' Interactive Time Series Charting Library. R package (2016)
7. Kaiser, S., Santamaria, R., Khamiakova, T., Sill, M., Theron, R., Quintales, L., Leisch, F., DeTroyer, E.: Biclust: BiCluster Algorithms. R package version 1.2.0 (2015)
8. Cheng, Y., Church, G.: Biclustering of expression data. In: Proceedings of the Eighth International Conference on Intelligent Systems for Molecular Biology, pp. 93–103 (2000)

9. Madeira, S., Oliveira, A.: Biclustering algorithms for biological data analysis: a survey. J. IEEE/ACM Trans. Comput. Biol. Bioinform. **1**, 24–45 (2004)
10. Cachucho, R., Nijssen, S., Liu, K., Knobbe, A.: Bipeline: a web-based visualization tool for biclustering of multivariate time series. In: Berendt, B., Bringmann, B., Fromont, E. (eds.) ECML PKDD 2016, Part III. LNCS(LNAI), vol. 9853. pp. 12–16. Springer, Heidelberg (2016)
11. Sugiyama, M., Kanamori, T., Suzuki, T., Plessis, M., Liu, S., Takeuchi, I.: Density-difference estimation. In: Proceedings of NIPS, pp. 683–691 (2012)
12. http://fr.liacs.nl:7000

h(odor): Interactive Discovery of Hypotheses on the Structure-Odor Relationship in Neuroscience

Guillaume Bosc[1]([⊠]), Marc Plantevit[1], Jean-François Boulicaut[1],
Moustafa Bensafi[2], and Mehdi Kaytoue[1]

[1] Université de Lyon, CNRS, INSA-Lyon, LIRIS, UMR5205, 69621 Lyon, France
{guillaume.bosc,marc.plantevit,
jean-francois.boulicaut,mehdi.kaytoue}@liris.cnrs.fr
[2] Université de Lyon, CNRS, CRNL, UMR5292, INSERM U1028 Lyon, Lyon, France
moustafa.bensafi@liris.cnrs.fr

Abstract. From a molecule to the brain perception, olfaction is a complex phenomenon that remains to be fully understood in neuroscience. Latest studies reveal that the physico-chemical properties of volatile molecules can partly explain the odor perception. Neuroscientists are then looking for new hypotheses to guide their research: physico-chemical descriptors distinguishing a subset of perceived odors. To answer this problem, we present the platform *h(odor)* that implements descriptive rule discovery algorithms suited for this task. Most importantly, the olfaction experts can interact with the discovery algorithm to guide the search in a huge description space w.r.t their non-formalized background knowledge thanks to an ergonomic user interface.

1   Introduction

Olfaction, or the ability to perceive odors, was acknowledged as an object of science (Nobel prize 2004 [2]). The olfactory percept encoded in odorant chemicals contributes to our emotional balance and well-being. It is indeed agreed that the physico-chemical characteristics of odorants affect the olfactory percept [6], but no simple and/or universal rule governing this Structure Odor Relationship (SOR) has yet been identified. Why does this odorant smell of roses and that one of lemon? As only a part of the odorant message is encoded in the chemical structure, chemists and neuro-scientists are interested in eliciting hypotheses for the SOR problem under the form of human-readable descriptive rules: for example, $\langle MolecularWeight \leq 151.28, 23 \leq \#atoms \rangle \rightarrow \{Honey, Vanillin\}$. The discovery of such rules should bring new insights in the understanding of olfaction and has applications for Healthcare and the perfume and flavor industries.

Subgroup Discovery algorithms are able to discover such rules [7]. As olfaction datasets are composed of thousands of attributes, multi-labeled with a highly skewed distribution, an interactive mining of rules is interesting for experts that cannot formalize their domain knowledge, neither their mining preferences. Existing interactive subgroup discovery tools [3–5] can thus not be directly used due to the specificity of olfactory datasets. As such, we propose an original

© Springer International Publishing AG 2016
B. Berendt et al. (Eds.): ECML PKDD 2016, Part III, LNAI 9853, pp. 17–21, 2016.
DOI: 10.1007/978-3-319-46131-1_4

platform, $h(odor)$, that enables to extract descriptive rules on physicochemical properties that distinguish odors through an interactive process between the algorithm and the neuroscientists.

2   System Overview

Input data and desired output. Our demo olfaction dataset is composed of 1,700 odorant molecules (objects) described by $1,500$ physicochemical descriptors [1] and are associated to several olfactory qualities (odors) among 74 odors given by scent experts. The data are represented in a tabular format (several CSV files). The physicochemical properties are numeric attributes and each olfactory quality is boolean. The goal is to extract subgroups $s = (d, L)$, i.e., descriptive rules, that covers a subset of molecules $(supp(s))$ where the description d over the physicochemical descriptors distinguishes a subset of odors L.

Algorithm sketch. The search space of subgroups is a lattice based on both the attribute space and the target space. The child $s' = (d', L')$ of a subgroup $s = (d, L)$ of the lattice is a specialization of s. This specialization consists of (i) restricting the interval of a descriptor in d, or (ii) adding a new odor to L. Since the search space growths exponentially with the number of descriptors and labels, a naive exploration (DFS or BFS) is not suitable. For that, we use the beam-search heuristic (BS). BS enables to proceed to a restricted BFS, i.e., for each level of the search space only a part of the subgroups are kept and put into the beam. Only the subgroups in the beam of the current level are explored in the next one [4]. The quality of a subgroup is evaluated by a measure. It adapts the F1-score by taking into account the label distribution for weighting the precision and recall.

System architecture. A core module (server) is contacted by a client (Web interface) to initiate the mining algorithm with the given parameters. This core module allows the user to interact/guide the algorithm exploration based on the likes/dislikes of the user (Fig. 1).

Core Module. This is the back-end of the $h(odor)$ application. Based on *NodeJS*, the *Core Module* is the gateway between the user and the algorithm: it is in charge of the interaction. For that, JSON data are sent to and received from the *SD Algorithm* through sockets thanks to a dedicated communication process. Moreover, this module controls the UI to display results extracted from the *SD Algorithm* and collects the user preferences (like/dislike).

User Interface (UI). The front-end of the application, based on *Bootstrap* and *AngularJS*, enables the user to select the parameters of the *SD Algorithm* and to run it. Once the subgroups of the first level of the *beam search* are extracted (the algorithm

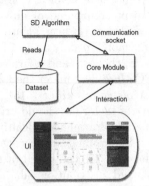

Fig. 1. System architecture

is paused waiting the user preferences), the UI displays these subgroups and the user can like/dislike some of them: the liked subgroups are forced to be within the beam for the next step, and the disliked subgroups are removed from the beam. When the algorithm finishes, the UI displays the results.

3   Use Case: Eliciting Hypotheses for the Musk Odor

We develop a use case of the application as an end user, typically a neuroscientist or a odor-chemist that seeks to extract descriptive rules to study the Structure-Odor Relationships. The application is available online with a video tutorial supporting this use case http://liris.cnrs.fr/dm2l/hodor/. In this scenario, we proceed in the following steps, knowing that the expert wishes to discover rules involving at least the *musk* odor.

1- Algorithms, parameters and dataset selection. In the *Algorithms* section of the left hand side menu, the user can choose the exploration method and its parameters. In this use case, we consider the ELMMUT algorithm. This algorithm implements a beam search strategy to extract subgroups based on a quality measure. We plan to add new/existing algorithms and subgroup quality measures. Once the exploration method is chosen, we have to select the olfactory dataset as introduced in the previous section, and choose to focus on the *musk* odor. Considering our use case, we decided to set the size of the beam to 50 (the exploration is quite large enough) and the minimum support threshold to 15 (since $|supp(Musk)| = 52$, at least the subgroups have to cover 30 % of the musk odorants). Other parameters are fixed to their default value.

2- Interactive running steps. When the datasets and the parameters have been fixed, the user can launch the mining task clicking on the *Start mining* button. When the first step of the *beam search* is finished, the *SD Algorithm* is paused and the subgroups obtained at this step are displayed to the user. The interaction view in the front-end presents the olfactory qualities involved at this level of the exploration (see Fig. 2). Each subgroup is displayed in a white box with the current descriptive rule on the physicochemical descriptors and some quantitative measures. For each subgroup box, the user can select in the top right corner if he likes/dislikes this subgroup. For example, at the first step, the application displays the subgroups extracted at the first level for the *Musk* odor. As it is a known fact in chemistry that the *musk* odor involves large molecules, we *like* the subgroup which description is $d = [238.46 \leq MW \leq 270.41]$. After that, we keep on exploring by clicking the *Next* button. Another interactive step begins, but the expert has no particular opinion so he can jump to the *next* level. Once the algorithm finished (the quality measures cannot improve), we can study the table of results. For example, the description of one of the best extracted subgroups s is: $[238.46 \leq MW \leq 270.41][-0.931 \leq Hy \leq -0.621][2.714 \leq MLOGP \leq 4.817][384.96 \leq SAtot \leq 447.506][0 \leq nR07 \leq 0][0 \leq ARR \leq 0.316][1 \leq nCsp2 \leq 7]$ that involves large odorants. Moreover, according to the experts, this latter topological descriptor is consistent with the presence of double bonds (or so-called sp2 carbon atoms) within most musky chemical structure, that provides them with a certain hydrophilicity. The goal of the *h(odor)* application is to confirm knowledge and to elicit new

Fig. 2. The interaction view of the application. For each step of the beam search, the algorithm waits for the user's preferences (like/dislike). The subgroups are displayed into white boxes. On the right part, complementary information is displayed: part of value domain of a chosen restriction on a descriptor, and parameters of the run.

hypotheses for the SOR problem. In the case of s, the neuroscientists are interested in understanding why these descriptors (excepted the Molecular Weight) are involved in the $Musk$ odor.

Learning user preferences. Besides, the $h(odor)$ application enables to save all the choices taken by the different users. Indeed, the application archived all the actions the users did into log files. The goal here is to use these log files to learn user preferences, not only for a single run of the algorithm [3] but for all experiments performed by the users. This kind data (choices made by experts) is hard to collect by simply asking experts and will be explored in future work.

Acknowledgments. The authors thank Florian Paturaux, Sylvio Menubarbe and Pierre Houdyer for helping developing the prototype. This research is partially supported by the *Institut rhônalpin des systémes complexes* (IXXI) and by the *Centre National de Recheche Scientifique* (Préfute PEPS FASCIDO, CNRS).

References

1. Arctander, S.: Perfume and Flavor Materials of Natural Origin, vol. 2. Allured Publishing Corp., USA (1994)
2. Buck, L., Axel, R.: A novel multigene family may encode odorant receptors: a molecular basis for odor recognition. Cell **65**(1), 175–187 (1991)
3. Dzyuba, V., van Leeuwen, M., Nijssen, S., Raedt, L.D.: Interactive learning of pattern rankings. Int. J. Artif. Intell. Tools **23**(6), 1–31 (2014)
4. Galbrun, E., Miettinen, P.: Siren: an interactive tool for mining and visualizing geospatial redescriptions. In: KDD, pp. 1544–1547 (2012)

5. Goethals, B., Moens, S., Vreeken, J.: MIME: a framework for interactive visual pattern mining. In: Gunopulos, D., Hofmann, T., Malerba, D., Vazirgiannis, M. (eds.) ECML PKDD 2011. LNCS (LNAI), pp. 634–637. Springer, Heidelberg (2011). doi:10.1007/978-3-642-23808-6_45

6. de March, C.A., Ryu, S., Sicard, G., Moon, C., Golebiowski, J.: Structure-odour relationships reviewed in the postgenomic era. Flavour Fragrance J. **30**(5), 342–361 (2015). http://dx.doi.org/10.1002/ffj.3249

7. Novak, P.K., Lavrač, N., Webb, G.I.: Supervised descriptive rule discovery: A unifying survey of contrast set, emerging pattern and subgroup mining. J. Mach. Learn. Res. **10**, 377–403 (2009)

INSIGHT: Dynamic Traffic Management Using Heterogeneous Urban Data

Nikolaos Panagiotou[1], Nikolas Zygouras[1(✉)], Ioannis Katakis[1],
Dimitrios Gunopulos[1], Nikos Zacheilas[2], Ioannis Boutsis[2], Vana Kalogeraki[2],
Stephen Lynch[3], Brendan O'Brien[3], Dermot Kinane[3], Jakub Mareček[4],
Jia Yuan Yu[4], Rudi Verago[4], Elizabeth Daly[4], Nico Piatkowski[5],
Thomas Liebig[5], Christian Bockermann[5], Katharina Morik[5],
Francois Schnitzler[6], Matthias Weidlich[7], Avigdor Gal[8], Shie Mannor[8],
Hendrik Stange[9], Werner Halft[9], and Gennady Andrienko[9]

[1] National and Kapodistrian University of Athens, Athens, Greece
nzygouras@di.uoa.gr
[2] Athens University of Economics and Business, Athens, Greece
[3] Dublin City Council, Dublin, Ireland
[4] IBM Research, Dublin, Ireland
[5] AI Group, TU Dortmund, Dortmund, Germany
[6] Technicolor, Paris, France
[7] Humboldt-Universität zu Berlin, Berlin, Germany
[8] Technion - Israel Institude of Technology, Haifa, Israel
[9] Fraunhofer IAIS, Sankt Augustin, Germany

Abstract. In this demo we present INSIGHT, a system that provides traffic event detection in Dublin by exploiting Big Data and Crowdsourcing techniques. Our system is able to process and analyze input from multiple heterogeneous urban data sources.

1 Introduction

In this demo we present a traffic monitoring system that is currently *deployed* in Dublin and utilized for city event detection. The purpose of this demo is to show, that, using novel data mining techniques we are able to monitor diverse data coming from city-wide infrastructures and extract useful information to present to the city operators. We collaborated with Dublin City Council (DCC) and designed a system that is able to process *real-time* data from diverse input sources such as sensors mounted on top of buses, traffic sensors embedded in street intersections or even citizen's tweets. INSIGHT identifies events of interest such as traffic congestion, construction works and accidents [1].

Although sensor data are available to smart city authorities, it is very difficult for human operators to monitor the vast amount of information. Figure 1 shows the DCC control center, where one of the screens displays INSIGHT[1]. Our system identifies events and aids operators to react in a timely manner by providing tools

---

[1] http://www.insight-ict.eu/ - INSIGHT was funded by an FP7 EU grant.

© Springer International Publishing AG 2016
B. Berendt et al. (Eds.): ECML PKDD 2016, Part III, LNAI 9853, pp. 22–26, 2016.
DOI: 10.1007/978-3-319-46131-1_5

that: (1) Automatically identify the locations where abnormal sensor behaviour occurs, (2) Analyze historical data and build models that capture the sensors' normal behaviour, and (3) Exploit ubiquitous human sensors to complement the existing data sources.

2 System Description

Our system, as described in Fig. 3 consists of multiple components; each is responsible for monitoring an individual data source. Data are generated from a set of heterogeneous traffic sensor networks in Dublin. Each stream is processed by an *Intelligent Sensor Agent (ISA)* that identifies *anomalies*. The detected anomalies are fused to infer *events* (see Round Table) and are presented to the traffic operators. In case of uncertainty, Crowdsourcing tasks are initiated.

Data Sources. Our system analyzes heterogeneous streaming data from the following input sources: (i) *Buses* that transmit their GPS location and information regarding their route. (ii) *SCATS* sensors that are deployed at intersections and transmit traffic flow measurements and the degree of saturation. (iii) *Tweets* that are posted by users in the Dublin area. (iv) *Crowdsourcing* input from users that provide feedback by reporting an event or answering to dynamic system queries using the CrowdAlert app from their mobile devices (see Fig. 2).

Fig. 1. DCC's traffic control center. INSIGHT system is running at the top monitor

Architecture. Our system receives raw data from the heterogeneous data sources and channels processed information to the corresponding *ISA*[2]. The *SCATS ISA* checks whether the SCATS sensor's behaviour deviates significantly from its neighbours behaviour, using a multivariate ARIMA model.

The *Bus ISA* is responsible to detect anomalies, monitoring the bus data. This component exploits the Lambda architecture, using Storm[3], to transmit data to concurrently running Complex Event Processing (CEP) engines. These engines monitor whether the streaming data trigger the set up rules. The *Bus ISA*

---

[2] http://www.insight-ict.eu/sites/default/files/deliverables/D2-1.pdf.
[3] http://storm.apache.org/.

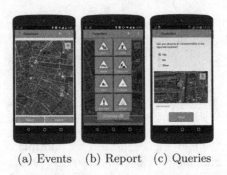

(a) Events   (b) Report   (c) Queries

Fig. 2. CrowdAlert app

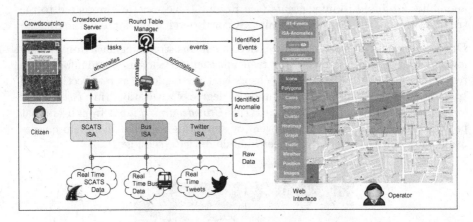

Fig. 3. Overview of the INSIGHT system showing the inputs, the interface to the operators and citizens, and the connectivity between the components.

enables the automatic adjustment of needed resources (elasticity), estimating the system load in upcoming time windows, using Gaussian Processes [4].

The Twitter ISA analyzes tweets that are geo-located in the Dublin area and identifies those that describe an event (traffic, flood, or fire), using a SVM text classifier. The Twitter API sets a number of constraints in terms of how many users, keywords or locations can be tracked at the same time. INSIGHT utilizes a dynamic filtering approach that continuously evaluates candidate keywords, users, and locations and selects an optimal subset.

The output of the ISAs is integrated in the *Round Table* component, which aggregates information [3], infers about what is happening and forwards the detected events either to the DCC operators through a web interface or to the crowdsourcing users requesting clarification. The *Crowdsourcing Server* acts as a middleware among the system and the users. Its main responsibilities are: (i) to keep track of the active users that are able to participate in the crowdsourcing tasks, (ii) to assign tasks to the Crowdsourcing users, in order to resolve possible ambiguities [2], (iii) to receive reports asynchronously from the users when an event occurs, and (iv) to provide feedback to the users regarding ongoing events near their location. The *Crowdsourcing App* is an Android application, called CrowdAlert[4] (illustrated in Fig. 2) that enables the human crowd to monitor the ongoing traffic events and provide feedback to the authorities.

User Evaluation. INSIGHT has been evaluated by the Traffic Management team at DCC. A set of employees with diverse responsibilities were selected, evaluated the system and filled a questionnaire. The questions regarded the usefulness and accuracy of the system and how it affected their workflow. The team members were quite happy with the system that provided them accurate events in real time and a set of tools that visualize real-time and historical data[5].

3   Demo Description

We will demonstrate the web interface and the CrowdAlert app that visualize the detected events, from the different *ISAs*. During the demonstration, both, real-time, and historical data will be available, to highlight multiple aspects of the system. An example of an identified fire event is shown in Fig. 4.

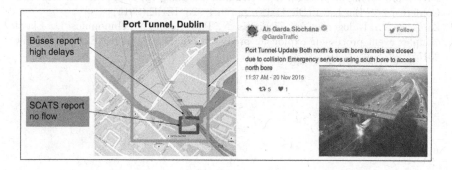

Fig. 4. Visualization of a fire event in Port Tunnel, Dublin

Equipment. We will be using a laptop in order to demonstrate the front-end of INSIGHT, which is currently used by DCC. We will also be using Android devices to demonstrate the CrowdAlert app.

Demo Plan. The users will be able to view and interact with the system that is currently deployed and running at DCC. They will be able to overview information about the anomalies and events identified by the analysis components in real time. Moreover, the users will be able to select past time periods, redefine the analysis thresholds and parameters and re-run the system. Users will be able to generate plots that visualize data originating from the Buses and

---

[4] http://crowdalert.aueb.gr/.

[5] www.insight-ict.eu/sites/default/files/deliverables/D6-2.pdf.

the SCATS sensors. Finally, we will also demonstrate the CrowdAlert app using smartphones and tablets. The users will be able to use CrowdAlert and interact with our system. Thus, they will be able to observe events that take place in Dublin in real-time as well as to report events and provide feedback through crowdsourcing tasks.

Acknowledgments. This research has been financed by the European Union through the FP7 ERC IDEAS 308019 NGHCS project, the Horizon2020 688380 VaVeL project and a Yahoo Faculty award.

References

1. Artikis, A., Weidlich, M., Schnitzler, F., Boutsis, I., Liebig, T., Piatkowski, N., Bockermann, C., Morik, K., Kalogeraki, V., Marecek, J., Gal, A., Mannor, S., Kinane, D., Gunopulos, D.: Heterogeneous stream processing and crowdsourcing for urban traffic management. In: EDBT (2014)
2. Boutsis, I., Kalogeraki, V.: On task assignment for real-time reliable crowdsourcing. In: ICDCS, pp. 1–10, Madrid, Spain, June 2014
3. Schnizler, F., Liebig, T., Marmor, S., Souto, G., Bothe, S., Stange, H.: Heterogeneous stream processing for disaster detection and alarming. In: 2014 IEEE International Conference on Big Data (Big Data), pp. 914–923. IEEE (2014)
4. Zacheilas, N., Kalogeraki, V., Zygouras, N., Panagiotou, N., Gunopulos, D.: Elastic complex event processing exploiting prediction. In: Big Data. IEEE, USA, October 2015

Coordinate Transformations for Characterization and Cluster Analysis of Spatial Configurations in Football

Gennady Andrienko[1,2(✉)], Natalia Andrienko[1,2], Guido Budziak[3],
Tatiana von Landesberger[4], and Hendrik Weber[5]

[1] Fraunhofer Institute IAIS, Sankt Augustin, Germany
{gennady.andrienko,natalia.andrienko}@iais.fraunhofer.de
[2] City University London, London, UK
[3] TU Eindhoven, Eindhoven, The Netherlands
[4] TU Darmstadt, Darmstadt, Germany
[5] DFL Deutsche Fussball Liga GmbH, Frankfurt, Germany

Abstract. Current technologies allow movements of the players and the ball in football matches to be tracked and recorded with high accuracy and temporal frequency. We demonstrate an approach to analyzing football data with the aim to find typical patterns of spatial arrangement of the field players. It involves transformation of original coordinates to relative positions of the players and the ball with respect to the center and attack vector of each team. From these relative positions, we derive features for characterizing spatial configurations in different time steps during a football game. We apply clustering to these features, which groups the spatial configurations by similarity. By summarizing groups of similar configurations, we obtain representation of spatial arrangement patterns practiced by each team. The patterns are represented visually by density maps built in the teams' relative coordinate systems. Using additional displays, we can investigate under what conditions each pattern was applied.

1 Introduction

Current tracking technologies enable measuring and recording of the spatial positions and movements of the players and the ball in football (a.k.a. soccer) games with high accuracy and temporal frequency. Analysis of the resulting trajectories can bring valuable knowledge about the movement behaviors of the players and teams and interactions between the players and between the teams. A lot of research has been done on analyzing various aspects of a football game, such as players' performance (e.g. [4]), passes (e.g. [3]) and pass opportunities (e.g. [5]), team formations (e.g. [2]), and others. Analysis of formations mostly focuses on identifying long- and short-term roles of individual players [2] or typical geometric configurations of tactical groups [6].

We demonstrate a novel approach to analyzing collective movement behaviors of football teams with the aim to find typical patterns of spatial arrangement of all field players within their teams and in relation to the opponent teams. For this purpose, we transform the positions of the players in the field into their relative positions with

B. Berendt et al. (Eds.): ECML PKDD 2016, Part III, LNAI 9853, pp. 27–31, 2016.
DOI: 10.1007/978-3-319-46131-1_6

respect to the centers and attack vectors of the teams and analyze the distributions of the players in these "team spaces" (similarly to [1]).

2   Characterization and Analysis of Spatial Configurations

We use the term *spatial configuration* for the relative arrangement of the field players in one time moment. A spatial configuration can be characterized by the coordinates of the players in "team spaces". The team space of one team is defined in each moment by the team centroid (i.e., the mean position of all field players of this team) and the direction towards the opponents' goal (attack direction). The team centroid is taken as the origin of the coordinates. The vertical axis of the team space corresponds to the attack direction, and the horizontal axis is perpendicular to it. Such coordinate system is created for each team. For each moment of the game, the positions of the players of both teams and the positions of the ball are transformed to the coordinate system of each team. The horizontal coordinate shows the relative position in the left or right side of the team, and the vertical coordinate shows the relative position in the back or front side of the team.

The transformation is illustrated in Fig. 1 by example of trajectories of two players from opposing teams (black and green). Figure 1A shows the original trajectories in the space of the pitch. The goal of the black team is on the left and the goal of the green team is on the right. The attack directions of the teams are indicated by colored arrows at the bottom of the image. In Fig. 1B, the trajectories are composed from the relative positions of the players in their own teams. The image represents simultaneously the spaces of both teams, i.e., their coordinate systems are aligned. We see that, despite quite strong separation of the two trajectories in the field space (Fig. 1A), they cover similar regions in their team spaces, i.e., they have similar roles in their teams. Image 1C represents the space of the black team. The green trajectory shows how the green team player was positioned in relation to the opponent team. Likewise, image 1D represents the space of the green team and the positioning of the black team player with respect to the opponents.

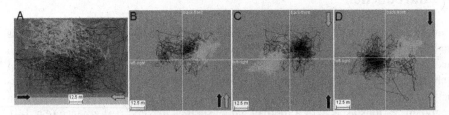

Fig. 1. Transformation of coordinates is shown by example of trajectories of two players from opposing teams. A: original trajectories in the space of the pitch; B, C, D: the trajectories transformed to the spaces of the own teams (B), black team (C) and green team (D). (Color figure online)

To find typical patterns of spatial arrangement, which is the goal of our analysis, we need to group the spatial configurations that occurred throughout the game by similarity. This can be achieved by means of clustering. To apply clustering, we need to represent the spatial configurations by appropriate features. We construct feature vectors of the spatial configurations by putting the players' relative positions in the team spaces in the order of decreasing vertical coordinates, i.e., from front to back. To reduce the sensitivity of the descriptions of the spatial configurations to substitutions of players and changes of players' roles, we apply ordering from left to right when the vertical difference is below a threshold (e.g., 2 m).

The full feature vector of a spatial configuration consists of the coordinates of the field players of each team in their own team space and in the team space of the opponents plus the coordinates of the ball in the spaces of both teams. On demand, a subset of features can be used for clustering, e.g., features of only one team.

By summarizing the clusters of spatial configurations produced by a clustering algorithm (e.g., k-means or EM), we obtain a representation of typical spatial arrangement patterns. Since each spatial configuration is basically a set of points (which represent players' positions), we summarize a cluster of configurations by computing joint point density fields in the team and field spaces using kernel density estimation.

In the demonstration, we show interactive clustering of spatial configurations and visual exploration of spatial arrangement patterns supported by interactive tools.

3   Example

In this example, we use data collected during the German Bundesliga game of Borussia Dortmund against VfL Wolfsburg on 10/12/2015. Our analysis focus is the arrangement patterns of the field players of Dortmund in relation to the opponents. For clustering, we use a subset of features consisting of the relative positions of the Dortmund's players in the space of Wolfsburg. Figure 2 shows the results for k-means clustering with k = 8. The density maps represent the Dortmund's configuration patterns in the Wolfsburg's team space. Here we use a color scale from light blue for low densities to red for high densities. The movements of the ball during the times when the configurations took place are represented by semi-transparent black lines.

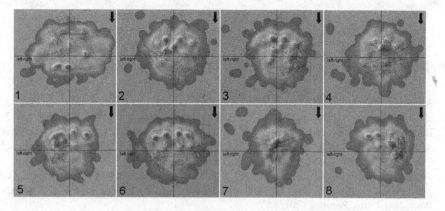

Fig. 2. Density maps represent clustered arrangements of the players of Dortmund in relation to the opponent team (i.e., in the team space of Wolfsburg). (Color figure online)

To investigate in what conditions these different arrangement patterns were applied, we use additional displays. Thus, Fig. 3 shows how the patterns are related to the ball possession. The lengths of the green and red bars represent the numbers of time moments when the ball was possessed by Wolfsburg (green) and Dortmund (black). The uppermost pair of bars correspond to the whole game, excluding the time when the ball was out of play. The remaining rows correspond to the clusters of the spatial configurations. We see that the pattern of cluster 1 was used almost exclusively under ball possession by Dortmund, whereas the patterns of clusters 5 and 6 were mostly applied under ball possession by Wolfsburg. The remaining patterns are not so clearly related to the ball possession by either of the teams. In Fig. 4, we see how the patterns are related to the positions of the Dortmund's players and the ball in the field.

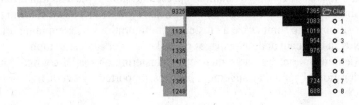

Fig. 3. Distribution of the time moments with ball possession by VfL Wolfsburg (green) and Borussia Dortmund (black) across the clusters of the spatial configurations. (Color figure online)

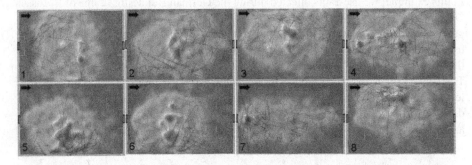

Fig. 4. Relation of the configuration patterns to the players' positions in the field.

References

1. Andrienko, N., Andrienko, G., Barrett, L., Dostie, M., Henzi, P.: Space transformation for understanding group movement. IEEE Trans. Vis. Comput. Graph. **19**(12), 2169–2178 (2013)
2. Bialkowski, A., Lucey, P., Carr, P., Yue, Y., Sridharan, S., Matthews, I.A.: Large-scale analysis of soccer matches using spatiotemporal tracking data. In: IEEE ICDM 2014 (International Conference on Data Mining), pp. 725–730. IEEE (2014)
3. Cintia, P., Rinzivillo, S., Pappalardo, L.: A network-based approach to evaluate the performance of football teams. In: Machine Learning and Data Mining for Sports Analytics Workshop, Porto, Portugal (2015)

4. Di Salvo, V., Baron, R., Tschan, H., Calderon Montero, F., Bachl, N., Pigozzi, F.: Performance characteristics according to playing position in elite soccer. Int. J. Sports Med. **28**(3), 222–227 (2007)
5. Gudmundsson, J., Wolle, T.: Football analysis using spatio-temporal tools. Comput. Environ. Urban Syst. **47**, 16–27 (2014)
6. Perl, J., Grunz, A., Memmert, D.: Tactics analysis in soccer: an advanced approach. Int. J. Comput. Sci. Sport **12**(1), 33–44 (2013)

Leveraging Spatial Abstraction in Traffic Analysis and Forecasting with Visual Analytics

Natalia Andrienko[1,2(✉)], Gennady Andrienko[1,2], and Salvatore Rinzivillo[3]

[1] Fraunhofer Institute IAIS, Sankt Augustin, Germany
{natalia.andrienko,gennady.andrienko}@iais.fraunhofer.de
[2] City University London, London, UK
[3] Area della Ricerca CNR, Istituto di Scienza e Tecnologie dell'Informazione, Pisa, Italy
rinzivillo@isti.cnr.it

Abstract. By applying spatio-temporal aggregation to traffic data consisting of vehicle trajectories, we generate a spatially abstracted transportation network, which is a directed graph where nodes stand for territory compartments (areas in geographic space) and links (edges) are abstractions of the possible paths between neighboring areas. From time series of traffic characteristics obtained for the links, we reconstruct mathematical models of the interdependencies between the traffic intensity (a.k.a. traffic flow or flux) and mean velocity. Graphical representations of these interdependencies have the same shape as the fundamental diagram of traffic flow through a physical street segment, which is known in transportation science. This key finding substantiates our approach to traffic analysis, forecasting, and simulation leveraging spatial abstraction. We present the process of data-driven generation of traffic forecasting and simulation models, in which each step is supported by visual analytics techniques.

1 Introduction

The topic of this presentation, based on [4], is derivation of traffic forecasting and simulation models from traffic data. Traffic data in the form of trajectories of vehicles are currently collected in great amounts, but their potential remains largely underexploited. By means of visual analytics methods, we discovered fundamental patterns of traffic flow dynamics that are common for different areas and spatial scales. On this basis, we created interactive visual interfaces for representing these patterns by mathematical models and devised a lightweight traffic forecasting and simulation algorithm that exploits these models. We developed interactive visual embedding for defining initial conditions, running simulations, and analyzing the outcomes. Since simulations could be prepared and performed very fast, thus allowing interactive operation, our tools allow the users to imitate various interventions altering network properties and/or traffic routes and investigate their impacts on the traffic situation development, including comparative analysis of various "what if" scenarios.

2 Approach Summary

Given a set of trajectories, we apply a method [2] that derives an abstracted network consisting of territory compartments (further called cells) and links between them. In

© Springer International Publishing AG 2016
B. Berendt et al. (Eds.): ECML PKDD 2016, Part III, LNAI 9853, pp. 32–35, 2016.
DOI: 10.1007/978-3-319-46131-1_7

brief, the method organizes points sampled from the trajectories into groups fitting in circles of a user-specified maximal radius. The medoids of the groups are taken as generating seeds for Voronoi tessellation. Smaller or larger cells (Voronoi polygons) can be generated by varying the maximal circle radius, thus allowing traffic analysis at a chosen spatial scale. Moreover, it is possible to vary the spatial scale across the territory depending on the data density [1]. Next, the trajectories are transformed into *flows* (aggregate movements) between the cells by time intervals. For each pair of neighboring cells (C_i, C_j) and each time interval T_k, the flow is an aggregate of all moves from C_i to C_j that ended within the interval T_k and started within either T_k or T_{k-1}. The flow is characterized by the number of moves and the mean speed (velocity) of the movement. The number of moves (traffic volume) per time interval is called *traffic intensity* (a.k.a. *traffic flow* or *flux*). Since available trajectories typically cover only a sample of vehicles that move within a network and not the entire population, the computed traffic intensities need to be appropriately scaled, to approximate real intensities. Appropriate scaling parameters or functions can be derived by comparing the computed vehicle counts with measured counts obtained from traffic sensors [7].

To study and quantify the relationships between the traffic intensities and mean speeds, the data are further transformed in the following way. Let A and B be two time-dependent attributes associated with the same link and defined for the same time steps. The value range of attribute A, which is taken as an independent variable, is divided into intervals. For each value interval, all time steps in which values from this interval occur are found, and all values of attribute B occurring in the same time steps are collected. From these values of B, summary statistics are computed: quartiles, 9th decile, and maximum. For each statistical measure, a sequence of values of B corresponding to the value intervals of A is constructed. These sequences are called *dependency series*. We first take the traffic intensity as the independent variable and derive dependency series of the mean speed. Then we take the mean speed as the independent variable and derive dependency series of the traffic intensity. Dependency series may be derived using either the absolute or relative traffic intensities, the latter being the ratios or percentages of the absolute intensities to the maximal intensities attained on the same links.

In Fig. 1, two maps on the left represent abstracted transportation networks of Milan with different levels of spatial abstraction. Curved lines in the upper map and half-arrow symbols in the lower map represent the network links. On the right of each map, the upper graph shows the dependencies of the mean speed on the relative traffic intensity. The horizontal axis corresponds to the traffic intensity and the vertical axis to the 9th decile of the mean speed (this statistical measure is less sensitive to outliers as the maximum). The lower graph shows the dependencies of the relative traffic intensity on the mean speed. The horizontal axis corresponds to the mean speed and the vertical axis to the maximal traffic intensity. The network links have been clustered by similarity of the speed-intensity dependencies. The coloring of the link symbols on the map and lines in the graphs represents the cluster membership. The shapes of the dependency lines are very similar to the curves in the fundamental diagram of traffic flow describing the relationship between the flow velocity and traffic flux [5, 6] in a physical transportation network consisting of street segments. We see that the same relationships exist also in a spatially abstracted network. Moreover, we

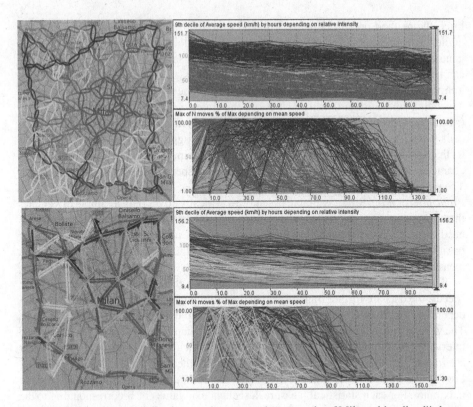

Fig. 1. The maps show spatially abstracted transportation networks of Milan with cell radii about 2 km (top) and 4 km (bottom). The graphs to the right of each map represent the dependencies between the relative traffic intensities and the mean speeds on the network links.

have found that the relationships conforming to the fundamental traffic diagram exist on different levels of spatial abstraction, as illustrated in Fig. 1.

We have developed interactive visual tools supporting derivation of formal models from the time series of flow characteristics and from the dependency series [3]. Models are built for clusters of links rather than individual links, to avoid over-fitting and reduce the impacts of noise and local outliers. Predictions made for link clusters are individually adjusted for each link based on the statistics of its original values [3]. We have also developed a novel traffic simulation algorithm that can directly work with the derived models. The main idea is following: for each link, the algorithm finds how many vehicles need to move through it in the current minute, determines the mean speed that is possible for this link load (using the dependency model from the traffic intensity to the mean speed), then determines how many vehicles will actually be able to move through the link in this minute (using the dependency model from the mean speed to the traffic intensity), and then promotes this number of vehicles to the end place of the link and suspends the remaining vehicles in the start place of the link (Fig. 2).

To perform a simulation, the analyst defines a scenario. A wizard guides the analyst through the required steps and providing visual feedback at each step. We describe the

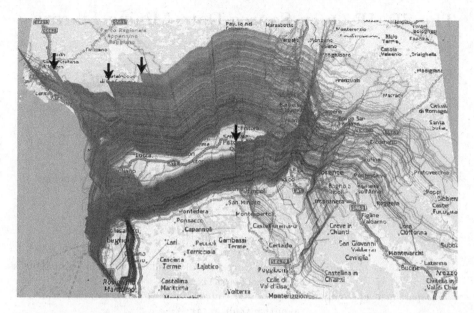

Fig. 2. For a scenario of mass evacuation from the coastal areas in Tuscany (Italy), simulated car trajectories are represented in a space-time cube, where two dimensions represent geographical time and one dimension time. The arrows point at the places of major traffic suspensions.

simulation of a scenario of mass evacuation from the coastal area in Tuscany (Italy). The appendix to the paper (http://geoanalytics.net/and/is2015/) includes a video demonstration of the process of model building, scenario definition, simulation, and exploration of results supported by interactive visual interfaces.

References

1. Andrienko, G., Andrienko, N., Bak, P., Keim, D., Wrobel, S.: Visual Analytics of Movement. Springer, Heidelberg (2013)
2. Andrienko, N., Andrienko, G.: Spatial generalization and aggregation of massive movement data. IEEE Trans. Visual. Comput. Graph. **17**(2), 205–219 (2011)
3. Andrienko, N., Andrienko, G.: A visual analytics framework for spatio-temporal analysis and modeling. Data Mining Knowl. Discovery **27**(1), 55–83 (2013)
4. Andrienko, N., Andrienko, G., Rinzivillo, S.: Leveraging spatial abstraction in traffic analysis and forecasting with visual analytics. Inf. Syst. **57**(1), 172–194 (2016). Appendix: http://geoanalytics.net/and/is2015/
5. Gazis, D.C.: Traffic Theory. Kluwer Academic, Boston (2002)
6. http://en.wikipedia.org/wiki/Fundamental_diagram_of_traffic_flow
7. Pappalardo, L., Rinzivillo, S., Qu, Z., Pedreschi, D., Giannotti, F.: Understanding the patterns of car travel. Eur. Phys. J. Special Topics **215**, 61–73 (2013)

The SPMF Open-Source Data Mining Library Version 2

Philippe Fournier-Viger[1]([✉]), Jerry Chun-Wei Lin[2], Antonio Gomariz[3],
Ted Gueniche[4], Azadeh Soltani[5], Zhihong Deng[6], and Hoang Thanh Lam[7]

[1] School of Natural Sciences and Humanities, Harbin Institute of Technology
Shenzhen Graduate School, Shenzhen, China
philfv8@yahoo.com
[2] School of Computer Science and Technology, Harbin Institute of Technology
Shenzhen Graduate School, Shenzhen, China
jerrylin@ieee.org
[3] Department of Information and Communication Engineering,
University of Murcia, Murcia, Spain
agomariz@gmail.com
[4] Department of Computer Science, University of Moncton, Moncton, Canada
ted.gueniche@gmail.com
[5] Department of Computer Engineering, University of Bojnord, Bojnord, Iran
a.soltani@ub.ac.ir
[6] School of Electronics Engineering and Computer Science, Peking University,
Beijing, China
zhdeng@cis.pku.edu.cn
[7] IBM Ireland Research Lab, Dublin, Ireland
t.l.hoang@ie.ibm.com

Abstract. SPMF is an open-source data mining library, specialized in
pattern mining, offering implementations of more than 120 data mining
algorithms. It has been used in more than 310 research papers to solve
applied problems in a wide range of domains from authorship attribution
to restaurant recommendation. Its implementations are also commonly
used as benchmarks in research papers, and it has also been integrated
in several data analysis software programs. After three years of development, this paper introduces the second major revision of the library,
named SPMF 2, which provides (1) more than 60 new algorithm implementations (including novel algorithms for sequence prediction), (2) an
improved user interface with pattern visualization (3) a novel plug-in
system, (4) improved performance, and (5) support for text mining.

Keywords: Open-source library · Data mining · Frequent pattern mining

1 Introduction

Several open-source general purpose data mining libraries or programs have been
developed such as Knime [2], Mahout [8], and Weka [9]. Although these software

B. Berendt et al. (Eds.): ECML PKDD 2016, Part III, LNAI 9853, pp. 36–40, 2016.
DOI: 10.1007/978-3-319-46131-1_8

programs provide algorithms for many data mining tasks, they provide very few algorithms for mining frequent patterns in databases, while hundreds of algorithms have been proposed in this field during the last twenty years [10]. Moreover, the majority of researchers in the field of frequent pattern mining do not share their implementation or source code online. As a result, a user who wants to apply specific algorithms from this field, providing some particular features required by an application, often needs to implement the algorithms again, which is time-consuming and requires programming knowledge. To address this issue, the SPMF (Sequential Pattern Mining Framework) [5] open-source library has been created in 2009. The goal is to provide a common library for sharing the source code of efficient implementations of frequent pattern mining algorithms to increase their use in real applications, and also to provide a set of reference implementations for researchers to compare algorithms. Initially, SPMF was designed as a library for mining frequent patterns [1] in sequences (hence its name). But over the years, it has evolved to include all kinds of pattern mining algorithms for discovering patterns such as itemsets and association rules, sequential patterns [1], periodic patterns [11], and high-utility patterns [10]. It also provides a simple user-interface for quick testing and a command-line interface for easy integration with other systems. In the past five years, SPMF has been used in more than 310 research papers to solve applied problems in a wide range of domains ranging from authorship attribution, retail forecasting, chemistry, music analysis to restaurant recommendation. The algorithm implementations of SPMF are optimized and commonly used as benchmarks in research papers. SPMF has also been integrated in several popular data analysis software programs such as ScaVis and MOA [3]. Nowadays, SPMF offers by far the largest library of pattern mining algorithms with over 120 algorithms. Moreover, it is open-source, it can be used in commercial projects, and it is an active project, unlike similar smaller projects such as Coron [4] and LUCS-PKDD [7]. Moreover, SPMF is lightweight as it has no dependencies to any other projects. The first major release of SPMF is version 0.94, released in 2013 [5]. This paper introduces the second major release of the library, named SPMF 2.

2   Novel Features

SPMF 2 introduces five major novelties. First, it offers about 60 novel algorithm implementations. Thus, the number of algorithms has doubled since the previous major release, offering a greater range of algorithms to users. In particular, a novel module has been integrated in SPMF offering seven state-of-the-art algorithms for sequence prediction named DG, LZ78, AKOM, TDAG, PPM, CPT and CPT+. Sequence prediction (predicting the next symbol of a sequence of symbols based on a set of training sequences) has wide-applications in many domains such as web page prefetching and path recommendation [6]. Moreover, SPMF now offers about 20 more algorithms for utility pattern mining [10], which is probably the most active research area in frequent pattern mining. Utility pattern mining consists of finding patterns that may not be frequent but have a high-utility, where utility can be defined for example as the products generating the highest profit in a transaction database.

Second, the user interface has been improved. The main window is shown in the left side of Fig. 1. It is designed as a minimalistic user interface that let the user choose an algorithm, set its parameters and choose and input and output file, to then launch the algorithm. But an important novelty in SPMF 2 is a new pattern visualization window that let the user explore the patterns found by any algorithm in a table view (right side of Fig. 1). Using that window, the user can browse patterns, search patterns, and apply complex filters with boolean conditions, sorts, and export the result of these operations to various formats such as text and CSV files. Thus, this window lets the user perform post-processing of the patterns found by the algorithms using various criteria.

Third, another important novel feature is a plug-in system. In SPMF 2, a user can implement new algorithms by sub-classing a class named *DescriptionOfAlgorithm*. SPMF can automatically detect algorithms sub-classing this class (which can be stored in another JAR file) and load the additional algorithms in its user-interface and show them in the same list as its built-in algorithms. This allow researchers to easily extend the software with new algorithms, and reuse the same user interface.

Fourth, in this new version of SPMF, many performance optimizations have been performed to increase the performance of the algorithms already offered in SPMF. For example, the performance of the new implementation of PrefixSpan introduced in SPMF 2 is up to 10 times faster and consumes up to twice less memory than the previous version. Extensive performance comparison of various versions of algorithms and optimizations in SPMF are not presented in this paper due to length limitations but can be found on the SPMF website at: http://www.philippe-fournier-viger.com/spmf/.

Fifth, support for additional input formats has been added to SPMF. In SPMF 2, mining patterns in text documents is now natively supported. Thus, algorithms for discovering patterns such as itemsets and sequential patterns can now be applied to files containing texts. This is a very important feature as SPMF has been used in many papers related to text mining but previous versions of SPMF required that the user preprocesses input files to convert them to the SPMF format, which was inconvenient. When the new version of SPMF is applied to a text document, each word is seen as a symbol, and each sentence is viewed as a transaction or sequence. The document is transformed to an internal representation used by the algorithms and the result is then transformed again to be displayed to the user.

3   Conclusion and Future Work

In this paper, we presented the second major release of the SPMF library (version 2), which offers many new algorithms, an improved user interface with pattern visualization, a novel plug-in system, many performance optimizations, as well as support for additional formats such as text files.

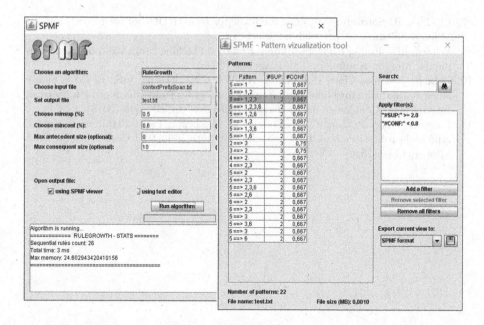

Fig. 1. The SPMF user interface

The SPMF library is an active project. Many contributors have provided algorithm implementations to the project from universities all around the world. The current development of SPMF is focused on providing more algorithms especially for discovering patterns in graphs and time-series, types of data that have not yet been considered in SPMF. Besides, an enhanced user interface for visually combining several algorithms in a workflow, and for interactive mining are currently planned for the next release.

References

1. Agrawal, R., Ramakrishnan, S.: Mining sequential patterns. In: Proceedings 11th International Conference on Data Engineering, pp. 3–14. IEEE (1995)
2. Berthold, M.R., et al.: KNIME - the Konstanz information miner: version 2.0 and beyond. SIGKDD Explor. **11**(1), 26–31 (2009)
3. Bifet, A., et al.: MOA: massive online analysis, a framework for stream classification and clustering. J. Mach. Learn. Res. (JMLR) **11**, 1601–1604 (2010)
4. Coron Software: http://coron.loria.fr/site/index.php
5. Fournier-Viger, P., Gomariz, A., Gueniche, T., Soltani, A., Wu, C., Tseng, V.S.: SPMF: a java open-source pattern mining library. J. Mach. Learn. Res. (JMLR) **15**, 3389–3393 (2014)
6. Gueniche, T., Fournier-Viger, P., Raman, R., Tseng, V.S.: CPT+: decreasing the time/space complexity of the compact prediction tree. In: Cao, T., Lim, E.-P., Zhou, Z.-H., Ho, T.-B., Cheung, D., Motoda, H. (eds.) PAKDD 2015. LNCS (LNAI), vol. 9078, pp. 625–636. Springer, Heidelberg (2015). doi:10.1007/978-3-319-18032-8_49

7. LUCS-KDD Software: http://cgi.csc.liv.ac.uk/frans/KDD/Software/
8. Mahout Software: http://mahout.apache.org/
9. Witten, I.H., Frank, E.: Data Mining: Practical Machine Learning Tools and Techniques. Morgan Kaufmann, San Francisco (2005)
10. Zida, S., Fournier-Viger, P., Lin, JC.-W., Wu, C.-W., Tseng, V.S.: EFIM: a highly efficient algorithm for high-utility itemset mining. In: Proceedings of 14th Mexican International Conference on Artificial Intelligence, pp. 530–546 (2015)
11. Fournier-Viger, P., Lin, C.W., Duong, Q.-H., Dam, T.-L.: PHM: mining periodic high-utility itemsets. In: Proceedings of 16th Industrial Conference on Data Mining, 15 p. (2016)

DANCer: Dynamic Attributed Network with Community Structure Generator

Oualid Benyahia[1], Christine Largeron[1(⊠)],
Baptiste Jeudy[1], and Osmar R. Zaïane[2]

[1] Univ Lyon, UJM-Saint-Etienne, CNRS, Institut d Optique Graduate School,
Laboratoire Hubert Curien UMR 5516, 42023 Saint-Etienne, France
{oualid.benyahia,christine.largeron,baptiste.jeudy}@univ-st-etienne.fr
[2] Department of Computer Science, University of Alberta, Edmonton, Canada
zaiane@cs.ualberta.ca

Abstract. We propose a new generator for dynamic attributed networks with community structure which follow the known properties of real-world networks such as preferential attachment, small world and homophily. After the generation, the different graphs forming the dynamic network as well as its evolution can be displayed in the interface. Several measures are also computed to evaluate the properties verified by each graph. Finally, the generated dynamic network, the parameters and the measures can be saved as a collection of files.

Keywords: Social network mining · Attributed graph · Synthetic data generator

1 Introduction

The proliferation of complex information networks in diverse fields of application has led to the proposal of a panoply of methods to analyze and discover relevant patterns in these networks. However, evaluating these methods and the comparison of the different approaches are not very easy due to the lack of large real networks with ground truth freely accessible to researchers. The alternative consists in using synthetic data provided by generators. There is a large bibliography regarding generation for static graphs, including the classic Erdős-Rényi (ER) model which generates random graphs or the Barabási-Albert (BA) model that generates random scale-free networks, but very few generators allow the construction of evolving graphs, exhibiting or not a community structure and, none of them takes into account the attribute values of the vertices. The interest of community detection, link prediction and more generally pattern discovery in dynamic networks where vertices are associated with attributes led us to develop the generator DANCer for attributed dynamic graphs with embedded community structure. This generator is an extended version of a previous generator, ANC dedicated to static graphs [1].

© Springer International Publishing AG 2016
B. Berendt et al. (Eds.): ECML PKDD 2016, Part III, LNAI 9853, pp. 41–44, 2016.
DOI: 10.1007/978-3-319-46131-1_9

2   Model

An attributed dynamic network generated by DANCer is represented by (1) a sequence of T attributed graphs $\mathcal{G}_i = (\mathcal{V}_i, \mathcal{E}_i)$, $i \in \{1, \ldots, T\}$, where $\mathcal{V}_i$ is a set of vertices, $\mathcal{E}_i$ a set of undirected edges and where for each vertex $v \in \mathcal{V}_i$ and each real attribute $A \in \mathcal{A}$, v_A denotes the attribute value of A assigned to vertex v and (2) a sequence of T partitions $\mathcal{P}_i$ of $\mathcal{V}_i$, $i \in \{1, \ldots, T\}$ which gives a community for each vertex in the corresponding graph $\mathcal{G}_i$, $i \in \{1, \ldots, T\}$. Each partition allows to define a community structure on a graph (i.e., the network at a single timestamp) in such a way that the nodes are grouped into sets densely connected and relatively homogeneous with regard to the attributes, while they are less connected to vertices belonging to other groups, and less similar with regard to their attributes.

The generation of the network is carried out in two phases. In phase one, an initial graph $\mathcal{G}_1 = (\mathcal{V}_1, \mathcal{E}_1)$ is built while respecting the well-known network properties such as preferential attachment, small world or homophily and, in the second phase, this initial graph is modified through two kinds of operations. The first set of operations, called micro operations, consist in removing or adding vertices and edges or updating their attributes whereas the second kind of operations is applied on the communities, i.e., at a macro level. They consist in (1) migrating members of a community to either a new community or an existing one, (2) splitting a community into two new sub-communities and (3) merging two existing communities into a single one.

3   Software Overview

The user interface has three panels as shown in Fig. 1. In the parameter panel, the user selects the dynamic generator parameters presented in Table 1[1]. Note that a seed is used for the random number generator. It can be saved to reproduce exactly the same network.

The visualization panel allows to display the generated network and its dynamic evolution. Each graph in the sequence can be selected with a timestamp scrollbar and viewed separately (Fig. 1). This panel can also display the size and the evolution of the different communities in the sequence of graphs according to the macro dynamic operations (split, merge and migrate) (see Fig. 2).

The sequence of attributed graphs is built while preserving properties of real networks and several measures, like modularity, clustering coefficient, diameter, expected and observed homophily or within inertia rate are computed on each graph of the dynamic network to describe its properties[2]. The changes in these different measures on the sequence of graphs are presented at the bottom of the interface in the measure panel (Fig. 3).

The bottom of the user interface includes also a panel displaying the distribution of vertex degrees on each graph of the sequence as shown in Fig. 4.

---

[1] The reader is referred to [1] for the static network generator parameters http://journals.plos.org/plosone/article?id=10.1371/journal.pone.0122777.

[2] See [1] for a more detailed presentation of the properties and corresponding measures.

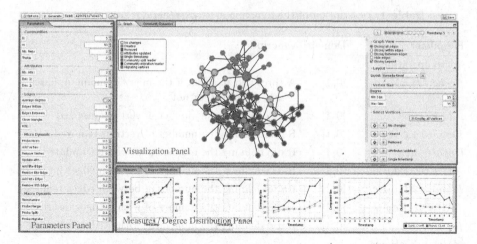

Fig. 1. User interface of the generator DANCer.

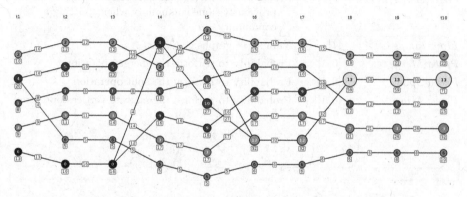

Fig. 2. Community dynamics display in visualization panel.

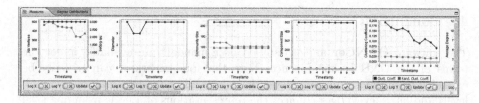

Fig. 3. Measures panel of the generator.

The generated dynamic network can be saved as a collection of files. For each graph of the sequence, a file indicates the composition of the graph (vertices and edges) and a "**parameters**" file enumerates all the parameters used by the generator. The graph measures and community dynamics can also be saved in separated files.

Table 1. Description of the dynamic network generator parameters

Parameter	Domain	Description
Micro operations		
Proba Micro	$[0,1]$	A threshold to select if the micro dynamic updates are performed or not
Add Vertex	$[0,1]$	Ratio defining the number of vertices inserted
Remove Vertex	$[0,1]$	Ratio defining the number of vertices removed
Update Attr.	$[0,1]$	Ratio defining the number of attributes updated
Add Btw. Edges	$[0,1]$	Ratio defining the number of between edges inserted
Remove Btw. Edges	$[0,1]$	Ratio defining the number of between edges removed
Add Wth. Edges	$[0,1]$	Ratio defining the number of within edges inserted
Remove Wth. Edges	$[0,1]$	Ratio defining the number of within edges removed
Macro operations		
$P_{removeEdgeSplit}$	$[0,1]$	Proba. to remove an edge between two vertices in the previously same community when splitting a community
Timestamps	$\mathbb{N}^{+}$	Number of graphs generated
Proba Merge	$[0,1]$	Probability to perform the merge operation
Proba Split	$[0,1]$	Probability to perform the split operation
Proba Migrate	$[0,1]$	Probability to perform the migrate vertices operation

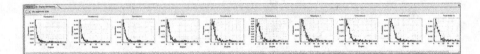

Fig. 4. Degree distribution panel.

4   Conclusion

The software **DANCer** and a detailed user manual[3] are available under the terms of the GNU Public Licence. Note that our generator can trivially be extended to produce multiplex networks, where all nodes are omnipresent in all levels and intra-level edges connect the representations of a node from one level to the other. This conversion is possible by simply converting each timestamp graph into a layer of the multiplex network and adding the necessary intra-level edges.

Reference

1. Largeron, C., Mougel, P.N., Rabbany, R., Zaïane, O.R.: Generating attributed networks with communities. PLoS ONE **10**(4), e0122777 (2015)

---

[3] http://perso.univ-st-etienne.fr/largeron/DANCer_Generator/.

Topy: Real-Time Story Tracking via Social Tags

Gevorg Poghosyan[✉], M. Atif Qureshi, and Georgiana Ifrim

Insight Centre for Data Analytics, University College Dublin, Dublin, Ireland
gevorg.poghosyan@insight-centre.org

Abstract. The *Topy* system automates real-time story tracking by utilizing crowd-sourced tagging on social media platforms. Topy employs a state-of-the-art Twitter hashtag recommender to continuously annotate news articles with hashtags, a rich meta-data source that allows connecting articles under drastically different timelines than typical keyword based story tracking systems. Employing social tags for story tracking has the following advantages: (1) social annotation of news enables the detection of emerging concepts and topic drift in a story; (2) hashtags go beyond topics by grouping articles based on connected themes (e.g., #rip, #blacklivesmatter, #icantbreath); (3) hashtags link articles that focus on subplots of the same story (e.g., #palmyra, #isis, #refugeecrisis).

Keywords: Story tracking · News · Social media · Social tags

1 Introduction

Although keyword and semantic-based matching of news have advanced considerably [1,4], the problem of automatically tracking story timelines and their evolution in real-time remains very challenging. A news story often discusses multiple related events, which take place in different time periods and may involve different entities. Some stories are relatively short-lived, for example, the 2016 Champions League final, and some others span many years and discuss multiple events, for example, the Ebola outbreak. For instance, the story of the Syrian war has evolved in time, by shifting the discussion **topic** (*Middle East, migration, human rights, politics*), the discussed **entities** (*Assad, ISIS, Putin, USA, Turkey, Belgium*) and the discussed **events** (*rebel uprising, destruction of Syria's chemical weapons, Yazidi massacres, camerawoman kicks a migrant*). Figure 1 illustrates this drift in the news article space projected on the topic-event-entity dimension. Stories may share articles, e.g., the article "Turkey carries out air strikes" may appear in several stories: *Syrian war, PKK in Syria, Turkey elections 2015.*

Topy takes a different approach to story tracking by building on crowd-sourced social tags and a hashtag recommender, to link news articles in complex story timelines. The choice of Twitter hashtags as rich social annotations is motivated by the following factors: (i) most stories have a lot of quality discussions centered on focused hashtags on Twitter, (ii) creation, popularity and abandonment of hashtags implicitly encode the concept drift in the story, (iii) hashtags

© Springer International Publishing AG 2016
B. Berendt et al. (Eds.): ECML PKDD 2016, Part III, LNAI 9853, pp. 45–49, 2016.
DOI: 10.1007/978-3-319-46131-1_10

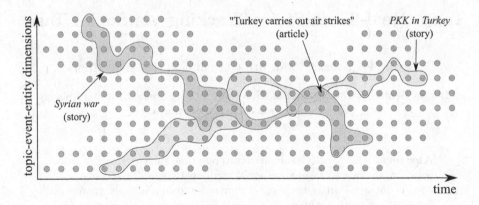

Fig. 1. Stories' drift in topic-event-entity-time space.

allow cross-platform multi-modal content linking (text, image, video). This approach is also consistent with recent trends in news media: (a) *The Guardian*[1], *Huffington Post*[2], *AJ+*, *BBC*, write articles about popular hashtags to inform and engage the public on discussion trends, (b) *The Sun* published a newspaper with a hashtag alongside an article to allow readers "to share their opinions and continue the story online"[3]. Most automated story tracking solutions are either limited in the number of events that can be tracked or are not real-time. Some news organizations have story-pages on their websites, i.e., curated collections of news articles that allow the reader to get an overview on particular events, e.g., referendums, elections, budgets. The Irish Times has dedicated story-pages for issues of relevance to the Irish society, e.g., the inquiry into the banking collapse[4] of 2008 (hashtag #bankinginquiry). Preparing these story-pages relies on prior agreement among journalists to manually tag all articles relevant to a set of stories, with the same set of tags. Once a decision is taken to create a story-page, those articles are continuously retrieved from the news archive via the manual tags. The problem with this approach is that it relies on foresight over which stories are worth covering and what is the right tag to use for those story-articles. Additionally, the manual process does not scale well on many stories that require in depth coverage, e.g., tools such as provided by www.newsdeeply.com although useful, update slowly and lack behind the fast pace of the real-world.

To the best of our knowledge there is no similar system that makes use of Twitter hashtags for story tracking. State-of-the-art systems rely on keyword/semantic matching and require often slow-to-change offline snapshots of knowledge bases [3] or need computationally expensive, complex clustering or semantic models, where parameters, such as number of topics [2], timespan of stories [1,4] and cluster sizes [5] significantly affect the system performance.

---

[1] www.theguardian.com/technology/hashtags.

[2] www.huffingtonpost.com/news/hashtags/.

[3] www.huffingtonpost.com/2014/03/26/sun-hashtag-newspaper-murdoch-british_n_5034639.html.

[4] www.irishtimes.com/news/banking-inquiry.

2   Topy System Overview

Topy maps news articles to stories in real-time by grouping articles with connected events, entities and topics that are discussed together on Twitter. Story tracking is formulated as a retrieval task with queries that allow mixing of keywords and hashtags. This allows tracking stories on-the-fly rather being restricted to pre-determined stories.

Real-time annotation of news articles with Twitter hashtags. We build on top of the Hashtagger infrastructure [6] for collecting, processing and storing news articles and Twitter data. The system architecture is illustrated in Fig. 2. An article is represented by its headline, subheadline, body, a set of summary keywords and a set of hashtags recommended to the article over a period of 24 h from the article publication time [6]. Hashtagger achieves Precision of more than 85 %. Around 70 % of processed news articles have at least one recommended hashtag.

Query-based story tracking. The hashtags of each article are binned into 20 confidence bins with ranges from $(0.975, 1.0]$ to $(0.5, 0.525]$ and indexed as child documents for the corresponding article documents. Parent-child relationship and the chosen mapping enable an efficient search on article fields with different weighing using the BM25 algorithm. A query is composed of (i) words $w_1, ..., w_n$, which are matched on article keywords, headline, subheadline and content with score boost of correspondingly $\times 4$, $\times 3$, $\times 2$ and $\times 1$, and (ii) hashtags $\#h_1, ..., \#h_m$, which are searched on $k = 10$ hashtag confidence bins with score boosting of $6 - \frac{(i+1) \times 2}{20}$ for a match on bin $1 < i < k$. To get the articles covering a certain story, we do a two-step retrieval over a time period given by the user. We first expand the query in the hashtag space using the recommended hashtags of the top-10 articles from the initial search. This forms what we call the story tracking query. The second retrieval with the expanded query returns up to 1,000 articles ranked by their relevance, which are presented to the user. Note that the method works even for cases where there are no hashtags recommended to an article.

The Web user interface. Allows saving and curating stories over time. The *Topy* page provides a search box for queries and a time period menu, from 3 days to a year in the past. Once a query is issued, the user gets a ranked list of relevant articles for that story, together with a list of relevant hashtags as shown in Fig. 3. The user can like or remove articles or hashtags from the story. The *MyStories* page shows a dashboard for tracked stories. When loading a saved story, the retrieval is triggered and the updated list of relevant articles is returned. The user can issue the same query over different time periods, each will be saved as a different story in the user dashboard. The system can be seen in action at http://ada.ucd.ie/tutorial_video/.

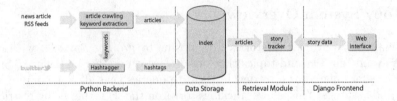

Fig. 2. System architecture.

3   Topy Use Case

We collect 27 RSS news feeds from 8 news organizations, starting from August 2015. This allows us to track stories that have started capturing the public attention almost a year ago. One such case is the complex refugee crisis that has developed in connection to war conflicts worldwide, Syria in particular. We show here related use cases where the user is interested in the story of "refugee crisis". The user issues the query "refugee crisis" with a time period of 1 year. This retrieves 19,881 ranked articles, grouped by news source. The retrieval also returns related hashtags: *#eu, #crisis, #refugee, #refugees*. For example, the article "Asylum seekers may receive funding for college" may not match any of the query terms but with Topy it is retrieved by matching the *#refugees* hashtag. The user is interested in searching for subplots of the story, hence issues a new query "#refugeeswelcome". The system retrieves 3,944 ranked articles with related hashtags: *#syria, #turkey, #refugeeswelcome, #refugeecrisis, #aylan, #syrianrefugees*. The user can observe the emphasis on the tragic death of Aylan Kurdi that triggered empathic reactions from EU citizens towards Syrian refugees. Similarly, the query "#pegida" focuses on an opposite subplot of the refugee story, that emphasizes negative reactions towards refugees, among the related hashtags *#bachmann* is discovered, who is the founder of this movement. Each hashtag can be clicked to get tweets with that hashtag, e.g., "*#German far-right #Pegida founder #Bachmann guilty of race charge* https://t.co/YFWYPlhLoP". The system can discover hashtags that may cause a

Fig. 3. System screenshot.

topical drift. These can be manually removed to direct the story towards the user's preference[5].

References

1. Conrad, J., Bender, M.: Semi-supervised events clustering in news retrieval. In: NewsIR (2016)
2. Hou, L., Li, J., Wang, Z., Tang, J., Zhang, P., Yang, R., Zheng, Q.: Newsminer: multifaceted news analysis for event search. Knowl. Based Syst. **76**, 17–29 (2015)
3. Kuzey, E., Weikum, G.: Evin: building a knowledge base of events. In: WWW (2014)
4. Leban, G., Fortuna, B., Grobelnik, M.: Using news articles for real-time cross-lingual event detection and filtering. In: NewsIR (2016)
5. Pouliquen, B., Steinberger, R., Deguernel, O.: Story tracking: linking similar news over time and across languages. In: MMIES (2008)
6. Shi, B., Ifrim, G., Hurley, N.: Learning-to-rank for real-time high-precision hashtag recommendation for streaming news. In: WWW (2016)

---

[5] This work was funded by Science Foundation Ireland (SFI) under grant number 12/RC/2289.

Ranking Researchers Through Collaboration Pattern Analysis

Mario Cataldi[1,2,3]([⊠]), Luigi Di Caro[1,2,3], and Claudio Schifanella[1,2,3]

[1] Université Paris 8, Paris, France
m.cataldi@iut.univ-paris8.fr
[2] University of Torino, Turin, Italy
dicaro@di.unito.it
[3] RAI Research Centre, Turin, Italy
claudio.schifanella@rai.it

Abstract. The academic world utterly relies on the concept of scientific collaboration. As in every collaborative network, however, the production of research articles follows hidden co-authoring principles as well as temporal dynamics which generate latent and complex collaboration patterns. In this paper, we present an online advanced tool for real-time rankings of computer scientists under these perspectives.

Keywords: Bibliometrics · Collaboration patterns · Authors ranking

1 Introduction

Scientists are object of evaluation for funding allocation and career promotions. The discovery of leading scientists is an important task that simple statistics over long publication records may miss. In this demo, we present an online tool for analyzing researchers under a *collaborative* perspective by studying and ranking their capacity to maintain the same quality/quantity levels in different research environments.

Bibliometric indicators are increasingly used to evaluate scientific careers based on personal publication records. The simple number of papers published by an author rather than the received citations are still common ways to capture both the quantity and the impact of a scientist's work. However, these measures represent only an evaluation of what is knowable from simple database searches. Still, these numbers actually make strong assumptions on the co-authorship of the research works in terms of how proportional the collaboration was among the co-authors. In a sense, scientists may look favorably good if working in a dynamic and active research environment. On the contrary, they may result unfairly below par due to modest research collaborators.

In literature, a number of related concepts have been presented, such as the *undeserved co-authorship* [4] and the *scientific relevance* [1]. Instead, our proposed application system is oriented to the study of what *collaboration* means. A research collaboration can be defined as a two-way process where individuals

© Springer International Publishing AG 2016
B. Berendt et al. (Eds.): ECML PKDD 2016, Part III, LNAI 9853, pp. 50–54, 2016.
DOI: 10.1007/978-3-319-46131-1_11

and/or organizations share learning, ideas and experiences to produce joint scientific outcomes. Collaborations are intrinsically necessary to the production of complete and groundbreaking research. In the light of this, one of the key aspect (and more demanded in recruitment scenarios) of a successful researcher is the development of a large and active network of collaborators that helps researchers bring new solutions within the research community.

The presented online tool is able to automatically compare scientists by deeply analyzing their local co-authorship networks and how they have been crucial in the production of research articles over time. Along the paper, we will present both the theoretical and algorithmic parts of the tool as well as the set of available features which can be freely tested on http://d-index.di.unito.it.

2 Background: Formalization of Scientific Collaborations

Based on the theoretical works proposed in [2,3], for this demo application, we make use of a formalization of the *co-authorship network* that represents the environment in which a researcher has produced his/her scientific outcomes. Given two collaborating researchers (also called authors), r_i, r_j and their common scientific network N_{r_i,r_j}^t, defined as the set of researchers who collaborated with them, the autonomy of their collaboration a_{r_i,r_j}^t at time t is calculated as:

$$a_{r_i,r_j}^t = \begin{cases} 0 & \text{if } N_{r_i,r_j}^t = \emptyset \\ \dfrac{1}{\sum_{r_k \in N_{r_i,r_j}^t} \left(\sum_{x=1}^{c(r_k, O_{r_i,r_j}^t)} \frac{1}{x} \right)} & \text{if } N_{r_i,r_j}^t \neq \emptyset \end{cases}$$

where the function $c(r_k, O_{r_i,r_j}^t)$ returns the number of times a researcher r_k co-authored a paper with both r_i and r_j at time t. The higher the autonomy the more independent the work of r_i and r_j is from their research environment. We then define the *dependence value* of r_i on the collaboration with r_j as $d_{r_i \rightarrow r_j}^t$ as

$$d_{r_i \rightarrow r_j}^t = \frac{p_{r_i,r_j}^t}{p_{r_i}^t} \times \frac{a_{r_i,r_j,N_{r_i}^t}^t + a_{r_j,\neg r_i,N_{r_i}^t}^t}{a_{r_i,r_j,N_{r_i}^t}^t + a_{r_j,\neg r_i,N_{r_i}^t}^t + a_{r_i,\neg r_j,N_{r_i}^t}^t},$$

where p is a productivity score (number of published works) of a is the autonomy score. The dependence value $d_{r_j \rightarrow r_i}^t$ ranges from 0 to 1; in particular, $d_{r_i \rightarrow r_j}^t \approx 0$ indicates that the dependence of r_i on r_j, at the time t, is negligible, while a $d_{r_i \rightarrow r_j}^t \approx 1$ highlights the contrary.

Thus, given the complete set of dependence values, for each year and relative to each co-author, we calculate the researcher's *dependence trajectory*, by calculating the standard deviation, along the time, of each dependence value, for each co-author, from the optimal attended value of 0 (which would mean a dependence score of 0; i.e., the production of the considered researcher is independent from the collaboration with the considered co-author). In a sense, we aim at evaluating the overall independence of a researcher from the surrounding community. More formally, given a researcher r_i, we define his/her dependence trajectory $\overrightarrow{d_{r_i}} = \{sd_{r_i}^t, sd_{r_i}^{t+1}, \cdots, sd_{r_i}^{t+n}\}$, where $sd_{r_i}^t$ is calculated as

$$sd_{r_i}^t = \sqrt{\frac{\sum_{r_k \in N_{r_i}} (d_{r_i \to r_k}^t)^2}{|N_{r_i}|}}.$$

In words, the system detects anomalies in the collaboration patterns with respect to the attended behavior. Researchers, in fact, are expected to increment their collaboration network over time becoming independent from their single collaborations.

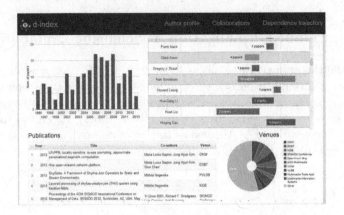

Fig. 1. Profile of a researcher in the presented application demo.

We can use these values to properly compare, and rank researchers with similar characteristics. More in detail, we provide a radar chart that can rank the independence performance of a considered researcher with respect to those who have (i) similar career length, (ii) similar number of publications, (iii) similar number of co-authors. We also provide a comparison with the active researchers and the whole community. Finally, we will integrate a feature to compare researchers with respect to topics automatically extracted from publication titles.

3   Application and Demo Scenario

In this section, we present our application, available at http://d-index.di. unito.it, for analyzing, comparing and ranking scientific collaboration patterns of researchers. As data input, we considered the DBLP data set[1], containing information about 1,717,211 authors and 3,268,812 scientific papers[2].

---

[1] http://dblp.uni-trier.de/db.
[2] Information updated at May 2016.

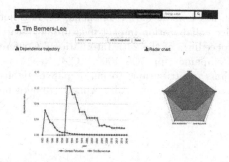

Fig. 2. Another screen-shot, taken from http://d-index.di.unito.it, that compares collaboration patters of Sir Tim Berners-Lee and Dr. Christos Faloutsos.

As shown in Fig. 1, the proposed application initially allows to search for any author indexed by DBLP and to analyze his/her scientific profile and her/his collaboration history over time (through several features and visualizations). Then, the online demo provides the following analyses:

– **Collaborations over time.** The user can analyze the evolution over time of each scientific collaboration for a searched researcher.
– **Collaboration Pattern Analysis.** The system can visualize the above-mentioned dependence patterns through a curve metaphor, mapping the evolution of the dependence of a researcher on the support of each co-author along the career. With this chart, it is also possible to select/deselect additional co-authors to make further analyses and comparisons.
– **Temporal Analysis.** The demo provides a dynamic visualization chart ("time-lapse") which allows the user to focus on a specific time interval and/or a subset of co-authors.
– **Ranking.** This tool also allows to compare and rank the overall independence of an author, along his/her whole career, with the whole research community (Fig. 2). This visualization permits to focus on how much the entire production of a researcher can be considered dependent on the interactions with her/his local community.

The presented demo can be used to analyze each researcher in the entire DBLP community by also considering similar profiles (with parameters such as number of papers, number of co-authors, and length of career).

References

1. Ausloos, M.: A scientometrics law about co-authors and their ranking: the co-author core. Scientometrics **95**(3), 895–909 (2013)
2. Caro, L.D., Cataldi, M., Schifanella, C.: The d-index: discovering dependences among scientific collaborators from their bibliographic data records. Scientometrics **93**(3), 583–607 (2012)

3. Di Caro, L., Cataldi, M., Lamolle, M., Schifanella, C.: It is not what but who you know: a time-sensitive collaboration impact measure of researchers in surrounding communities. In: Proceedings of the 24th International Conference on World Wide Web, WWW 2015 Companion, pp. 995–1000. ACM, New York (2015)
4. Slone, R.M.: Coauthors' contributions to major papers published in the ajr: frequency of undeserved coauthorship. AJR Am. J. Roentgenol. **167**(3), 571–579 (1996)

Learning Language Models from Images with ReGLL

Leonor Becerra-Bonache[1], Hendrik Blockeel[2], Maria Galván[1],
and François Jacquenet[1(✉)]

[1] Université de Lyon, UJM-Saint-Etienne, CNRS, Saint-Etienne, France
`Francois.Jacquenet@univ-st-etienne.fr`
[2] Department of Computer Science, KU Leuven, Leuven, Belgium

Abstract. In this demonstration, we present ReGLL, a system that is able to learn language models taking into account the perceptual context in which the sentences of the model are produced. Thus, ReGLL learns from pairs (Context, Sentence) where: Context is given in the form of an image whose objects have been identified, and Sentence gives a (partial) description of the image. ReGLL uses Inductive Logic Programming Techniques and learns some mappings between n-grams and first order representations of their meanings. The demonstration shows some applications of the language models learned, such as generating relevant sentences describing new images given by the user and translating some sentences from one language to another without the need of any parallel corpus.

1 Introduction

Learning language models has been a very active domain of research for a long time and Grammatical Inference, a subdomain of Machine Learning dedicated to that task, has produced a huge number of results in the literature. That has already led to the implementation of tools used in various applications (see [3] for an overview of the domain).

This research has mainly focused on learning language models from a syntactic point of view considering training sets only made up of strings of characters or sequences of words. Nevertheless, it seems obvious that human beings do not learn languages in that way. If we look at very young children starting to learn their native language, we can note that they are exposed to many sentences that refer to things in a perceptible scene. Thus, some work have been done to integrate some semantic information in the language learning process. The work from Chen et al. [2] is one example of this way of learning language models. Nevertheless, in this approach the meaning of each sentence has to be provided for each example of the training set. The ReGLL (Relational Grounded Language Learning) prototype proposes a different approach in which the meaning

H. Blockeel—Work supported by KU Leuven (sabbatical grant) and FWO-Vlaanderen (WOG "Declarative Methods in Informatics").

B. Berendt et al. (Eds.): ECML PKDD 2016, Part III, LNAI 9853, pp. 55–58, 2016.
DOI: 10.1007/978-3-319-46131-1_12

of each sentence is automatically discovered by the system, thanks to the context associated with it.

Some work has focused on learning to caption images using some deep learning approaches. The work from Karpathy et al. [4] is one example of such an approach. The main idea behind these approaches is to learn a function that ranks sequences of words given some images. It is different from the ReGLL prototype that is able to build a general semantic representation of the meaning of (part of) sentences. Doing in that way makes it possible later to use this representation to reason about the universe that has been described by the set of images and sentences of the training set.

2   The ReGLL prototype

Due to space limitations we cannot detail the theoretical and algorithmic aspects behind ReGLL, for that purpose, the reader may refer to [1].

The input of the system is a dataset D1 made up of pairs (I,S) where I is an image that has been built using a scene builder and S is a sentence that describes (part of) the image. The scene builder provides a set of cliparts and the user can drag and drop cliparts to design a new scene. A preprocessing step can then transform D1 in a dataset D2 made up of Prolog facts that provide pairs (C,L) where C is a set of grounded atoms that contains all the information about the objects of the image I and L is the list of words of the sentence S. During the learning step, the system takes the dataset D2 as an input and generates a language model. In the demo we provide three families of datasets where the sentences are written in English and Spanish. Thus the language models learned are subsets of language models for English and Spanish.

ReGLL is based on Inductive Logic Programming (ILP) techniques where each learning step uses the *least general generalization* (lgg) operator [5]. The basic idea behind the ReGLL engine is to traverse the training set and, given an n-gram NG ($1 \leq n \leq 8$) that appears in a sentence, generate a most specific generalization of all the contexts of NG in the training set. The process is iterated for each n-gram of the training set. During the learning process, the system learns the meaning of 1-grams (words) and then learns the meaning of n-grams ($n \geq 2$).

From an operational point of view, ReGLL is run through an interface that allows the user to act in various ways. One may: (i) load or design some training sets, (ii) learn some language models, (iii) load a language model to: (a) visualize the meaning of words, (b) generate relevant sentences given an image, (c) translate some sentences from a language L1 to a language L2 given the language models of L1 and L2.

3   Overview of the Demonstration

The demonstration is mainly based on the Abstract Scene Dataset built by Zitnick et al. [6] with a scene builder, nevertheless the attendees will be allowed to build some new datasets if they want to explore this functionality. It shows:

1. The way we can build a dataset from scratch using an abstract scene builder and then learn a language model from this dataset.
2. What can be done using a language model that has been learned.

Figure 1 shows two screenshots of the ReGLL system generating sentences describing images and showing the meaning of words it has discovered from a given dataset.

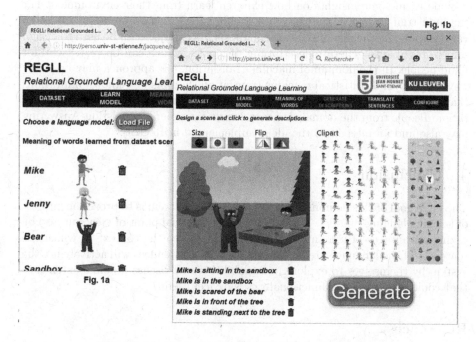

Fig. 1. Screenshots of the ReGLL interface.

Three functionalities are thus mainly demonstrated:

- *Visualizing the meaning of words.* We can visualize what are the meanings that have been learned by ReGLL from a training set. Each word that has an associated meaning is displayed. The user may choose to delete some incorrect associations in order to help the system to be more efficient on the two other functionalities (see Fig. 1a).
- *Describing images.* Given an image that is built by the user using an image builder, the system may generate all the relevant sentences (ordered by decreasing relevance) that describe this image (see Fig. 1b).
- *Translating sentences.* We show that the language models learned by the system can be used to translate sentences written in a language L1 to sentences written in a language L2, while preserving the meaning. The languages available at the moment for this demonstration are English and Spanish.

Of course, the demonstration will allow attendees to look "inside the machine". Indeed, it may be interesting for people familiar with Prolog to observe the code associated with the training sets, the main components of the learner and the language models.

We think this demonstration may be useful for people from both the academic and the industrial world working in the domain of natural language processing. For the academic audience, it may be interesting for people specialized in grammatical inference and people from the computation linguistic area. That can provide them some insights on how children learn from their environment. For the industrial audience, the core ideas behind ReGLL may be useful to design various tools. As the demonstration shows, such techniques can be used to design some tools able to generate descriptions of images, which can be very useful for blind people. In the domain of machine translation, our approach may be a new way to go beyond statistical machine translation that has well-known limitations that could be avoided by passing through a relational, more semantic representation. People from the domains of text summarization and Question-Answering may also find an interest in the ideas implemented in ReGLL.

4   Conclusion

The main goal of this demonstration of the ReGLL system is to prove the interest of learning language models not only from a syntactical point of view but also by taking advantages of semantic information related to the context in which the sentences of the language are produced. We expect attendees will actively use the system by themselves to explore its capabilities and discuss possible extensions that could integrate new functionalities they feel useful.

References

1. Becerra-Bonache, L., Blockeel, H., Galván, M., Jacquenet, F.: A first-order-logic based model for grounded language learning. In: Fromont, E., et al. (eds.) IDA 2015. LNCS, vol. 9385, pp. 49–60. Springer, Heidelberg (2015). doi:10.1007/978-3-319-24465-5_5
2. Chen, D.L., Kim, J., Mooney, R.J.: Training a multilingual sportscaster: using perceptual context to learn language. J. Art. Int. Res. **37**, 397–435 (2010)
3. de la Higuera, C.: Grammatical Inference, Learning Automata and Grammars. Cambridge University Press (2010)
4. Karpathy, A., Joulin, A., Li, F.: Deep fragment embeddings for bidirectional image sentence mapping. In: Proceedings of the 28th NIPS Conference, pp. 1889–1897 (2014)
5. Plotkin, G.D.: A note on inductive generalization. In: Machine Intelligence 5, pp. 153–163. Edinburgh University Press (1970)
6. Zitnick, C.L., Vedantam, R., Parikh, D.: Adopting abstract images for semantic scene understanding. IEEE TPAMI **38**(4), 627–638 (2016)

Exploratory Analysis of Text Collections Through Visualization and Hybrid Biclustering

Nicolas Médoc[1,2(✉)], Mohammad Ghoniem[2], and Mohamed Nadif[1]

[1] LIPADE, University of Paris Descartes,
45, rue des saints pères, 75006 Paris, France
mohamed.nadif@mi.parisdescartes.fr
[2] ERIN-eScience, Luxembourg Institute of Science and Technology,
41, rue du Brill, 4422 Belvaux, Luxembourg
{nicolas.medoc,mohammad.ghoniem}@list.lu

Abstract. We propose a visual analytics tool to support analytic journalists in the exploration of large text corpora. Our tool combines graph modularity-based diagonal biclustering to extract high-level topics with overlapping bi-clustering to elicit fine-grained topic variants. A hybrid topic treemap visualization gives the analyst an overview of all topics. Coordinated sunburst and heatmap visualizations let the analyst inspect and compare topic variants and access document content on demand.

1 Introduction

We present a visual analytics tool designed to help analytic journalists explore large text corpora. Analytic journalists typically start by getting an overview of the field under investigation, then focus on specific aspects to identify facts and viewpoints that verify, refine or refute their hypothesis. Text corpora are often modeled by *Term × Document* matrices, from which topics may be extracted using graph modularity-based diagonal biclustering [1]. Word cloud views are popular representations of individual topics and have been extended in many ways. In the considered use case, the journalist needs to grasp dozens of topics at a glance and appreciate topic importance. A good visualization may further ease this task by displaying topic relationships. Once the journalist has identified a topic of interest, his concern shifts to understanding topic variants and identifying distinctive documents and terms for each. The visualization of overlapping biclusters has been approached in various ways e.g. transparent overlapping hulls in node-link diagrams, matrix visualizations and parallel coordinates by Santamaría et al. [5]. BiSet [6] represents chained bipartite graphs enhanced with semantic bundles to represent chained bicluster relationships. These representations fail to convey an overview of a large number of overlapping biclusters while identifying common and distinctive terms and documents.

2 Tool Overview

To support the topic mapping task, we apply diagonal biclustering based on graph modularity [1] on the *Term × Document* matrix. The *Weighted Topic Map*

© Springer International Publishing AG 2016
B. Berendt et al. (Eds.): ECML PKDD 2016, Part III, LNAI 9853, pp. 59–62, 2016.
DOI: 10.1007/978-3-319-46131-1_13

visualization in Fig. 1 is a hybrid Treemap view where rectangular tiles represent individual topics, tile area encodes topic importance, while topic details are shown as a nested word cloud. Term size and color reflect its representativeness of the topic and the number of documents where it appears. An MDS projection computed from the similarity matrix of the diagonal biclusters generates 2D positions which are fed to the Weighted Map visualization algorithm [2]. This results in similar topics being placed in adjacent tiles. Jaccard similarity is used to display links to the five most similar topics when the analyst hovers over a topic, as shown in Fig. 1. Showing topic relationships aims to alleviate the hard partitioning due to the diagonal biclustering. This overview enables the analyst to discover the main topics and select one for further scrutiny.

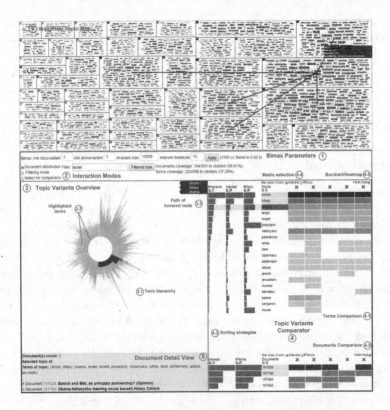

Fig. 1. The US presidential election topic is selected from 3,992 online news articles collected between Nov. 2nd and Nov. 16th, 2015. Five topic variants concerning Hillary Clinton have been sent for comparison (https://youtu.be/xY6mgZyg3jA).

When the analyst selects a topic, `Bimax` [4], a pattern-based overlapping biclustering, extracts the topic variants by identifying all maximal combinations of terms shared by a maximal set of documents. While the exhaustiveness of `Bimax` may serve the needs of the analyst, it produces a very large number of biclusters. To make sense of the numerous `Bimax` biclusters, we hierarchize them based on term overlaps using the `FPTree` algorithm [3]. The resulting term hierarchy is represented as a sunburst visualization (3.1 in Fig. 1). The most common terms have a higher overlap degree and appear closer to the root, while the most distinctive terms are placed further away. Each path, from root to leaf, represents a unique association of terms grouped by one bicluster. As we move away from the root along a given path, the word combination becomes more specific and retains fewer documents. At the leaf level, only the documents of one bicluster are retained. By exploring this view and the coordinated comparator view (4), the journalist can focus on a specific aspect of a topic and depict all document relationships to identify facts or viewpoints related to his hypotheses.

The text of the documents can be read in the Document Detail View. In addition, we provide multiple interaction modes illustrated in Fig. 2. Hovering over a term in the hierarchy colors all its occurrences in red (3.3 in Fig. 1) and shows the corresponding term sequence on the right (3.2). The comparator view allows to analyze the common and distinctive terms as well as the distribution of documents across the selected topic variants. Multiple sorting strategies are proposed to facilitate the identification of the most informative terms.

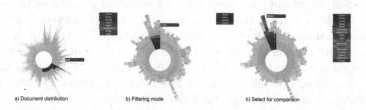

a) Document distribution   b) Filtering mode   b) Select for comparison

Fig. 2. Interaction Modes. (a) The orange biclusters contain any document selected by the clicked node "Israel". (b) The biclusters not matching the term "Israel" are filtered. (c) The bicluster colored in blue are sent to the topic variant comparator. (Color figure online)

3   Parameter Setting by the User

The number of `Bimax` biclusters increases with the size or the density of the diagonal bicluster blocks up to more than ten thousand biclusters. To reduce this number, we allow the user to modify the parameters of `Bimax`: the minimum number of terms or documents per bicluster ($MinT$, $MinD$) and the maximum number of biclusters ($MaxB$). As `Bimax` uses binary matrices, we also enable the user to change the binarization threshold (Thr) applied on the TF-IDF weights. Increasing the threshold selects, for each document, the most representative terms and reduces the density and the dimensions of the matrix.

In Fig. 3, we visualize the effect of varying each parameter separately on the term hierarchy built from the U.S. presidential elections topic. After each

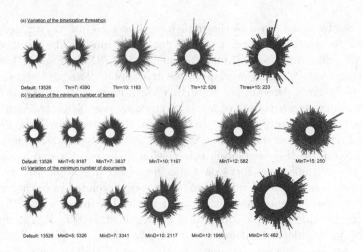

Fig. 3. Number of biclusters as the parameters of `Bimax` vary.

parameter variation, the root node "Obama" is clicked to highlight in orange the distribution of the selected documents. With the default parameters ($MinT = 3$, $MinD = 4$, $Thr = 5$), only the first levels of the 13,000 biclusters are visible in the sunburst visualization. Increasing both Thr and $MinT$ reduces the dispersion of the documents concerning "Obama", but the changes of Thr maintain the variety regarding the number of terms. As $MinD$ increases, the number of terms tends to be reduced but the documents selected by the node "Obama" remain largely dispersed in the biclusters until the node disappears.

References

1. Ailem, M., Role, F., Nadif, M.: Co-clustering document-term matrices by direct maximization of graph modularity. In: Proceedings of the 24th ACM International on CIKM, CIKM 2015, pp. 1807–1810. ACM, NY (2015)
2. Ghoniem, M., Cornil, M., Broeksema, B., Stefas, M., Otjacques, B.: Weighted maps: treemap visualization of geolocated quantitative data. In: IS&T/SPIE Electronic Imaging, p. 93970G–93970G. Int. Soc. for Optics and Photonics (2015)
3. Han, J., Pei, J., Yin, Y.: Mining frequent patterns without candidate generation. In: Proceedings of the 2000 ACM SIGMOD International Conference on Management of Data, SIGMOD 2000, pp. 1–12. ACM, NY (2000)
4. Prelić, A., Bleuler, S., Zimmermann, P., Wille, A., Bhlmann, P., Gruissem, W., Hennig, L., Thiele, L., Zitzler, E.: A systematic comparison and evaluation of biclustering methods for gene expression data. Bioinformatics **22**(9), 1122–1129 (2006)
5. Santamaría, R., Therón, R., Quintales, L.: A visual analytics approach for understanding biclustering results from microarray data. BMC Bioinform. **9**(1), 247 (2008)
6. Sun, M., Mi, P., North, C., Ramakrishnan, N., BiSet: semantic edge bundling with biclusters for sensemaking. IEEE TVCG **PP**(99), 1 (2015)

SITS-P2miner: Pattern-Based Mining of Satellite Image Time Series

Tuan Nguyen[1(⊠)], Nicolas Méger[1], Christophe Rigotti[2], Catherine Pothier[3], and Rémi Andreoli[4]

[1] Université Savoie Mont Blanc, Polytech Annecy-Chambéry, LISTIC,
74944 Annecy-le-vieux, France
{hoang-viet-tuan.nguyen,nicolas.meger}@univ-smb.fr
[2] Univ Lyon, INSA-Lyon, CNRS, Inria, LIRIS, UMR5205,
69621 Villeurbanne, France
[3] Univ Lyon, INSA-Lyon, SMS-ID, 69621 Villeurbanne, France
[4] Bluecham S.A.S., 98807 Nouméa, Nouvelle-Calédonie, France

Abstract. This paper presents a mining system for extracting patterns from Satellite Image Time Series. This system is a fully-fledged tool comprising four main modules for pre-processing, pattern extraction, pattern ranking and pattern visualization. It is based on the extraction of grouped frequent sequential patterns and on swap randomization.

1 Introduction

A Satellite Image Time Series (SITS) is a series of images covering a same area acquired by satellites over time. SITS analysis is a still growing research field, stimulated by the enhancement of the spatial resolution, the reduction of the time intervals between acquisitions and the development of new acquisition modes. Considering the large volume and the raw nature of such SITS, it is not possible to process them manually, and unsupervised mining techniques demonstrate their potential to describe and discover spatiotemporal phenomena in SITS. These techniques rely either on global models such as clustering (e.g. [2]) or on local patterns such as sequential patterns (e.g. [3]).

This paper presents *SITS-P2miner* (Pattern maP miner), a system that implements the pattern mining method introduced in [4] and the swap randomization ranking presented in [5], together with appropriated pre-processing and visualization tools. The salient features of the resulting system, with respect to other state-of-the-art methods, are: (1) its ability to process both optical and radar satellite images; and (2) its robustness against frequent quality degradation sources inherent to satellite images (atmospheric perturbations, missing values, sensor defects, irregular time spacing).

In a SITS, the covered area is represented as a grid of pixels, and, for each pixel, the SITS contains the sequence of values (integers or floating point numbers) acquired over time for that location. In *SITS-P2miner*, the input SITS is quantized in a pre-processing step to replace pixel values by symbols denoting discrete levels $(1, 2, 3, \ldots)$. This symbolic SITS is then mined to extract GFS-patterns [4], where a GFS-pattern is a sequential pattern [1] satisfying the two

B. Berendt et al. (Eds.): ECML PKDD 2016, Part III, LNAI 9853, pp. 63–66, 2016.
DOI: 10.1007/978-3-319-46131-1_14

following constraints. First, it must occur in a sufficient number of sequences (being frequent, in the usual sense). Secondly, the occurrences of the pattern have to be somehow coherent over space, i.e., if the pattern occurs in the sequence of values of a pixel, then it must also tend to occur in the spatial neighborhood of this pixel (but eventually with a shift in time).

2   System Description

The architecture of the system is presented in Fig. 1. As an input, it takes a SITS expressed as a single synthetic band of interest such as the ground motion magnitude in the line of sight of a radar satellite, or a vegetation index when dealing with optical images. First, the SITS is quantized by the *pre-processing* module, using one of the different available discretization strategies, to produce a *Symbolic SITS*. This symbolic SITS is in turn processed by the *pattern extraction* module to extract *maximal GFS-patterns*. These patterns are then assessed by swap randomization of the symbolic SITS and ranked using a normalized mutual information measure reflecting the impact of the randomization upon the pattern occurrences. Finally, different *maps* depicting the location in space and time of the occurrences of the top-ranked patterns are computed by the *visualization* module. The reader is referred to [4,5] for the complete definition of the patterns, the description of the extraction/ranking steps and the guidelines for parameter settings.

The whole process is driven by a single human readable parameter file. The output is stored in folders whose hierarchy is structured according to the processing steps, the parameter values and the execution time stamps. This output includes maps, patterns, intermediate ranking information as well as monitoring logfiles that are organized for quick result browsing and easy iterative mining.

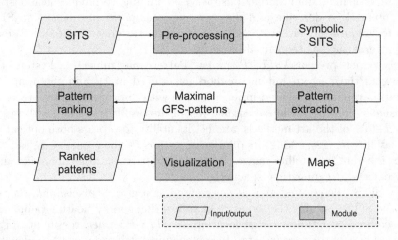

Fig. 1. System architecture.

The most resource consuming steps are the GFS-pattern extraction and the swap randomization. Therefore, the corresponding modules are implemented in C. The other ones are implemented in Python. All modules are chained in Python, which allows to add new modules simply. The system can be run on *Windows, Linux* and *Mac OS X* operating systems using a standard computing platform (e.g., single core on 2.7 GHz Intel Core i7, 8 GB memory).

3   Demonstration

During the demonstration, we will present the analysis of two real SITS. The first one (provided by Marie-Pierre Doin, ISTerre lab., CNRS), is an ENVISAT-based SITS covering Mount Etna (16 radar images 598×553 from 2003 to 2010). In this series, the effects of the stratified atmosphere have been corrected, but not the ones due to the turbulent atmosphere. The pixel values give ground motion magnitudes in the satellite line of sight. The other series is a Landsat 7 SITS (16 optical images 513×513 from 2004 to 2011) covering the area of Yaté in New Caledonia and containing values expressing the presence/absence of vegetation (NDVI index). The limited size of these series allows for live computation of the maps during the demonstration on a standard laptop. Four images of the second series are shown in Fig. 2, illustrating typical problems of satellite data such as missing values, artifacts, sensor defects, presence of clouds, etc.

Two examples of maps of occurrences of GFS-patterns, selected among the best ranked maps found in these series, are shown in Fig. 3. The colored pixels denote the locations of the occurrences in space while the colors correspond to the ending dates of the occurrences (middle of the SITS in blue and end of the SITS in pink). The map of Fig. 3a corresponds to GFS-pattern 1-2-2-2-2-2-3-3 over the Mount Etna motion series. It sketches a trend from low magnitude motion (symbol 1) to high magnitude motion (symbol 3). The upper part of the map exhibits a moving part of the volcano flank and the lower part unveils a fault system. Figure 3b shows the map of GFS-pattern 2-2-3-2-2-2-3 over the New Caledonia vegetation series. It denotes a cycling variation from normal vegetation index (symbol 2) to high vegetation index (symbol 3) and corresponds to the presence of maquis (evergreen vegetation). The best ranked maps over the New Caledonia vegetation series highlight various phenomena related not only to the vegetation

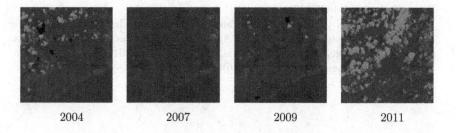

| 2004 | 2007 | 2009 | 2011 |

Fig. 2. Landsat 7 images (RGB color space). (Color figure online)

(a) Mount Etna   (b) New Caledonia

Fig. 3. Examples of top-ranked maps of GFS-patterns. (Color figure online)

but also to anthropic activities, and have been integrated as data layers in the *Qëhnelö environmental management platform* (http://www.yate.nc/).

The system used in this demonstration is available at: https://www.polytech. univ-savoie.fr/fileadmin/polytech_autres_sites/sites/listic/projets/sitsmining/ SITSP2MINER.zip

Acknowledgments. Funding for this project was provided by a grant from la Région Rhône-Alpes (Tuan Nguyen's grant). Other support: C. Rigotti and C. Pothier are members of LabEx IMU (ANR-10-LABX-0088).

References

1. Agrawal, R., Srikant, R.: Mining sequential patterns. In: Proceedings of Conference ICDE, pp. 3–14 (1995)
2. Gueguen, L., Datcu, M.: Image time-series data mining based on the information-bottleneck principle. IEEE Trans. Geosci. Remote Sens. **45**(4), 827–838 (2007)
3. Honda, R., Konishi, O.: Temporal rule discovery for time-series satellite images and integration with RDB. In: Siebes, A., De Raedt, L. (eds.) PKDD 2001. LNCS (LNAI), vol. 2168, pp. 204–215. Springer, Heidelberg (2001)
4. Julea, A., Méger, N., Bolon, P., Rigotti, C., Doin, M.P., Lasserre, C., Trouvé, E., Lăzărescu, V.N.: Unsupervised spatiotemporal mining of satellite image time series using grouped frequent sequential patterns. IEEE Trans. Geosci. Remote Sens. **49**(4), 1417–1430 (2011)
5. Méger, N., Rigotti, C., Pothier, C.: Swap randomization of bases of sequences for mining satellite image times series. In: Appice, A., Rodrigues, P.P., Santos Costa, V., Gama, J., Jorge, A., Soares, C. (eds.) ECML PKDD 2015. LNCS, vol. 9285, pp. 190–205. Springer, Heidelberg (2015)

Finding Incident-Related Social Media Messages for Emergency Awareness

Alexander Nieuwenhuijse[1(✉)], Jorn Bakker[1], and Mykola Pechenizkiy[2]

[1] Coosto, Fellenoord 35, 5612 Eindhoven, The Netherlands
{alexander.nieuwenhuijse,jorn.bakker}@coosto.com
[2] Eindhoven University of Technology,
PO. Box 513, 5600 Eindhoven, The Netherlands
m.pechenizkiy@tue.nl

Abstract. An information retrieval framework is proposed which searches for incident-related social media messages in an automated fashion. Using P2000 messages as an input for this framework and by extracting location information from text, using simple natural language processing techniques, a search for incident-related messages is conducted. A machine learned ranker is trained to create an ordering of the retrieved messages, based on their relevance. This provides an easy accessible interface for emergency response managers to aid them in their decision making process.

Keyword: Incident related social media monitoring

1 Introduction

With the ever growing social media networks, and the information shared on those networks, the opportunity for automated data analysis is increased with it. A large part of this data is generated by private users, posting about their everyday life and their surroundings. By combining the data generated by the users of these social media platforms new insights can be gathered concerning the geographic hotspots about which people talk. This information can help in the decision making process of emergency response units and their managers, who need to make decisions about sending aid to incident locations.

In the work of MacEachern et al. [3] a survey was conducted among emergency managers from International Association of Emergency Managers (IAEM) and FirstResponder.gov about their current and envisioned use of geovisual tools that support social media analysis. This survey shows that only 39.1 % of the participants use social media to gather information from the public. The most important feature (94.7 %) requested by the emergency managers is the inclusion of maps showing geographical information of an incident. The second most requested feature (71.1 %) is the option to search through photo and video collections relating to the incident, since these graphics allow a domain expert to assess a situation.

© Springer International Publishing AG 2016
B. Berendt et al. (Eds.): ECML PKDD 2016, Part III, LNAI 9853, pp. 67–70, 2016.
DOI: 10.1007/978-3-319-46131-1_15

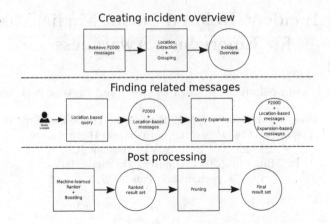

Fig. 1. Overview of the developed framework

These survey results show a relatively low use of social media data to gather information, due to the large quantity of social media messages and the time consuming process to analyse this data. As such, there is a demand for a more structured and more accessible representation of social media data. The framework described below is integrated within the Coosto solution and enables emergency response units to easily access these social media messages to aid them in their decision making process.

2   Framework

Figure 1 shows an overview of the developed framework. Only one input is required from the user: a selection of an incident for which to retrieve related social media messages. The other components of the framework operate fully automated.

An overview of active incidents is created from a set of P2000 messages. These messages are created by emergency response managers and are publicly broadcasted over the P2000 paging network to alert and dispatch emergency response units towards a reported incident. Due to the nature of these messages they are always highly reliable and contain exact location information of the reported incidents, thus creating a solid basis from which to start searching for other, non-P2000, messages.

Using the extracted location an initial search is conducted to retrieve social media messages mentioning the incident location. Even if these messages are not directly related to the incident they can be used to a create context, and also to get in contact with people at, or traveling towards, the incident location. To retrieve harder to find messages, not explicitly mentioning the incident location, a query expansion step is applied. This query expansion extracts relevant terms from the intermediate result set and uses these terms to construct a more

Fig. 2. Media objects related to the Chemelot fire. They inform about the location and scale of the incident, wind direction, shape and color of the smoke plume. (Color figure online)

complex query. This query expansion is tuned towards a high recall, to ensure that we retrieve as many relevant messages as possible.

Finally, to reduce the noise in the final result set, a machine learned ranking is applied to the results. To accomplish this we considered multiple ranking algorithms. Train and test sets were created by manually labeling messages in the result sets as either being *very relevant, somewhat relevant* or *irrelevant* and the performance of the rankers was measured using the Normalized Discounted Cumulative Gain [2]. From these experiments the RankBoost [1] algorithm yielded the best results and was adopted into the framework. The rank assigned by the machine learned ranker is also boosted, based on the occurence of terms retrieved from the query expansion step. This boosting is applied to compensate for the out-of-vocabulary words in the ranking model and to cover incident types that were not present in the training set of the machine learned ranker.

Combining all these steps yields the result set of social media messages, automatically retrieved and ranked, ready to be presented to the user.

3 Usecase - Chemelot

In the morning of November 9th 2015 a large fire developed at a chemical complex called Chemelot, located near Geleen, the Netherlands. Using the developed framework the media objects shown in Fig. 2, related to this fire, were obtained. As the idiom states: "A picture is worth a thousand words", these photos provide a lot of information for an emergency response manager, located in a response

Fig. 3. Concerned tweet asking about ammonia hazard.

center far away from the incident location. A fireman can use these photos to estimate how the fire is developing based on the smoke trail, color and direction. These initial pictures can help in the decision making process of what units to dispatch to the incident location.

As shown in Fig. 3, people are asking questions about toxicity of the fumes. These messages can be important for press officers and the related information management. Given that the fire was located at a chemical complex, and by analysing the color of the smoke, an alert can be given to the people living in the areas effected by the smoke, with the advice to stay indoors.

The overview created by the framework can easily give insights for the emergency response units and support in their decision making process. It allows for easy interaction with concerned people and informing the public, without it being a time consuming process.

4 Conclusion

The developed framework provides easy access to incident-related social media messages and helps providing emergency awareness. This application can be used by anyone. It is no longer a time consuming process to construct complex queries, and it does not require domain specific knowledge to operate. It enables emergency response managers in their decision making process, making social media more accessible.

References

1. Freund, Y., Iyer, R., Schapire, R.E., Singer, Y.: An efficient boosting algorithm for combining preferences. J. Mach. Learn. Res. **4**, 933–969 (2003)
2. Järvelin, K., Kekäläinen, J.: Cumulated gain-based evaluation of ir techniques. ACM Trans. Inf. Syst. **20**(4), 422–446 (2002)
3. MacEachren, A.M., Jaiswal, A., Robinson, A.C., Pezanowski, S., Savelyev, A., Mitra, P., Zhang, X., Blanford, J.: Senseplace2: geotwitter analytics support for situational awareness. In: 2011 IEEE Conference on Visual Analytics Science and Technology (VAST), pp. 181–190. IEEE (2011)

TwitterCracy: Exploratory Monitoring of Twitter Streams for the 2016 U.S. Presidential Election Cycle

M. Atif Qureshi[(✉)], Arjumand Younus, and Derek Greene

Insight Center for Data Analytics, University College Dublin, Dublin, Ireland
{muhammad.qureshi,arjumand.younus,derek.greene}@ucd.ie

Abstract. We present *TwitterCracy*, an exploratory search system that allows users to search and monitor across the Twitter streams of political entities. Its exploratory capabilities stem from the application of light-weight time-series based clustering together with biased PageRank to extract facets from tweets and presenting them in a manner that facilitates exploration.

1 Introduction

Twitter has established itself as an important medium for online political discourse, as evidenced during events such as the Arab Spring, Barack Obama's 2012 presidential campaign, and India's General Elections in 2014. This has subsequently led to the increased usage of the platform by politicians as a part of their campaign activities [4,6]. Following this trend, Fortune Magazine has termed the 2016 U.S Presidential Election as the *"social media election"* [1]. The research community has experienced a surge of interest in the analysis of political chatter over Twitter [5]. Much of the current focus lies in the prediction of election outcomes, with relatively few state-of-the-art studies [3,8] conducted on the analysis of political discussion by general users [5]. Despite the attention given to election predictions in the literature [9], such methods fail to empower the general public in the spirit of "democracy" and "voter empowerment".

The rising prominence of social media as a platform for political discourse has fundamentally altered the way in which candidates conduct election campaign [6]. It is therefore necessary for voters, analysts, and journalists to keep a close eye on the online activity of politicians. We believe such monitoring can help to increase political awareness among the general public, thereby enabling them to make informed choices in electing their representatives. This, in turn, dictates a clear need for analytical tools that can delve into the communication behaviors of politicians on social media.

Towards this end, we have created *TwitterCracy*, a system which aims to facilitate voters and analysts, by keeping them aware of the key agenda issues that are of interest to politicians, as reflected by their ongoing activity on Twitter. The core functionality of the system enables the exploration of various facets of these issues, via the extraction of keywords from politicians' tweets. Our

B. Berendt et al. (Eds.): ECML PKDD 2016, Part III, LNAI 9853, pp. 71–75, 2016.
DOI: 10.1007/978-3-319-46131-1_16

technique for exploratory analysis is based on the application of biased PageRank [2] to a graph of terms, mentions, and hashtags appearing in tweets. In line with the *TweetMotif* tool [7], our system allows a user to navigate via the extracted keywords and drill down into the data in more depth. However, unlike *TweetMotif*, which only operates on a static corpus, *TwitterCracy* indexes a live stream of tweets and extracts query-specific facets in real-time, while incorporating a light-weight time-series clustering mechanism for the efficient application of the PageRank model. Another novel aspect of *TwitterCracy* is the incorporation of valuable metrics based on theoretical constructs within relational sociology [3] to provide deeper insights into the communication patterns of politicians. To illustrate the use of *TwitterCracy*, we consider the 2016 U.S. Presidential Election as a case study, analyzing the activity of 635 relevant politicians and political organizations on Twitter during the campaign. click

2   TwitterCracy Architecture

In this section, we present an overview of the architecture of *TwitterCracy*, as illustrated in Fig. 1. The user, who is central to the system, issues a "query"[1], which is processed by the query module to produce a ranked list of relevant tweets. This ranked list then passes through various components of our processing pipeline: (1) clustering and compression module, (2) facet extraction module, (3) social extraction module and finally, (4) rendering module. We now explain the first three modules in the following sub-sections, as these represent the key system components, while the rendering module simply produces the HTML output. Separately, the crawler module is responsible for back-end data acquisition, continuously collecting from the live stream of politicians' tweets and matching them with the user metadata. This data stream is immediately indexed to provide the user with real-time updates.

2.1   Key Components

Clustering and compression module: This module is responsible for reducing the large, dense graph of terms, mentions, and hashtags into a relatively small, sparse graph for efficient computation of PageRank. First, we apply cost-effective, time-series based clustering to the ranked list of tweets. Based on the assumption that bursts of tweets are likely to indicate significant events [10], we apply k-means clustering over the timestamps of the retrieved tweets to cluster bursts of tweets together. From these clusters, we then pick the top retrieved tweets, in proportion to the size of each cluster. This reduces the full stream to a representative sub-sample of tweets prior to the application of PageRank in the next stage of the processing pipeline.

---

[1] Note a query can be a phrase entered by the user or the live stream depicting last 15 min.

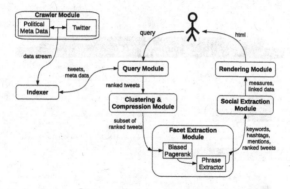

Fig. 1. Architecture of the *TwitterCracy* system.

Facet extraction module: This module extracts various facets[2] from the retrieved tweets by applying biased PageRank. In the graph, the nodes are terms extracted from retrieved tweets, and edges connect pairs of terms that occur together in a tweet. The weight on an edge is the relevance score of the tweet relative to the original query. The biasing of PageRank vector is explained as follows:

- The terms in retrieved tweets are biased in proportion to the amount of their significance calculated by chi-square test of independence.
- The named entities in retrieved tweets are biased in proportion to their correlation with an event where the correlation is calculated by means of their document frequencies in retrieved tweets.

Finally, we merge single terms identified by biased PageRank to extract longer keywords as facets[3]. To achieve this, we add the individual PageRank scores of the co-occurring terms according to their probability of co-occurrence. This means that sets of terms with high PageRank scores and that co-occur frequently are extracted as facets, and appear in the exploratory search interface (see Fig. 2)

Social extraction module: This module applies theoretical measures from relational sociology to quantify various aspects of online conversational practices of politicians. More specifically, we make use of three measures introduced by Lietz et al. [3]: cultural similarity, cultural focus, and cultural reproduction. The level of similarity between the stances of political parties (e.g. Democrats and Republicans) in relation to various issues is measured by means of cultural similarity. The stability of a political party's ideology can be quantified by both cultural focus and cultural reproduction.

3 Case Study: 2016 U.S. Presidential Election

To illustrate the use of *TwitterCracy*, we consider the 2016 U.S. Presidential Election as a case study, analyzing the activity of 635 relevant politicians and

---

[2] Facets here are keywords, mentions and hashtags.
[3] Note that we restrict this extraction to bigrams as tweets are short.

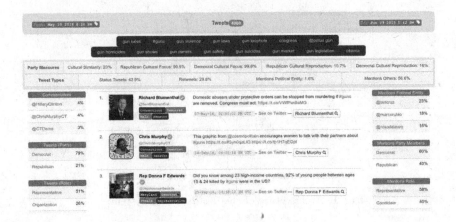

Fig. 2. *TwitterCracy* user interface showing results for a sample query "guns". Identified facets include "gun violence" and "gun legislation", which can be explored in more detail.

political organizations on Twitter during the campaign. The dataset contains 1,473,514 number of tweets (from 3 June 2008 to 11 May 2016) and it is still growing. A video demonstrating the system can be accessed at http://mlg.ucd.ie/twittercracy. A query such as "guns" can reveal significant insights (see Fig. 2): we observe the low level of cultural similarity between parties, while aspects like "gun sales", "gun violence", and "gun legislation" highlight various facets within this topic which the user can navigate for further exploration. Together with the various insights from theoretical measures, these facets help uncover various issues of U.S. politics that may concern the voter. Three further examples are: (1) the different facets evident between the parties for the query "abortion", (2) the high level of cultural similarity between parties on matters of foreign policy, such as "Israel" and "Syria", (3) the low level of cultural similarity between parties on matters of domestic policy such as "drugs".

Acknowledgments. This publication has emanated from research conducted with the support of Science Foundation Ireland (SFI) under Grant Number SFI/12/RC/2289.

References

1. This is why social media will decide the 2016 election. http://fortune.com/2015/12/01/social-media-2016-election/, December 2015
2. Haveliwala, T.H.: Topic-sensitive pagerank. In: WWW (2005)
3. Lietz, H., Wagner, C., Bleier, A., Strohmaier, M.: When politicians talk: Assessing online conversational practices of political parties on twitter. In: ICWSM (2014)
4. Mejova, Y., Srinivasan, P., Boynton, B.: Gop primary season on twitter: popular political sentiment in social media. In: WSDM (2013)

5. Mejova, Y., Weber, I., Macy, M.W.: Twitter: A Digital Socioscope. Cambridge University Press, New York (2015)
6. Nulty, P., Theocharis, Y., Popa, S.A., Parnet, O., Benoit, K.: Social Media and Political Communication in the 2014 Elections to the European Parliament (2015)
7. O'Connor, B., Krieger, M., Ahn, D.: Tweetmotif: Exploratory search and topic summarization for twitter. In: ICWSM (2010)
8. Schweitzer, E.J.: Normalization 2.0: a longitudinal analysis of german online campaigns in the national elections 2002–9. Eur. J. Commun. **26**(4), 310–327 (2011)
9. Tumasjan, A., Sprenger, T.O., Sandner, P.G., Welpe, I.M.: Predicting elections with twitter: What 140 characters reveal about political sentiment. ICWSM **10**, 178–185 (2010)
10. Weng, J., Lee, B.-S.: Event detection in twitter. In: ICWSM (2011)

Industrial Track Contributions

Using Social Media to Promote STEM Education: Matching College Students with Role Models

Ling He[(✉)], Lee Murphy, and Jiebo Luo

Goergen Institute for Data Science, University of Rochester,
Rochester, NY 14627, USA
{lhe4,lmurp14}@u.rochester.edu, jiebo.luo@gmail.com

Abstract. STEM (Science, Technology, Engineering, and Mathematics) fields have become increasingly central to U.S. economic competitiveness and growth. The shortage in the STEM workforce has brought promoting STEM education upfront. The rapid growth of social media usage provides a unique opportunity to predict users' real-life identities and interests from online texts and photos. In this paper, we propose an innovative approach by leveraging social media to promote STEM education: matching Twitter college student users with diverse LinkedIn STEM professionals using a ranking algorithm based on the similarities of their demographics and interests. We share the belief that increasing STEM presence in the form of introducing career role models who share similar interests and demographics will inspire students to develop interests in STEM related fields and emulate their models. Our evaluation on 2,000 real college students demonstrated the accuracy of our ranking algorithm. We also design a novel implementation that recommends matched role models to the students.

Keywords: STEM · Recommendation systems · Social media · Text mining

1 Introduction

The importance of the STEM industry to the development of our nation cannot be understated. As the world becomes more technology-oriented, there is a necessity for a continued increase in the STEM workforce. However, the U.S. has been experiencing the opposite. In the United States, 200,000 engineering positions go unfilled every year, largely due to the fact that only about 60,000 students are graduating with STEM degrees in the United States annually [17]. Another obvious indication is the relatively fast growth in wages in most STEM-oriented occupations: for computer workers alone, there are around 40,000 computer science bachelors degree earners each year, but roughly 4 million job vacancies [29]. Therefore, our motivation is to solve this problem of STEM workforce shortage by promoting STEM education and careers to college students so as to so to increase

© Springer International Publishing AG 2016
B. Berendt et al. (Eds.): ECML PKDD 2016, Part III, LNAI 9853, pp. 79–95, 2016.
DOI: 10.1007/978-3-319-46131-1_17

the number of people who are interested in pursuing STEM majors in college or STEM careers after graduation.

In this paper, we present an innovative approach to promote STEM education and careers using social media in the form of introducing STEM role models to college students. We chose college students as our target population since they are at a life stage where role models are important and may influence their career decision-making [15]. Social media is useful for our study in the following two ways: (1) the massive amount of personal data on social media enables us to predict users real life identities and interests so we can identify college students and role models from mainstream social networking websites such as the microblogging website Twitter and professional networking website LinkedIn; (2) social media itself also can serve as a *natural and effective platform* by which we can connect students with people already in STEM industries (Fig. 1).

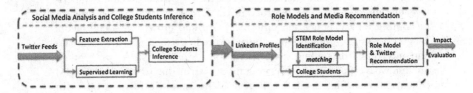

Fig. 1. The framework for promoting STEM education and careers using social media to match college students with STEM role models.

Our approach is effective in the following three ways. First, increasing STEM presence will inspire students to develop interests in STEM fields [18]. Second, the exposure of career STEM role models that students can identify with will have positive influence on students, as strongly supported by previous studies [12]. Finally, as a form of altruism, accomplished people are likely to help young people [6,11] and people who resemble them when they were young [21]. More importantly, social learning theory [1,2], psychological studies, and empirical research have suggested that students prefer to have role models whose race and gender are the same as their own [12,15,30] as well as who share similar demographics [7] and interests [16]. Motivated and supported by the findings of these related studies, we select *gender, race, geographic location, and interests* as the four attributes that we will use for matching the students with STEM role models. In addition, similar interests and close location will further facilitate the potential *personal connection* between the students and role models.

In particular, we first use social media as a tool to identify college students and STEM role models using the data mined from Twitter and LinkedIn. As a popular online network, on the average, Twitter has over 350,000 tweets sent per minute [27]. Moreover, in 2014, 37 % social media users within the age range of 18–29 use Twitter [5]. This suggests a large population of college users on Twitter. In contrast, as worlds largest professional network, LinkedIn only has roughly 10 % college users out of more than 400 million members [25], but has a rich population of professional users. Part of its mission is to connect the world's professionals and provide a platform to get access to people and insights that help its users [14]. Our goal, to connect college students with role models, is *organically consistent with LinkedIn's mission and business model.*

Specifically, we train a reliable classifier to identify college student users on Twitter, and we build a program that finds STEM role models on LinkedIn. We employ various methods to extract gender, race, geographic location and interests from college students and STEM role models based on their respective social media public profiles and feeds. We then develop a ranking algorithm that ranks the top-5 STEM role models for each college student based on the similarities of their attributes. We evaluated our ranking algorithm on 2,000 college students from the 297 most populated cities in the United States, and our results have shown that around half of the students are correctly matched with at least one STEM role model from the same city. If we expand our geographic location standard to the state-level, this percentage increases by 13 %; if we look at the college students who are from the top 10 cities that our STEM role models come from separately, this percentage increases by 33 %.

Our objective is to do social good, and we expect to promote STEM education and careers to real and diverse student population. In order to make a real life impact on the college students after we obtain the matches from the ranking algorithm, we design an implementation to help establish connections between the students and STEM role models using social media as the platform. For each student, we generate a personalized webpage with his top-5 ranked STEM role models' LinkedIn public profile links as well as a feedback survey, and recommend the webpage to the student via Twitter. Ultimately, it is entirely up to the student and the role models if they would like to get connected via LinkedIn or other ways, and we believe these connections are beneficial for increasing interest in STEM fields. It is noteworthy that *LinkedIn has already implemented a suite of mechanisms to make connection recommendations,* even though none of which is intended to promote STEM career specifically. Figure 2 illustrates how our implementation naturally fits into the work flow and business model of LinkedIn.

Our study has many advantages. Leveraging existing social media ensures that we are able to retrieve a large scale of sampling users and thus our implementation is able to influence a large scale of students. Also, due to available APIs

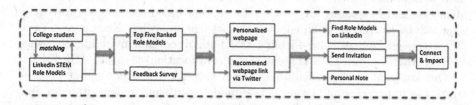

Fig. 2. The framework of our implementation to help establish the connections between the college students and the STEM role models.

and existing social media infrastructures, our data collection and our implementation are low cost or virtually free. More importantly, unlike some traditional intervention methods, we recommend STEM role models to college students in a *non-intrusive* way. We tweet at a student with the link of his personalized webpage, and it depends on himself if he wants to take actions afterwards. Finally, our approach is failure-safe in delivery. If there are some Twitter users that are classified incorrectly as college students, it has no harmful impact on them even if we promote STEM education to them.

The major contributions made in this study are fourfold. First, we take advantage of social media to do social good in solving a problem of paramount national interest. Second, we take advantage of human psychology, motivation, and altruism. That people are more likely to be inspired by models who are like them, and people who are accomplished are likely to help young people who share similarities with them. Third, we have developed a simple yet effective ranking algorithm to achieve our goal and verified its effectiveness using real students. Lastly, we design an implementation that seamlessly mashes up with the natural work flow and business model of LinkedIn to establish the connections between students and role models.

2   Related Work

STEM workforce is significant to our nation, and the shortage in such fields makes promoting STEM education and careers indispensable. We review the existing methods of promoting STEM education and build on previous research in both computer science and human psychology.

Previous effort has been made to promote STEM education. Most existing intervention methods focus on promoting through school educators [19], external STEM workshops [26], and public events such as conferences [22]. However, very little evidence has shown that these strategies were effective. On the other hand, while none of the methods has utilized the rich database and powerful networking ability of social media, social media-driven approaches have succeeded in many applications, such as health promotion and behavior change [33].

The abundance of social media data has attracted researchers from various fields. We benefit the most from studies that related to age prediction and user interest discovery. Nguyan et al. [20] studied various features for age prediction from tweets, and guided our feature selection for identifying college students. Michelson and Macskassy [31] proposed a concept-based user interest discovery approach by leveraging Wikipedia as a knowledge base while Xu, Lu, and Yang [32], Ramage, Dumais, and Liebling [23] both discovered user interest using methods that built on LDA (Latent Dirichlet Allocation) [3] or TF-IDF [24].

Our study also adopts knowledge from psychological studies that demonstrate the importance of having a role model with similar demographics and interests. Karunanayake [12] discussed the positive effect of having role models with the same race, and it holds across different races; Weber and Lockwood [30] discovered that female students are more likely to be inspired by female role models; Ensher and Murphy [7] indicated that liking, satisfaction, and contact with role models are higher when students perceive themselves to be more similar to them in demographics; and Lydon et al. [16] suggested that people are attracted to people who share similar interests. These studies help determine the attributes that we selected to match the students with role models.

3   Data

We used the REST API to retrieve Twitter data. Instead of directly searching for college Twitter users among all the general users, we focused on the followers of 112 U.S. college Twitter accounts since there is higher percentage of college students among these users. In total, we successfully retrieved more than 90,000 followers. For each user, we extracted the entities of his most recent 200 tweets (if a user has fewer than 200 tweets, all his tweets were extracted) and his user profile information, which includes geographic location, profile photo URL, and bio. After we filtered out API failures, duplicates, and users with zero tweet or empty profile, we are left with 8,688,638 tweets from 62,445 distinct users.

Due to the limited information that LinkedIn API allows us to retrieve, we employed web crawling techniques to obtain the desired information directly from the webpage. We built a program that does automated LinkedIn public people search and used it to search users based on the most common 1,000 surnames for Asians, Blacks, and Hispanic, and more than 5,000 common American given names[1]. Despite some overlapping surnames, the large number of names we searched is still able to ensure the diversity of the potential role models, and our results confirmed that. For each search, the maximum number of users returned is 25, and we collected the public profile URLs of all the returned users. After we deleted the duplicates, we retained 182,016 distinct LinkedIn users.

4   Identifying Twitter College Student Users

We employed machine learning techniques to identify Twitter college student users (i.e. from incoming freshmen to seniors). First, we labeled our training set. We used regular expression techniques to label college student users and non-college student users. Specifically, we studied patterns in users' tweets and bio, and constructed 45 different regular expressions for string matching. For example, expressions such as "I'm going to college", "#finalsweek", or "university'19" are used to label college students; and expressions such as "professor of", "manager of", or "father" are used to label non-college students. If a user's tweets and bio do not contain any of the 45 expressions, the user is unlabeled. We then manually checked and only counted the correctly labeled users. In the end, we are left with 2,413 labeled users, where 1,103 are college students and 1,310 are non-college students, as well as 60,032 unlabeled users.

---

[1] All the names were retrieved from http://names.mongabay.com.

Second, we trained our labeled data set to develop a reliable classifier using the LIBSVM Library [4] in WEKA [8]. We chose SVM for our binary classification because it is efficient for the size of our data set. We learned from Nguyan et al.'s study of language and age in tweets [20] that the usage of emoji, hashtag, and capitalized expressions such as "HAHA" and "LOL" are good age indicators. We built on their study and took a step further to use these three features for differentiating college students (i.e. specific age group) from general users. We were also curious about whether re-tweet would be another good age indicator, so we also extracted this feature. For each user, each feature is represented by its *relative frequency* among the user's tweets:

$$\frac{\# \text{ of tweets that contain this feature}}{\text{total } \# \text{ of tweets}} \tag{1}$$

Since relative frequencies are continuous, we discretized them into 10 bins with an equal width of 0.1 and assigned them with ordinal integer values for classification.

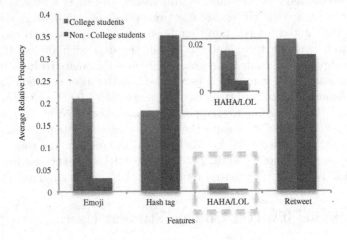

Fig. 3. Average usages of the four features for *college* and *non-college student users* among labeled Twitter users.

Figure 3 demonstrates our analysis of the four features. On average, college student users use emojis and HAHA/LOL more frequently while non-college student users use hashtags more frequently. We note that these results are consistent with the conclusions of a previous study [20]. However, there is not much difference in re-tweet between these two groups. We experimented training the classifier with and without re-tweet, our 10-folds cross-validation results showed that including re-tweet actually slightly lowers the accuracy of the classifier. Thus, we confirmed that re-tweet is a noise and does not help us to differentiate college student users. Our final classifier trained from the other three features achieves a high accuracy of 84 %. We then used this trained classifier to infer college student users among the unlabeled users. We further labeled 18,351 users as college students, and with our manually labeled college student users, together we have labeled 19,454 college student users in total.

5   Finding LinkedIn STEM Role Models

Our goal is to find diverse STEM role models from LinkedIn in terms of geographic locations and industries. While the definition of a role model is subjective to an individual student, we take an objective view and consider people who have received STEM education and work in STEM-related industries or have a career in STEM industries as role models.

We first filtered out users who are outside of the United States and then built a *Role Model Identification* program to find STEM role models. The program takes in a user's profile URL, crawls the contents in "industry" and "education" fields on the user's profile and only outputs the URL if the user is a STEM role model. Specifically, we divided all 147 LinkedIn industries into three groups, "non-STEM", "STEM", and "STEM-related". For example, "Biotechnology" and "Computer Software" are "STEM", "Music" and "Restaurants" are "Non-STEM", and "Financial services" and "Management consulting" are "STEM-related". We only consider those users who are under "STEM" or under "STEM-related" with a degree in STEM majors as role models. We used the 38 STEM majors offered at our University as our standard.

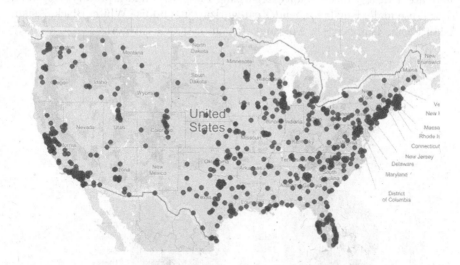

Fig. 4. The geographic location distribution of *STEM role models*. Darker color indicates higher density.

After we obtained the profile URLs of STEM role models, we crawled their entire profiles using the URLs. We successfully found 25,637 STEM role models from 2,022 distinct locations in the United States, including some places in

Hawaii and Alaska. Figure 4 shows a rough visualization of the diverse geographic locations the STEM role models come from. The top-10 cities that role models come from are, not surprisingly, San Francisco, New York City, Atlanta, Los Angeles, Dallas, Chicago, Washington D.C., Boston, Seattle and Houston.

6 Matching College Students with Role Models

This section presents the methods we employed to extract the gender, race, geographic location and interests from college students and STEM role models as well as our ranking algorithm that matches them based on the similarities of these attributes. We reiterate that our selection of attributes are supported by a variety of previous related studies. These factors can make the most influential pairing because they ensure that a student gets a mentor with a similar background for affinity. Moreover, close geographic location and similar interests are valuable for potential real life interaction between the students and role models.

6.1 Gender and Race Extraction

We extracted race and gender from both textual and visual features, namely the users' names and profile photos. We recognize that there are people who identify themselves with genders other than male and female; we also recognize that there are a variety of ways for categorizing races. To build a prototype system, we will use male, female for gender categorization, and use White, Black, Asian, Asian Pacific Islander (i.e. Api) and Hispanic for race categorization.

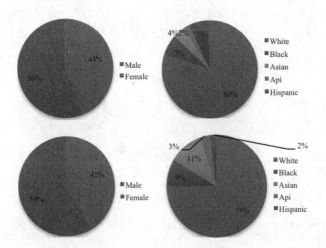

Fig. 5. Distribution of gender and race. Top row: college students; Bottom row: STEM role models.

In particular, we used Genderize.io[2], Face++[3], and Demographics[4] to extract these two attributes. Genderize.io and Demographics predict gender or both gender and race based on the users given name or full name while Face++ predicts both using the user's profile photo. In total, we obtained three gender predictions and two race predictions for each user. Each prediction is returned with an accuracy, and in the case the tool fails to predict, the prediction will be null. We picked the gender and race predictions with the highest accuracy.

As a result, we extracted the gender of 80 % college students and 97 % role models, and the race of 46 % college students and 92 % role models. Almost all role models have both attributes since we used their LinkedIn profiles, where the profile photos are usually high quality and the names are usually real. In contrast, Twitter profiles sometimes can contain profile photos with random objects and invented names. Figure 5 shows the make-up of those college students and STEM role models whose gender and race were successfully extracted.

6.2 Location and Interests Extraction

We directly extracted geographic locations from the "location" field in Twitter and LinkedIn profiles. The interests extraction is less straightforward and we used other features as proxies for this attribute.

We were able to extract the locations of all STEM role models since LinkedIn requires users to have a valid geographic location on their profiles. These locations usually contain the city and the state that role models work in. However, Twitter does not have this requirement, and we noticed that not every college student has filled the location field on his profile and some of the filled locations are not valid. In fact, 34 % Twitter users either did not fill the "location" field or provided fake geographic locations; among those valid locations, roughly 65 % are at city-level [11]. In addition, we observed that many students use the name of their educational institutions as locations, and some locations are not correctly spelled or formatted. For example, a student's location is "mcallentx", which refers to the city McAllen in Texas, but not a place called "mcallentx".

Due to the difference in the nature of LinkedIn and Twitter, we selected different proxies as interests for role models and college students. For role models, we directly extracted the contents in "interests" and "skills" fields as their interests because skills such as "Web Development" can also be an interest, and people usually are good at things that they are interested in. For college students, we extracted hashtags (excluding prefix "#") as interests. A hashtag is a user-defined, specially designated word in a tweet, prefixed with a "#" [31]. Originally, we experimented LDA topic modeling to discover topics of interests from all college students' tweets and intended to use these to define each student's interests. However, due to the noise and non-interest related terms in tweets

---

[2] A database that contains 216,286 distinct names across 79 countries.

[3] A face detection service that detects 83 points of the face and analyze features such as age, gender, and race.

[4] A database that contains U.S. census for demographics.

(excluding stop words and non-English words), most of the terms generated are too generic to be defined as topics of interests. Therefore, we extracted one's unique hashtags as proxy for interests. Hashtags have been used in characterizing topics in tweets [23] and have shown to be interest-related to a decent extent [32]. Although high-frequency hashtags are intuitively better representations of one's interests, including all unique hashtags allows us to extract a wilder range of interests. After we extracted interests from both students and role models, we stored everyone's interests as a set which we call *interest set*. The size of the set varies from user to user depending on the number of interests of that user.

6.3   Ranking Algorithm

We rank all STEM role models for each student based on the similarities of their attributes. Specifically, for each comparison of a student and a role model, we calculate the similarity of each attribute, and rank the role model based on the arithmetic average across similarities of all attributes. We will now explain our methods used for each comparison.

For gender and race, we simply compared if the two people have the same string for gender or race. In our case, there are two strings for gender, "female" and "male", and five strings for race, "White", "Black", "Asian", "Api", and "Hispanic". Therefore, the *gender similarity* is either 1 or 0 because two people either have the same gender or not, and the same went for *race similarity*.

For geographic locations, we used string comparison method to measure the similarity of two locations. Originally, we experimented two ways to calculate it: the actual distance between two locations based on their latitudes and longitudes, and the Levenshtein distance between the two strings that represent the two locations. Due the variety of possible expressions of the same location, traditional tool such as geocoder[5] can only correctly convert well-formatted locations that do not contain non-letter characters. For example, a real college student has location "buffalo state'18 psych majorr" and it cannot be successfully converted into coordinates using geocoder, but clearly that the student studies in buffalo. Since our objective is to be able to compare as many locations as possible, we decided to use string comparison, which allows the flexibility of using various location representations for the same place. Specifically, we employed Levenshtein distance[6] [13] to calculate the distance between two strings, and the *Levenshtein-based similarity* (a ratio between 0 and 1) is defined as:

$$\frac{length(S_1) + length(S_2) - Levenshtein\ distance(S_1, S_2)}{length(S_1) + length(S_2)} \quad (2)$$

where in the case of location, $S_1 (S_2)$ is the string of the student's (role model's) location, and we then have our *location similarity*. A minor problem of this similarity measure is that two geographically different locations might contain similar

---

[5] https://github.com/geopy/geopy.

[6] Levenshtein distance is the minimum number of single-character edits required to change one string into the other, and it is applicable to strings with different lengths. https://github.com/miohtama/python-Levenshtein.

words and have a high similarity, such as "Washington D.C." and "Washington State". But this happens relatively rare only if there are enough people from one of the location or both.

We used Jaccard coefficient [10] combined with *Levenshtein-based similarity* to compute the similarity of two *interest sets*. Hashtags are often not real words but a combination of words without spaces. While a real student's hashtag "computersciencelife" and a real role model's interest "computer science" clearly refer to the same interest in the field of computer science, the two strings are different and have a *Levenshtein-based similarity* of 0.86. Therefore, in order to capture the overlapping interests between two *interest sets*, we need a threshold for *Levenshtein-based similarity* that decides whether two strings refer to the same interest. After extensive experimenting with real data, we chose our threshold to be 0.8. Our *interest similarity* is then defined as:

$$\frac{|I_1 \bigcap I_2|}{|I_1 \bigcup I_2|} = \frac{\# \text{ of overlapping interests}}{|I_1| + |I_2| - \# \text{ of overlapping interests}} \tag{3}$$

where $I_1(I_2)$ is the student's (role model's) *interest set*. A potential problem is that since our measurement is string-based but not concept-based, it might not capture the synonymous of interests as overlapping interests.

After we calculated the similarities of all four attributes, we combined them by taking the arithmetic average and used that to rank the role models. In the cases of missing values, any unlabeled attributes is not taken into account. For instance, if a student does not have gender information, the arithmetic average will entirely depend on the similarities of his other three attributes.

6.4   Evaluation

In this section, we verified our ranking algorithm on 2,000 college students from the 297 most populated cities[7] in the United States [28]. All these students are randomly selected from our database. We manually evaluated their top-5 ranked role models, and we also recommended these role models to them via Twitter.

Although it is desirable to evaluate the ultimate impact of our study, we recognize that this would require tracking the subjects of the study over their career of substantial length (e.g., over 10 years). Therefore, it is beyond the scope of this study, and we decided to use matching accuracy as the performance measure, which is defined as:

$$\frac{\# \text{ of students that were } \textit{correctly} \text{ matched with n role models(s)}}{\# \text{ of total students}} \tag{4}$$

where n is the second metric, the specific number of role models out of the top-5 that are correctly matched with the student. It represents the granularity level of matching. We consider a student is *correctly* matched with a role model if the LinkedIn user is indeed a STEM role model and has the same gender, race

---

[7] With a population of at least 100,000.

and geographic location as the student. We did not evaluate interests since they are often not explicitly stated in social media and it would be too difficult to discover every student's real interests by reading his tweets.

We took a careful effort to manually evaluate the matching results of these 2,000 college students by checking their Twitter profile pages and the LinkedIn profile pages of their top-5 ranked STEM role models. We utilized all the information on their respective social media profiles to determine their gender, race, and geographic location. In order to determine if someone is indeed a STEM role model, we make our best judgment, as a career counselor would, based on the entire LinkedIn profile, which usually includes demographic background, personal summary, industry, education, working experience and skills.

If we failed to determine any of the three attributes of a student, we will have to consider that he is not correctly matched with any role model because we are unable to conduct the evaluation. Consequently, for Twitter public accounts and students with unlabeled gender, race, or invalid location, they all receive zero correctly matched role models. Location should not have been a problem since we selected these students by their locations, but we found that a handful of students have removed or changed their locations after we collected the data.

Table 1. Top: top-5 role models for a White, male student from "Atlanta, Georgia"; Bottom: top-5 role models for an Asian, female student from "Round Rock, TX"

	Gender	Race	Location	Current Occupation	Industry
#1 RM	Male	White	Atlanta, Georgia	Co-Founder/Managing Partner of a mobile application	Internet
#2 RM	Male	White	Atlanta, Georgia	Co-Founder & COO at Yik Yak, Inc.	Computer Software
#3 RM	Male	White	Atlanta, Georgia	Product Manager at Yik Yak	Computer Software
#4 RM	Male	White	Greater Atlanta Area	Entrepreneur, former Chief Strategy Officer at TeraData	Computer Software
#5 RM	Male	White	Atlanta, Georgia	Math Professor at Georgia Institute of Technology	Higher Education

	Gender	Race	Location	Current Occupation	Industry
#1 RM	Female	Asian	Lubbock, Texas	Enrollment Specialist at Virginia College	Higher Education
#2 RM	Female	Black	Houston, Texas	Design Consultant (Architectural Engineering Major)	Internet
#3 RM	Female	Asian	Houston, Texas Area	Registered Nurse	Hospital & Health Care
#4 RM	Female	Asian	Houston, Texas	Director, Nursing Knowledge Management	Hospital & Health Care
#5 RM	Female	White	Houston, Texas	Business Development Manager	Information Technology

Table 1 shows two randomly selected representative examples of the matching results for two students. We consider that the student in the top table was correctly matched with all five role models and the student in the bottom table was only correctly matched with #3 and #4 role models at state-level because #1 role model is not in STEM-related occupation and #2 and #5 are not Asian. None of the role models was correctly matched at city-level.

Taking into consideration that our limited database of STEM role models may have an impact on the performance of the ranking algorithm, we conducted evaluation in four levels: city-level for 297 cities, state-level for top 297 cities, city-level for top-10 cities, and state-level for top-10 cities. Among the 2,000 selected students, about a quarter of the selected students are from the top 10 cities. Intuitively, we expect more students to be correctly matched with role models at the state-level than city-level. Also, we expect students from the top-10 cities to be correctly matched with more STEM role models because there should be more diverse role models in these cities.

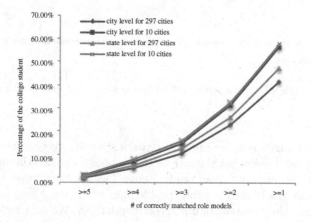

Fig. 6. The results of matching accuracy of 2,000 students from 297 cities in four different levels.

In Fig. 6 we show the overall matching accuracy in the four levels. We first look at our baseline, the city-level for 297 cities. 42 % of the college students were correctly matched with at least one role model. We noticed that around half of them was not matched with any role model and this is partly due to those college students with unlabeled gender and race information.

We then noticed that our ranking algorithm performs better in the 10 cities than in the 297 cities for both city and state levels. Numerically, the difference increases as the minimum number of correctly matched role models decreases. If we look at students who were correctly matched with at least one role model, for both city and state levels, the top-10 cities outperforms the 297 cities by 33 % and 21 %, respectively; the ranking algorithm achieves a decent accuracy of 57 % in both city and state levels for the 10 cities. Also, our ranking algorithm performs better in the state-level than in city-level for the 297 cities. With students who were at least correctly matched with one role model, the difference is 13 %, which is smaller but still very significant. However, there is almost no difference in state and city levels for the top-10 cities. A possible explanation is that because there are more STEM role models of various types in the top-10 cities, the student can usually get matched with STEM role models who are from the exact same city.

During our evaluation, we are encouraged to see that there is a good variety of STEM role models in different industries even for students with the same demographic background. We think this is a positive indicator that the attribute, interests, in fact contributes to our ranking algorithm.

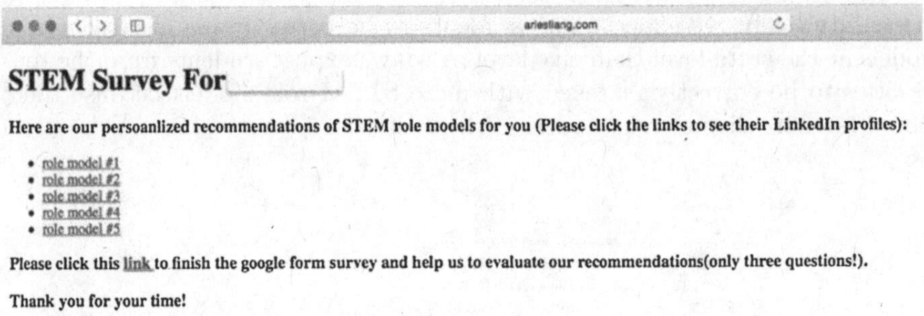

Fig. 7. An example of the personalized webpage for a real college student user on Twitter.

In order to make a real-life impact, for each student, we generated a personalized webpage and delivered the link of the webpage via tweeting at him from the official Twitter account of our study. Figure 7 shows an example of such webpage. It contains the LinkedIn public profile links of his top-5 role models and a survey regarding the accuracy of our recommendations. We only received a very small number of responses and conducted preliminary analysis. All responses indicated that they are indeed currently college students, a third agree that the recommendations are good and a third indicated that they would be more interested in STEM majors/careers if they had role models in STEM fields. We would need more responses to validate our implementation, and a potential way to do so is to cooperate with our university, apply the ranking algorithms on students who are Twitter users and ask for responses.

7   Conclusion and Future Work

In this paper, we present an innovative social media-based approach to promote STEM education by matching college students on Twitter with STEM role models from LinkedIn. Our ranking algorithm achieves a decent accuracy of 57 % in the city-level for the top-10 cities that the STEM role models come from. We also design a novel implementation that recommends the matched role models to the students. To achieve this, we identified college students from Twitter and STEM role models from LinkedIn, extracted race, gender, geographic location and interests from their social media profiles, and developed a ranking algorithm to rank the top-5 ranked STEM role models for each student. We then created a personalized webpage with the student' role models and recommended the webpage to the student via Twitter.

Our recommendation is not imposed on either side. It is the students' choice if they want to initiate the connection with the role models via LinkedIn or other methods; and it is for the role models to decide if they want to accept their LinkedIn invitations or other forms of communication. In the case of LinkedIn, note that if a student decides to approaches a potential role model, he can express why he would like to get connected (e.g., interest in STEM fields), and the role model can make his own judgment. One may worry that our implementation of recommendations may be considered a form of spamming on students, however, our intention is clearly to help their careers, and *not to profit* from them.

There are several possible extensions of our study in the future. Our approach might have a reduced effect for college seniors since it is more difficult for them to switch majors. However, it is not uncommon that students change their career paths after graduation, and in the future we could recommend role models with similar experiences specifically to seniors. We could also expand our target population to high school students or focus on promoting STEM education specifically to minority college students. In addition, we could classify STEM role models into specific groups such as current STEM major college students and experienced STEM role models since students might feel more comfortable reaching out to their peers. Finally, we could design an application based on our implementation to achieve real-time matching, where a college student could log into our application using their Twitter account, and we could collect their data, extract their attributes, and give them STEM role model recommendations in real-time. This application could also be generalized to other social media since many methods we used are compatible with other platforms.

We hope this study can serve as a starting point to make use of the rich data and powerful networking ability of social media "by the people" in order to promote STEM education and build positive influence "for the people".

Acknowledgments. This work was supported in part by Xerox Foundation, and New York State through the Goergen Institute for Data Science at the University of Rochester. We thank all anonymous subjects for contributing to the evaluation of our system.

References

1. Bandura, A.: Self-efficacy: toward a unifying theory of behavioral change. Psychol. Rev. **84**, 191–215 (1977)
2. Bandura, A.: Social Foundations of Thought and Action. Prentice Hall, Englewood Cliffs (1986)
3. Blei, D.M., Ng, A.Y., Jordan, M.I.: Latent Dirichlet Allocation. J. Mach. Learn. Res. **3**, 993–1022 (2003)
4. Chang, C.-C., Lin, C.-J.: LIBSVM: a library for support vector machines (2001)
5. Duggan, M., Ellison, N.B., Lampe, C., Lenhart, A., Madden, M.: Demographics of key social networking platforms, January 2015
6. Emmerik, H.V., Baugh, S.G., Euwema, M.C.: Who wants to be a mentor? An examination of attitudinal, instrumental, and social motivational components. Career Dev. Int. **10**(4), 310–340 (2005)
7. Ensher, E.A., Murphy, S.E.: Effects of race, gender, perceived similarity, and contact on mentor relationships. J. Vocat. Behav. **50**(3), 460–481 (1997)
8. Hall, M., Frank, E., Holmes, G., Pfahringer, B., Reutemann, P., Witten, I.H.: The WEKA data mining software: an update. SIGKDD Explor. **11**(1), 10–18 (2009)

9. Hecht, B., Hong, L., Suh, B., Chi, E.H.: Tweets from Justin Biebers heart: the dynamics of the location field in user profiles. In: Proceedings of ACM CHI Conference on Human Factors in Computing Systems (2011)
10. Jaccard, P.: Etude comparative de la distribution orale dans une portion des alpes et des jura. Bulletin del la Socit Vaudoise des Sciences Naturelles **37**, 547–579 (1901)
11. Judge, T.A., Kammeyer-Mueller, J.D.: On the value of aiming high: the causes and consequences of ambition. Appl. Psychol. **97**(4), 758–775 (2012)
12. Karunanayake, D., Nauta, M.M.: The relationship between race and students' identified career role models and perceived role model influence. Career Dev. Q. **52**(3), 225–234 (2004)
13. Levenshtein, V.I.: Binary codes capable of correcting deletions, insertions, and reversals. Sov. Phys. Dokl. **10**(8), 707–710 (1966)
14. LinkedIn Mission. https://www.linkedin.com/about-us
15. Lockwood, P.: Someone like me can be successful: do college students need same-gender role models? Psychol. Women Q. **30**, 36–46 (2006)
16. Lydon, J.E., Jamieson, D.W., Zanna, M.P.: Interpersonal similarity and the social and intellectual dimensions of first impressions. Soc. Cognit. **6**(4), 269–286 (1988)
17. Machi, E.: Improving U.S. Competitiveness with K-12 STEM education and training (SR 57). A Report on the STEM Education and National Security Conference 21–23 (2008), October 2008
18. Merrill, C., Custers, R.L., Daugherty, J., Westrick, M., Zeng, Y.: Delivering core engineering concepts to secondary level students. J. Technol. Educ. **20**(1), 48–64 (2008)
19. Metz, S.: Promoting STEM careers starts in the K12 classroom. ASCD Express **6**, 24 (2011)
20. Nguyen, D., Gravel, R., Trieschnigg, D., Meder, T.: How old do you think i am?: a study of language and age in Twitter. In: Proceedings of International Conference on Weblogs and Social Media, pp. 439–448 (2013)
21. Owen, C.J., Solomon, L.Z.: The importance of interpersonal similarities in the teacher mentor/prot relationship. Soc. Psychol. Educ. **9**(1), 83–89 (2006)
22. Promoting STEM education: a communications toolkit. https://www.iteea.org/File.aspx?id=40511&v=3cef257a
23. Ramage, D., Dumais, S., Liebling, D.: Characterizing microblogs with topic models. In: Proceedings of International Conference on Weblogs and Social Media (2010)
24. Salton, G., Buckley, C.: Term-weighting approaches in automatic text retrieval. Inf. Process. Manage. **24**(5), 513–523 (1988)
25. About LinkedIn. https://press.linkedin.com/about-linkedin
26. Tsui, L.: Effective strategies to increase diversity in STEM fields: a review of the research literature. J. Negro Educ. **76**(4), 555–581 (2007)
27. Twitter usage statistics (2015). http://www.internetlivestats.com/twitter-statistics/
28. U.S. Census Bureau, Population Division (2014). http://factfinder.census.gov/faces/tableservices/jsf/pages/productview.xhtml?src=bkmk
29. US News. Short on STEM talent. http://www.usnews.com/opinion/articles/2014/09/15/the-stem-worker-shortage-is-real
30. Weber, K.: Role models and informal STEM-related activities positively impact female interest in STEM. Technol. Eng. Teacher **71**(3), 18 (2011)
31. Michelson, M., Macskassy, S.A.: Discovering users' topics of interest on Twitter: a first look. In: Proceedings of the Fourth Workshop on Analytics for Noisy Unstructured Text Data (2010)

32. Xu, Z.H., Lu, R., Xiang, L., Yang, Q.: Discovering user interest on Twitter with a modified author-topic model. In: Proceedings of the 2011 IEEE/WIC/ACM International Conferences on Web Intelligence and Intelligent Agent Technology (2011)

33. Heldman, A.B., Schindelar, J., Weaver, J.B.: Social media engagement and public health communication: implications for public health organizations being truly "social". Publ. Health Rev. **35**, 1–18 (2013)

34. Hadgu, A.T., Jaschke, R.: Identifying and analyzing researchers on Twitter. In: Proceedings of ACM Conference on Web Science, pp. 23–32 (2014)

35. Mislove, A., Lehmann, S., Ahn, Y., Onnela, J., Rosenquist, J.N.: Understanding the demographics of Twitter users. In: Proceedings of Fifth International AAAI Conference on Weblogs and Social Media, pp. 554–557. AAAI Press (2012)

Concept Neurons – Handling Drift Issues for Real-Time Industrial Data Mining

Luis Moreira-Matias[1]($\boxtimes$), João Gama[2,4], and João Mendes-Moreira[2,3]

[1] NEC Laboratories Europe, Kurfürsten-Anlage 36, 69115 Heidelberg, Germany
luis.moreira.matias@gmail.com
[2] Faculdade de Economia, University of Porto, 4200-465 Porto, Portugal
{jgama,joao.mendes.moreira}@inescporto.pt
[3] LIAAD-INESC TEC, 4200-465 Porto, Portugal
[4] DEI-FEUP, University of Porto, 4200-465 Porto, Portugal

Abstract. Learning from data streams is a challenge faced by data science professionals from multiple industries. Most of them struggle hardly on applying traditional Machine Learning algorithms to solve these problems. It happens so due to their high availability on ready-to-use software libraries on big data technologies (e.g. SparkML). Nevertheless, most of them cannot cope with the key characteristics of this type of data such as high arrival rate and/or non-stationary distributions. In this paper, we introduce a generic and yet simplistic framework to fill this gap denominated Concept Neurons. It leverages on a combination of continuous inspection schemas and residual-based updates over the model parameters and/or the model output. Such framework can empower the resistance of most of induction learning algorithms to concept drifts. Two distinct and hence closely related flavors are introduced to handle different drift types. Experimental results on successful distinct applications on different domains along transportation industry are presented to uncover the hidden potential of this methodology.

Keywords: Supervised learning · Online learning · Concept drift · Perceptron · Stochastic gradient descent · Regression · Residuals · Transportation

1 Introduction

Today's hype around big data technologies floods the market of professionals with distinct backgrounds and yet a common job role: data scientist. Typically, they are actually very experienced on one of data science related fields (e.g. software engineering). However, they also commonly lack on the theoretical background required to adequately use more than off the shelf Machine Learning techniques and/or methodologies on their daily tasks.

The requirements for a more advanced framework varies naturally from task to task. Hitherto, this issue is more evident when a data mining (DM) task requires **real-time learning.** There are two key issues that empower such fact

© Springer International Publishing AG 2016
B. Berendt et al. (Eds.): ECML PKDD 2016, Part III, LNAI 9853, pp. 96–111, 2016.
DOI: 10.1007/978-3-319-46131-1_18

on these problems: (i) the high sample arrival rate and the constant and/or (ii) bursty drifts on the underlying probability distributions. These characteristics typically disallow the application of most of the traditional Machine Learning techniques (which assume finite training sets and/or stationary distributions) [1].

Recently, some simple approaches to handle this phenomenon have been scatterly introduced on different industries. Two of the most common ones are (i) windowing [2] and (ii) weight-based model selection [3]. The first approach consists into updating our model constantly based on every single arrived sample (i.e. incremental learning) or bunch of the most recent ones. The second one consists on combining multiple models through an weighted average of their outputs based on their recent performance. Although there is a growing interest for this type of methods followed by successful examples of their inclusion on modern large-scale Machine Learning libraries such as *Mahout* and *SparkML* (e.g. alternating least squares using stochastic gradient descent) – this movement is not certainly keeping up with the explosively increasing speed of industries needs to answer this particular problem. The most well-known exception is from the recommender systems area and, namely, the winning approach of NetFlix competition: Koren [4] pointed the temporal dynamics and concept drift as one key core ideas of their solution.

This paper intends to fill this gap by promoting a simplistic and yet effective framework that can handle drift on regression problems. Hereby, it is named Concept Neurons. The intuition behind its name comes from the need of a learning schema that can resist to concept drift and/or, *in extremis*, to a total absence of concept (i.e. bursty changes). In the context of predictive modeling in data streams, we have a two-stage (i.e. *predict and correct*) context-aware model [5]: firstly, a predition is made using a given offline/online learner. Secondly, the residuals distribution is monitored with a continuous inspection schema of interest. If a drift alarm is triggered, the prediction's residual is used to update the model whenever possible. Alternatively, we can also update directly the model output for a more bursty reaction to drift.

This schema can cope with most of traditional Machine Learning and/or time series forecasting methods. It was purposely designed on a simple fashion, targeting professionals who have not a strong background on fundamental statistical learning and/or optimization theory. By doing so, we aim to enlarge the pool of practitioners, increase the level of the results of their work as well as the quality of industrial DM pratices in general. Although not bringing a fundamental theoretical contribution, this paper proposes a fully functional idea, simple to understand, to use and with a tremendous applicational potential across industries. Besides the formal description of the present framework, this paper includes two concrete successful examples of their application on the transportation industry a field where the drift issues are classical problems – including operational control of taxis [6,7] and of highway networks [8]. Consequently, our contributions are two-fold: (1) to uncover applications of Supervised Learning with drift-handling mechanisms with real-world impact while (2) generalizing a framework that can be adopted by any practitioner on similar problems (from a fundamental point of view), regardless of his/her level of expertise or applicational domain.

The remaining of the manuscript is structured as follows: Sect. 2 depicts a problem illustration, as well as a brief overview on the related work. The third section formally describes our approach while Sect. 4 describe the two approach real-world case studies with distinct concept drift natures. Fifth Section describes our experimental test-bed and the obtained results on the abovementioned problems. Finally, conclusions are drawn.

2   Issues on Learning from Non-Stationary Distributions

Real-time DM involves to learn with one (or more) data sources providing samples in a sequential fashion. Typically, this type of data possesses unique and complex characteristics to deal with on carrying out Supervised Learning tasks. Some classical examples are high arrival rate, high labeling cost (e.g. [9]) and particularly, **non-stationary distributions**.

The non-stationarity phenomena can be translated in multiple ways. A common scenario is on dealing with datasets containing samples generated from multiple single/joint distributions. Although it is an issue for a vast majority of real-world DM problems (and datasets), it can also be neglected on most of the times to simplify potential paths to their solutions. On the top of the traditional stationarity assumption, many learning algorithms go one step further by assuming a functional form of the dependences and/or a particular residual's distribution (e.g. Gaussian Mixture Models with Expectation-Maximization for clustering; Ordinary Least Squares for regression). Albeit these facts, industrial practitioners rarely test the validity of these assumptions before applying these off-the-shelf Supervised Learning methods. It happens so because this approximation is fairly good for most of the traditional DM problems. Moreover, the trade-off between the time invested on getting alternative solutions and the performance gains often does not pay the effort back. Consequently, a question arises: why should we care about non-stationarity on real-time DM problems?

The main reason to focus on this issue lies on its **timewise** definition. Gama et al. [1] characterizes concept drift into four categories: (i) abrupt, (ii) incremental, (iii) gradual and (iv) recurrent. Fig. 1 illustrates a clear example of the latter one using time series data of integers (i.e. highway flow counts). In this particular example, it is somehow safe to assume that the underlying distribution, i.e. $p(y|x)$ is gaussian but for particular days/timespans (e.g. peak hours). This phenomenon is triggered by some sort of exogenous event (e.g. (iv) excessive demand load, (ii) car breakdown or (i) fast weather change) which is unexpected, absent of our data or somehow difficult to model and/or detect beforehand. In many applications, these time periods are actually the critical ones from a business perspective (e.g. peak-hours on transportation, prime time on media, happy hour/discounts on sales/retail).

Three of the most traditional techniques to deal with drift on DM tasks can be enumerated as follows: (1) dynamic model selection (i.e. meta-learning), (2) windowing and (3) re-training. In (1) model selection, we basically have a bucket of models which are combined dynamically along the time. Two common approaches of this type are weighting models [3] or categorizing samples using a meta-classifier [5]. The first one is simple to understand and to implement as well, being a good answer to (ii) incremental drifts. However, it can arguably deal with (i) abrupt drifts because, typically, the models in the bucket are only periodically updated. A meta-classifer one can handle either (i) abrupt or (iv) recurrent drifts by modeling samples into categories (which have associated labels). Nevertheless, an high level of expertise is required to put such learner in place. On the other hand, (2) windowing can help on dealing either with (ii) incremental and/or (iii) gradual drifts. It consists on considering one or just a bunch of the most recent samples to learn the models [2]. Although being quite simple, this approach is pointed by Gama et al. [1] to be slow on detecting (i) abrupt drifts. Model re-training (3) is the most simplistic approach to this problem and one of the most used among industry (e.g. wind power forecasting [10]). Often, it is combined with windowing for engineering-related purposes (e.g., see [11]). Although being pratical and require almost no tuning effort besides the window size, its *blind* reaction to drift – as the model update occurs independently on the samples content – represents a major drawback, thus resulting in a considerable probability of under/overfitting issues.

Our learning schema aims to combine the best of the abovementioned pratices on a simple fashion. The intuition behind it is to provide a very practical mechanism that can be build upon existing and somehow reliable Knowledge Discovery pipelines with proven results to improve their performance even further. The first big advantage on doing so is to re-use the existing DM frameworks (proprietary or not), avoiding costly re-engeneering tasks. By leveraging on the existing infrastructure (both physical and intellectual), this framework is easily adoptable by any industrial practitioners facing problems with similar drift-related issues.

3   Concept Neurons

From a high-level perspective, our algorithm operates in two stages: firstly, the residuals distribution produced by a given predictor is monitored by a continuous inspection schema of interest for drift detection purposes. This step aims to assess if the assumptions (here denominated as *Concept*) used to learn it (e.g. stationarity) are being violated. Secondly, a residual-based version of the parameter's inverse gradient is used to update the model whenever possible and/or directly its output. The second stage is only performed whenever an alarm is triggered on the first one, thus activating these updates (here conceptually denoted as *Neuron*).

The present methodology comes in two flavors: (I) asynchronous and (II) synchronous. The first aims on (I-2) (re-)training offline a near-*optimal* explanatory model at regular time intervals and (I-2) keep updating it incrementally in a stochastic fashion using the produced residuals. By extending the offline learning process through an incremental one, we purposely skip the monitoring stage by *blindly* assuming that the drift is **constantly happening**. It aims to handle (ii) incremental and (iii) gradual drifts. The second one consists on assuming

(II-1) an explanatory learning model (learned either offline or online) to be in place. Then, (II-2) a continuous inspection schema is used to monitor the recent residual's distribution (i.e. windowing) and trigger alarms. Whenever an alarm is triggered, a corrective neuron is activated to start adding up small percentages (i.e. learning rate) of the prediction's **residuals** to our model's output. This rate can be increased as novel alarms are triggered or *deactivated* instead in absence of an alarm for a long period (i.e. here denoted *activation* period). This mechanism aims to handle (iv) recurrent drifts which are limited in time or even bursty ones (when coping with an online learning model). This section describes this methodology fundamentally, departing from its roots in optimization theory till its practical application to Supervised Learning problems.

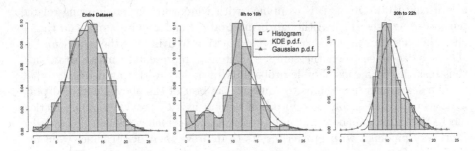

Fig. 1. Timewise drift illustration on a highway flow count data using kernel density estimation (KDE). Globally, the samples approximate a theoretical gaussian density curve. However, this is not true for some day periods due to drifts.

3.1 Stochastic Learning from Gradients

Let $y_1, ..., y_t : y_i \in \mathbb{R}, \forall i \in \{1..t\}$ denote the values of target variable of interest Y observed till current time t, e.g. train passenger load, and $x_1, ..., x_t : x_i \in \mathbb{R}^n, \forall i \in \{1..t\}$ be the values of an n-dimensional feature matrix $X \in \mathbb{R}^{n \times t}$. Regression problems aim to infer the following function:

$$\hat{f} : x_i, \theta \to \mathbb{R} \text{ such that } \hat{f}(x, \theta) = f(x_i) = y_i, \forall x_i \in X, y_i \in Y \qquad (1)$$

where $f(x_i)$ denotes the true unknown function which is generating the samples' target variable and $\hat{f}(x_i, \theta) = \hat{y}_i$ be an approximation dependent on the feature vector x_i and an unknown parameter vector $\theta \in \mathbb{R}^n$ (given by some induction model M). Typically, M determines the functional form of $\hat{f}(x_i, \theta)$ as well as the values of θ by formulating a data-driven optimization problem as

$$\hat{f}(x_i, \theta) = \underset{\hat{f}, \theta}{\arg\min} \sum\nolimits_{i=1}^{t} J(\theta, \hat{f}, x_i, y_i) \qquad (2)$$

where J denotes a cost function of interest and t the number of samples in the dataset. Standard gradient descent is a classical solver. Lets assume that we depart from a given (e.g. random) initialization of our parameter set, i.e. θ_0. The method updates θ iteratively until a certain stopping convergence criteria is met (e.g. ϵ where $\nabla_\theta > \epsilon$) as follows

$$\theta = \theta - \eta \nabla_\theta E[J(\theta, \hat{f}, X, Y)] \tag{3}$$

where the above expectation is computed with respect to the abovementioned cost and η denotes our constant learning rate (i.e. an user-defined parameter).

By doing so, we expect to converge to a local minima close enough to our global optimum. Obviously, this does not cope well with an infinite stream of data as our own (i.e. t is being constantly increased $\rightarrow t = +\infty$). A common way to handle this issue is with a stochastic learning (as known as SGD - Stochastic Gradient Descent) of θ. Instead of computing the expectation iteratively, we compute the inverse gradient, i.e. ∇_θ with respect to the most recent labeled sample (x_{t-1}, y_{t-1}), thus redefining recursively the Eq. 3 as follows

$$\theta_i = \theta_{i-1} - \eta \nabla_{\theta_{i-1}} J(\theta_{i-1}, \hat{f}, x_{i-1}, y_{i-1}) \tag{4}$$

The cost function most commonly used for regression problems is the well-known l_2 loss. If it is assumed to be in place and for a linear[1] $\hat{f}$, we obtain:

$$J(\theta_i, \hat{f}, x_i, y_i) = L_2(\theta_i, \hat{f}, x_i, y_i) = L_2(\hat{y}_i, y_i) = \frac{1}{2}(y_i - \hat{y}_i)^2 \tag{5}$$

$$\theta_i = \theta_{i-1} - \eta(y_{i-1} - \hat{y}_{i-1})x^T\theta_{i-1} = \theta_{i-1} \cdot \left(1 - \eta(r_{i-1}) \cdot x^T\right) \tag{6}$$

where r_i denotes the prediction's **residual** for sample (x_i, y_i) at time i.

3.2 Asynchronous Concept Neurons

In a real-time context, the simple computation of the ∇_{θ_i} can be problematic (e.g.: missing feature values, noise, $n >> 0$). Therefore, we propose a more naive approach by putting in place the following assumption:

Assumption 1. *Convergence is still possible at a smaller rate when done independently of X for a sufficiently small value of η and an adequate M.*

By doing so, we assume that most of the error is somehow proportional to the values of the parameter set. Formally, we transform Eq. 6 as follows:

$$\theta_i = \theta_{i-1}(1 - \eta(r_{i-1})) \tag{7}$$

One of the assumptions of SGD is that samples are drawn independently and are identically distributed (i.i.d.). From a theoretical point of view, drift is a violation of it. One way of circumventing this issue is to not keep a static learning rate but rather a time-variant one (i.e., $\eta(t)$;, e.g. [12]). The main intuition behind

---

[1] Despite the linear assumption (introduced for demonstrative purposes), SGD can also work on non-linear problems departing from a convex loss.

this idea is that the distribution is stationary through a *limited* period of time. Therefore, we can speed up/slow down convergence momentum according to our present learning context.

Departing from this intuition, we introduce a very simple idea in the Algorithm 1 based on three simple stages: (1) firstly, learn offline (using M) a model $\hat{f}(\theta, X)$ based on the samples obtained on recent window of time T; (2) Update θ incrementally using the model residuals; (3) re-compute $\hat{f}(\theta, X)$ after T_u periods. T, T_u and η are user-defined parameters and must be tuned for each particular application. Naturally, this approach is expected to handle poorly recurrent and/or bursty drifts as there is no drift detection mechanism embedded.

3.3   Synchronous Concept Neurons

To handle recurrent and/or abrupt drifts, we propose a slight change of the presented learning schema. Intuitively, the idea is that if the concept is dramatically *different*, we do not have time to learn it yet (and consequently, our current model approximation to the target function is quite *poor*). Let $A(R, \phi, \vec{\delta}, t) \in \{0, 1\}$ be a drift detection algorithm of interest where $R = r_1, ..., r_t$ denotes the set of residuals, ϕ denotes a sliding window size, $\vec{\delta}$ stands for generic user parameter set of interest specific for each possible type of A and t the current timestamp. Whenever $A = 1$, the model's output is *corrected* by re-engineering Eq. 7 as

$$\hat{y}_i = \hat{f}(x_i, \theta_i) - \eta_i(r_{i-1}) \tag{8}$$

where η is time-dependent from now on, i.e. η_i. If $a_i = 1$, then the learning rate is initialized as $\eta_i = \eta_0$ where η_0 is an initial learning rate set by the user. At this point, we are not fully *trusting* on what $\hat{f}$ is producing as outputs. For most of applications, it is recommended a conservative approach on the definition of η_i, i.e. $\eta_i << 0$.

Input: M - offline induction method, T - training window size, T_u - statonarity cyclic period; η - learning rate, X, Y - dataset;
Output: $\hat{f}$ - approximation function, θ_i - parameter vector
$W \leftarrow \emptyset$; //Initialization
foreach $i \leftarrow 1..t$ **do**
$\quad$ $W \leftarrow W \cup (x_i, y_i)$; // builds offline training set
$\quad$ **if** *(T_u mod $i == 0$)* **then**
$\quad\quad$ $\hat{f}, \theta_i \leftarrow M(W)$; // learns $\hat{f}$ and the parameter set θ_i from data
$\quad$ **end**
$\quad$ **if** *($T_u >= t \wedge T_u$ mod $i > 0$)* **then**
$\quad\quad$ $\theta_i = \theta_{i-1} - \eta(r_{i-1})\theta_{i-1}$; //update parameter set
$\quad\quad$ drop an element from the tail of W; //forgets outdated samples
$\quad$ **end**
end

Algorithm 1. Pseudocode for Asynchronous Concept Neurons (ACN).

Whenever a novel drift occurs, η_i is updated exponentially as $\eta_i = \eta_{i-1}(1+\gamma)$ where $\gamma \in [0,1]$ denotes a reactivability rate defined by the user. This methodology is not designed to update our models with respect to the observed drift - but rather to *handle* it instead. Intuitevely, in real-time DM problems, if we are facing a recurrent drift, M will be likely to still be useful in the future (as the validity of our current underlying distribution is limited in time). If facing an abrupt drift, this schema will slow down the performance deterioration of the model produced by M but it will not avoid standalone a further (re-)training stage. Consequently, we assume these drifts as time-limited phenomenons. Therefore, the decrease of η_i is operated abruptly as:

$$\sum_{i=t-\beta}^{t} A(R, \phi, \vec{\delta}, i) = 0 \tag{9}$$

In the present context, M can either be an offline or an online induction model. Algorithm (2) depicts the entire schema.

4 Case Studies

Hereby, we approach two different case studies in transportation industry: (A) demand prediction for taxi networks and (B) road traffic congestion prediction in highway networks. The target clients of the (A) are taxi dispatcher's and/or self-organized operators in the sector while (B) targets transit authorities and their road traffic management centers.

Case Study A is focused on predicting taxi-passenger demand for short-term horizons of $P-$minutes in a real-time setting [6,7]. The key idea is to improve the taxi driver's mobility intelligence through a live decision support system advising on best passenger-finding strategy to adopt in each moment (e.g. which is the stand/street/city area that he/she should head to in order to pick up the next passenger).

Case Study B is focused on predicting road Traffic congestion (i.e. incidents). It is possible to divide congestion in two types [8]: (i) *recurrent*, which happens on a regular basis within a given periodicity, e.g. peak hours on every Friday's evening, and a (ii) *stochastic* one, which is provoked by an external event, e.g.: car accidents. The problem is to predict the flow count (number of vehicles that traversed a sensor per unit of time) and occupancy rate (percentage of the time period that a car is over a sensor) on a short-term horizon of $P-$minutes. Then, a scenario-based threshold is considered to transform those discrete signals into binary ones (i.e. congestion/no congestion).

Brief summaries of the datasets are provided below. Additional details about preprocessing tasks conducted over these datasets can be found in Sects. 3.2 and 4 of [6,8] for case studies A and B, respectively.

4.1 (A) Taxi-Passenger Demand Prediction

Our data samples are a stream of timespamped location of events (e.g. pick-up, drop-off) obtained from taxi company (which runs 441 vehicles) operating in

Input: $\hat{f}$ - approximation function, θ - parameter set, ϕ - monitoring window size, η_0 - initial learning rate, γ - constant reactivability rate, β - activation period, A - drift detection algorithm; X, Y-dataset
Output: $\hat{y}$ - corrected predicted outputs
$W \leftarrow \emptyset$ and $\eta_1 \leftarrow 0$;
foreach $i \leftarrow 1..t$ **do**
  $\eta_i \leftarrow \eta_{i-1}$;
  if $(A(R, T, \delta, i) == 1)$ **then**
    if $(\eta_i > 0)$ **then**
      | $\eta_i \leftarrow \eta_{i-1}(1 + \gamma)$; //increase the learning rate
    else
      | $\eta_i \leftarrow \eta_0$; // activate the prediction corrections
    end
  end
  $\hat{y}_i \leftarrow \hat{f}(x_i, \theta_i) - \eta_i(r_{i-1})$; // correct our prediction output
  $W \leftarrow W \cup (r_{i-1})$; // add elements to the head of W
  if $(|W| == T)$ **then**
    | drop an element from the tail of W;
  end
  if $(\sum_{j=i-\beta}^{i} A(R, \phi, \delta, j) == 0)$ **then**
    | $\eta_i \leftarrow 0$; // deactivate the prediction corrections
  end
end

Algorithm 2. Pseudocode for Synchronous Concept Neurons (SCN).

Porto, Portugal between August 2011 and April 2012. This city is the center of a medium size urban area with 1.3 million habitants (see Fig. 2).

The drivers operate in 8 h shifts: midnight to 8am, 8am–4pm and 4pm to midnight. Each sample arrives has six attributes: (1) TYPE relative to the type of event reported and has four possible values: *busy* - the driver picked-up a passenger; *assign* the dispatch central assigned a service previously demanded; *free* the driver dropped-off a passenger and *park* - the driver parked at a taxi stand. The (2) STOP attribute is an integer with the ID of the related taxi stand. The (3) TIMESTAMP attribute is the date/time in seconds of the event and the (4) TAXI attribute is the driver code; attributes (5) and (6) refer to the LATITUDE and LONGITUDE corresponding to the acquired GPS position.

Table 1 details the number of taxi services demanded per daily shift and day type. Additionally, we can state that the central service assignment is 24 % of the total service (*versus* the 76 % of the one demanded directly in the street), while 77 % of the service demanded directly is dispatched in a stand (and 23 % is assigned in cruising time). The average driver waiting time in a stand is 42 min while the average cruising time for a service is only ~ 12 min.

4.2 (B) Highway Congestion Prediction

This dataset was collected through a traffic monitoring system of a major free-way deployed in an Asian country. The studied system broadcasts a stream of

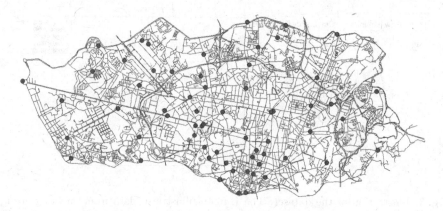

Fig. 2. The spatial distribution of the 63 taxi stands used by this fleet in Porto.

Table 1. Taxi services volume (Per Daytype/Daily Shift)

Daytype Group	Total Services Emerged	Averaged Service Demand per Daily Shift		
		0am to 8am	8am to 4pm	4pm to 0am
Workdays	957265	935	2055	1422
Weekends	226504	947	2411	1909
All Daytypes	1380153	1029	2023	1503

traffic-based measurements in real-time with distinct temporal granularities (depending on the type of sensor's installed on each lane). Each sensor measures traffic flow, lane occupancy rate and instantaneous vehicle's speed. The largest time granularity ($p = 5$ min) was used to normalize all the collected time series into a standard granularity level.

This network is composed by 106 sensors including both freeway's traffic directions. The covered segment's length is ~ 20 km while the sensor's sections are deployed each 500 m. Data was collected through 3 non-consecutive weeks.

Figure 3 depicts an illustration of the dataset. The (B)-figure contains one day of data from a particular section. Conversely, the other chart displays five sample-based p.d.f. obtained using a (gaussian) kernel density estimator over all the flow measurements available – one global and four specific for each of the considered timespans (divided by Periods I–IV, identified by the same display order as Fig. 3 legend). Table 2 details descriptive statistics. The top 10 sensors regarding the number of observed incidents are highlighted. As it is observable, the occupancy rate is higher in these sensors. Not surprisingly. the most critical period is the morning peak (P. II), comprised between 6:40 and 13:20.

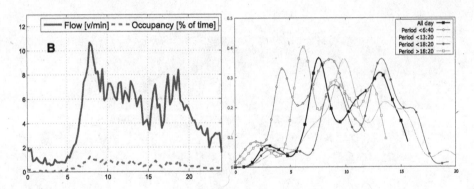

Fig. 3. Illustration of the dataset. The B-figure illustrates data from one-section on one particular day, while the other one depicts a flow-based p.d.f. estimation using all available data.

Table 2. Descriptive Statistics on all sensors *vs.* top-10 incindent ones.

Quantity	Flow				Occupancy			
	Mean	SD	Min.	Max.	Mean	SD	Min.	Max.
All day	9.9486	3.6514	0.1811	24.4104	2.1409	1.3276	0.0726	13.9967
Top 10, all day	8.2397	3.7001	0.1600	36.7667	3.4303	4.0282	0.0700	23.5567

5   Experiments

On both case studies, we assumed statistical independence among different taxi stands and road sections, respectively. Problem A consists into forecasting one term ahead (i.e. passenger demand count on a specific stand during the next P-minutes). To do it so, we chose a classical method M: Auto Regressive Integrated Moving Averages (ARIMA). For each stand, the ARIMA model was firstly set (and updated each 24 h) by detecting the underlying model in place from the time series of each stand during the recent $T = 15$ days (i.e. namely, the corresponding $15 \times 2 \times 24 = 1440$ periods). For that purpose, an automatic time series function was employed, i.e. auto-arima [13].

The parameters for each model are generally fit for each period/prediction using a generalized least squares (GLS) solver. Even considering that ARIMA use just a few bunch of recent samples T and low-dimensionality models (i.e. small n), the optimal fitting of its parameters can represent an unnecessary time-consuming process, i.e. $O(N^2)$. In problem A, we can be handling with hundreds of requests on a short amount of time (e.g. 4 different drivers dropping-off a passenger in an interval of two minutes will generate requests to process a total of 252 predictions/GLS – which is equivalent of doing roughly 2,1 model fittings per second on a single CPU) – which will raise undesired scalalibility issues. On the top of such computation issues, as the time series are bounded to the granularity of our forecasting horizon, we have to adapt them in order to obtain the right aggregation level for each on-demand prediction. One way to do it so is to maintain a *newly* calculated discrete time series each τ-minutes where $\tau << P$. By doing so, we can leverage on the additive properties of the series bins (similar to the ones exhibitted by histograms, e.g. [14]) to *roll* our time series into the desired bin positions (e.g. switch from 9:00, 9:30, 10:00, ... to 9:10, 9:40, 10:10, ...)–, e.g. as proposed by Moreira-Matias *et al.* [6].

To reduce the pratical computational time, we propose to replace GLS by Asynchronous Concept Neurons (ACN). The optimal parameter set θ is fit together with the model estimation stage (i.e. each $T_u = 96$ periods). Then, it is updated as depicted in Algorithm 1. The η value was tuned throughout a grid search procedure in $\{0.01k, \forall k \in \{1..20\}\}$ using a validation set with data collected on a previous time period.

To approach problem B, we departed from an online learning model which was composed of three main components: (a) an ARIMA-based model, (b) an Exponential Smoothing (ETS) model and (c) an online weighting model to combine both (i.e. ensemble). Similarly to the previous case study, the ARIMA prediction is also performed using a auto-arima+GLS+ACN procedure using $T = 2$ days. However, we are assuming here this schema as a fully incremental method for sake of simplicity. In this case, we decided to test the application of SCN to face the bursty nature of the non-recurrent traffic incidents (e.g. car accidents).

The parameter set θ is composed by both the ARIMA and the ETS model weights, as well as the two ensemble weights of each model. The online weighting ensemble is monitoring their performance over a sliding window of H-periods. The drift detection algorithm used was the Page-Hinkley (PH) test, an incremental inspection schema to detect drift [1] (consequently, $\phi = \infty$). The PH test depends on two parameters (i.e. $|\delta| = 2$). In our case, as we are monitoring two series of values (flow and occupancy), we have 4. Their values were set following traffic expert's suggestions. The remaining parameters of this framework η_0, β, γ and also H were tuned using another grid search procedure conducted over six of the 106 sensors of this case study. The full parameter setting employed in our experiments is summarized in Table 3.

5.1   Evaluation

In case study A, we compared traditional ARIMA trained with GLS (ARIGLS) with our ACN using the first as offline baseline (i.e. M). As test set, we considered the last 4 weeks of our data set. Experiments aimed to compare the model's error on it as well the computational time. As evaluation metric, we used an laplacian version of the Symmetric Mean Percentage Error averaged by all the taxi stands. The resulting metric (ASMAPE) is obtained as follows:

$$ASMAPE = \frac{1}{\Upsilon}\sum_{j=1}^{S}\sum_{i=1}^{t}\psi_j \frac{|y_{j,i} - \hat{y}_{j,i}|}{R_{j,i} + X_{j,i} + 1} : \psi_j = \sum_{i=1}^{t} y_{j,i}; \Upsilon = \sum_{j=1}^{S}\psi_j \quad (10)$$

where S denotes the total number of taxi stands.

In problem B, we compared three distinct online predictive methods: ARIMA (ARI), ETS and the hereby proposed SCN over an online weighted ensemble of both. On the top of the abovementioned sensor selection, we also assumed statistical independence between the data of each one of the three weeks (as they are non-consecutive). Consequently, it resulted on a total of 300 experiments (i.e. 100 sensors x 3 weeks by excluding the 6 sensors used in hyperparameter tuning).

The evaluation of these experiments were performed on two distinct dimensions: (1) numerical prediction and (2) event detection. In (1), we used *Root Mean Squared Error* (RMSE) and *Mean Absolute Error (MAE)* as evaluation metrics. On (2), we picked *Precision* (PRE) and *Recall* (REC). Similarly to A, the results were aggregated using an weighted average of these metrics, where each sensor's weight is given by the total number of incidents occurred.

5.2 Results

The evaluation of two models in case study A are displayed in Table 5. It is possible to observe than, despite their fundamental differences, their performance does not differ significantly. In terms of computational time, ARIGLS took 1.58 s to process each individual prediction while ACN took solely 0.60s (in average).

The results for experiments in case B are presented in three distinct folds: Table 4 presents the aggregated results. Left-hand side of Fig. 4 introduces an time-evolving evaluation in terms of RMSE produced by the three flow prediction methods hereby presented. The drift detection (i.e. *neuron activation*) and incident's boolean states are also exhibited on this chart. It is possible to observe that the SCN error is always lower than the one obtained from other methods. On the other hand, we can also conclude than the drift detection is not always necessarily correlated with an incident. The right-hand side of same Fig. 4 llustrates the prediction behavior along sensor with an increasing incident rate (on x-axis). The recall values are averaged using a sliding window considering just the recall values for the latest ten sensors with respect of the current one. By doing so, it is possible to conclude that the our method performance increases along with the number of incidents observed in each sensor.

5.3 Discussion

At a first glance, the high number of hyperparameters may appear a major drawback of our methodology. However, as we could demonstrate, they can be relatively easely tuned with the a validation set. From our experiments, we can sustain that the parameters related with the learning rate (e.g. η in ACN; η_0, β, γ in SCN) are the ones which provoke more variance on the target output. However, it is difficult to assess the framework's sensitivity to the parameter set without a careful evaluation procedure.

In case study A, the results illustrate the computational savings obtained by doing incremental approximations of the optimal model to deal with *soft* drift pheonomenas. In B, the high recall rates are illustrative of the potential of this framework on dealing with either bursty or recurrent drifts.

A work closely related to this one are the Kalman Filters. They are focused on signal processing problems, where our samples are simply a bunch of continuous measurements over time. Conceptually, it also relies on some sort of uncertain estimate/prediction of the series expected value and co-variance to then update it using the residuals co-variance. Formally, we can say that $f(x) = \hat{f}(x, \theta) + v$. Commonly, Kalman Filters assume stationarity on the residuals as $v \sim \mathcal{N}(0, \sigma^2)$. Conversely, our approach is fully non-parametric as **it makes no assumption on the residual's distribution**.

In this work, we end up using only linear induction methods as baseline learners for either ACN and SCN. However, **the authors want to highlight that this framework can be built upon non-linear learners as well** - see, for instance, the usage of SCN with decision trees for short-term bus travel time prediction [15]. By being generic and simple to understand as well as to put in pratice, this framework represents a pratical and yet inexpensive alternative to deal with drift on real-world Supervised Learning problems.

Table 3. Parameter Setting used in the experiments.

		Value	Description
A	P	30	forecasting horizon (in minutes)
	T	1440	training data size (i.e. 15 days)
	T_u	96	size of stationarity cycle period (i.e. 24 h)
	τ	5	minimum aggregation level (i.e. minutes)
	η	0.01	learning rate
	M	auto-arima + GLS	induction learner
	θ	arima model weights	model's parameter set
	H	4	sliding window size to compute our ensemble
B	A	Page-Hinkley test	drift inspection schema
	P	15	forecasting horizon (in minutes)
	ϕ	∞	drift monitoring window size
	δ_1^f	1.0	max. admissible flow prediction's residual for PH
	δ_2^o	0.1	max. admissible occupancy prediction's residual for PH
	δ_3^f	20	cumulative flow-based threshold to trigger PH alarm
	δ_4^o	4	cumulative occupancy-based threshold to trigger PH alarm
	η_0	0.3	initial learning rate
	β	6	activation period
	γ	0.2	reactivitability rate
	φ_f	10	flow-based min. threshold to trigger an incident
	φ_o	5	occupancy-based max. threshold to trigger an incident

6   Final Remarks

Today, experience on Data Science is one of most requested disciplines on job postings across different industries. The lack of qualified professionals on this

Table 4. Results on comparing SCN with ARIMA and ETS in case B.

Method	Week	Flow Prediction		Occ. Prediction		Event Detection	
		RMSE	MAE	RMSE	MAE	PREC	REC
ARI	ALL	1.6875	1.0743	2.1088	1.2939	0.8002	0.2823
ETS	ALL	1.7280	1.0765	2.3111	1.3057	0.8116	0.3000
SCN	ALL	**1.6389**	**1.0379**	**1.8151**	**1.0730**	**0.8199**	**0.3719**

Table 5. Error Comparison on the two Learning Models in A using ASMAPE.

Method	Periods			
	00 h–08 h	08 h–16 h	16 h–00 h	24 h
ACN	28.47 %	24.80 %	25.60 %	26.21 %
ARIGLS	28.23 %	24.70 %	24.93 %	25.80 %

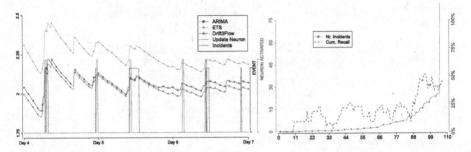

Fig. 4. Illustration of SCN Results: on left-hand side, we have a time-evolving flow-based evaluation on the top-event sensor using RMSE. The right-hand side depicts the average recall for all sensors (on x-axis) ordered by their number of incidents. Note SCN behavior.

area with respect to the number of vacancies is biasing companies towards hiring experienced programmers. Then, they are incited to use off-the-shelf libraries to do *magic* with little developping effort. Hitherto, the availability of drift-aware tools for real-time DM tasks on modern Big Data platforms is scarse. This scenario leads to the misusage of the available tools, poor performance and, ultimately, to reduced business value propositions.

This paper proposes a simple method for handling drift on real-time regression learning problems. It is designed generically, to run on the top of the Supervised Learning schemas popularly employed on modern industrial knowledge discovery pipelines. This two stage framework operates continousily by inspecting the residual's distributions without any predefined assumption on their functional form. Results conducted on real-world trials from the transportation domain demonstrated the potential of this method on reducing computational effort as well as to increase the regressor's generalization error.

As future work, we propose to conduct a sensitivity analysis on the parameter setting, as well as to generalize it even more this by introducing an inspection schema able not only to detect drift, but also to categorize its **nature**.

References

1. Gama, J., Žliobaitė, I., Bifet, A., Pechenizkiy, M., Bouchachia, A.: A survey on concept drift adaptation. ACM Comput. Surv. **46**(4), 44:1–44:37 (2014)
2. Widmer, G., Kubat, M.: Learning in the presence of concept drift and hidden contexts. Mach. Learn. **23**(1), 69–101 (1996)
3. Wang, H., Fan, W., Yu, P., Han, J.: Mining concept-drifting data streams using ensemble classifiers. In: Proceedings of the ninth ACM SIGKDD international conference on Knowledge discovery and data mining, pp. 226–235. ACM (2003)
4. Koren, Y.: Collaborative filtering with temporal dynamics. Commun. ACM **53**(4), 89–97 (2010)
5. Žliobaitė, I., Bakker, J., Pechenizkiy, M.: Beating the baseline prediction in food sales: how intelligent an intelligent predictor is? Expert Syst. Appl. **39**(1), 806–815 (2012)
6. Moreira-Matias, L., Gama, J., Ferreira, M., Mendes-Moreira, J., Damas, L.: On predicting the taxi-passenger demand: a real-time approach. In: Correia, L., Reis, L.P., Cascalho, J. (eds.) EPIA 2013. LNCS, vol. 8154, pp. 54–65. Springer, Heidelberg (2013)
7. Moreira-Matias, L., Gama, J., Ferreira, M., Mendes-Moreira, J., Damas, L.: Predicting taxi-passenger demand using streaming data. IEEE Trans. Intell. Transp. Syst. **14**(3), 1393–1402 (2013)
8. Moreira-Matias, L., Alesiani, F.: Drift3flow: freeway-incident prediction using real-time learning. In: IEEE 18th International Conference on Intelligent Transportation Systems (ITSC), pp. 566–571, September 2015
9. Žliobaitė, I., Bifet, A., Pfahringer, B., Holmes, G.: Active learning with drifting streaming data. IEEE Trans. Neural Network Learn. Syst. **25**(1), 27–39 (2014)
10. Monteiro, C., Bessa, R., Miranda, V., Botterud, A., Wang, J., Conzelmann, G., et al.: Wind power forecasting: state-of-the-art 2009. Technical report, Argonne National Laboratory (ANL) (2009)
11. Mendes-Moreira, J., Jorge, A., de Sousa, J., Soares, C.: Comparing state-of-the-art regression methods for long term travel time prediction. Intell. Data Anal. **16**(3), 427–449 (2012)
12. Ikonomovska, E., Gama, J., Džeroski, S.: Learning model trees from evolving data streams. Data Mining Knowl. Discov. **23**(1), 128–168 (2011)
13. Hyndman, R., Koehler, A., Snyder, R., Grose, S.: A state space framework for automatic forecasting using exponential smoothing methods. Int. J. Forecast. **18**(3), 439–454 (2002)
14. Gama, J., Pinto, C.: Discretization from data streams: applications to histograms and data mining. In: Proceedings of the 2006 ACM Symposium on Applied Computing, pp. 662–667. ACM (2006)
15. Moreira-Matias, L., Gama, J., Mendes-Moreira, J., Freire de Sousa, J.: An incremental probabilistic model to predict bus bunching in real-time. In: Blockeel, H., van Leeuwen, M., Vinciotti, V. (eds.) IDA 2014. LNCS, vol. 8819, pp. 227–238. Springer, Heidelberg (2014)

PULSE: A Real Time System for Crowd Flow Prediction at Metropolitan Subway Stations

Ermal Toto[1]([⊠]), Elke A. Rundensteiner[1], Yanhua Li[1], Richard Jordan[2], Mariya Ishutkina[2], Kajal Claypool[2], Jun Luo[3,4], and Fan Zhang[3]

[1] Worcester Polytechnic Institute, Worcester, USA
toto@wpi.edu
[2] MIT Lincoln Laboratory, Lexington, USA
[3] Shenzhen Institutes of Advanced Technology,
Chinese Academy of Sciences, Chengdu, China
[4] Lenovo Group Limited, Hong Kong SAR, China

Abstract. The fast pace of urbanization has given rise to complex transportation networks, such as subway systems, that deploy smart card readers generating detailed transactions of mobility. Predictions of human movement based on these transaction streams represents tremendous new opportunities from optimizing fleet allocation of on-demand transportation such as UBER and LYFT to dynamic pricing of services. However, transportation research thus far has primarily focused on tackling other challenges from traffic congestion to network capacity. To take on this new opportunity, we propose a real-time framework, called PULSE (**P**rediction Framework For **U**sage **L**oad on Subway Syst**E**ms), that offers accurate multi-granular arrival crowd flow prediction at subway stations. PULSE extracts and employs two types of features such as streaming features and station profile features. Streaming features are time-variant features including time, weather, and historical traffic at subway stations (as time-series of arrival/departure streams), where station profile features capture the time-invariant unique characteristics of stations, including each station's peak hour crowd flow, remoteness from the downtown area, and mean flow. Then, given a future prediction interval, we design novel stream feature selection and model selection algorithms to select the most appropriate machine learning models for each target station and tune that model by choosing an optimal subset of stream traffic features from other stations. We evaluate our PULSE framework using real transaction data of 11 million passengers from a subway system in Shenzhen, China. The results demonstrate that PULSE

This work is sponsored by the Department of Air Force under Air Force Contract FA 8722-05-C-0002. Opinions, interpretations, conclusions and recommendations are those of the authors and not necessarily endorsed by the United States Government. Yanhua Li is partly supported by a gift funding from Pitney Bowes, Inc. Jun Luo is partly supported by the National Natural Science Foundation of China (Grant No. 11271351). Prof. Rundensteiner also thanks Dept of Education for Phd student support on GAANN grant (P200A150306) for Big Data Computing Research supporting Ermal Toto and other Phd students. Prof. Rundensteiner thanks NSF for grants IIS-1018443 and CRI-1305258.

© Springer International Publishing AG 2016
B. Berendt et al. (Eds.): ECML PKDD 2016, Part III, LNAI 9853, pp. 112–128, 2016.
DOI: 10.1007/978-3-319-46131-1_19

greatly improves the accuracy of predictions at all subway stations by up to 49 % over baseline algorithms.

1   Introduction

Background. Subway systems provide unobstructed transit throughout an urban area. Starting in the early 90s, in order to streamline fare collection, subway authorities have implemented smart card enabled entry and exit systems [21]. These widely adopted systems generate a large amount of fine-grained data about passengers' mobility throughout the transportation network. Offering new opportunities in gaining in-depth insights into the performance and effectiveness of the system as well as the passenger mobility patterns.

Motivation. However a recent survey of smart card transaction usage [21] found that current research is limited to simple post-hoc analysis of generalized mobility patterns, thus risks missing potentially valuable opportunities for new mobility-related services. Predictions of crowd flow arriving at subway stations based on fine-grained smart card transaction streams open tremendous new opportunities for novel services, including optimizing fleet allocation and introducing dynamic fares in on-demand systems [20,22]. In addition, traditional transportation modes such as buses would also benefit from mobility prediction capabilities that would allow them to dynamically adjust stop frequency and routes [10,12]. These new classes of services increase quality of service and reduce emissions.

Limitations of the State of Art. In the literature, traffic prediction on road networks has been studied extensively, and many prediction models have been applied and developed [8,13,15,25,26,28,31]. However, when applying these methods directly on solving the arrival crowd flow prediction at subway stations, they fail to achieve high prediction accuracy, because these (general) methods do not explicitly take into account the unique features and characteristics of subways systems, such as the pairwise crowd flow between stations, attrition rate of subway stations, etc. Such arrival crowd flow prediction problem is challenging in practice. Figure 1(a) shows that the arrival crowd flows at different stations exhibit completely different time-series patterns, while Fig. 1(b) shows that for the same station, the arrival crowd flow changes its pattern over different days.

Our Proposed Approach. Given these challenges, in this paper, we make the first attempt to study the crowd flow prediction problem at subway stations. We propose a novel real-time framework, called PULSE (**P**rediction Framework For **U**sage **L**oad on Subway Syst**E**ms), that offers accurate multi-granular arrival crowd flow prediction at subway stations. Below we summarize our main contributions in this paper.

- PULSE extracts two types of features for the arrival crowd flow prediction, i.e., streaming features and station profile features. Streaming features are time-variant features including time, weather, and historical traffic at subway stations (as time-series of arrival/departure streams), where station profile

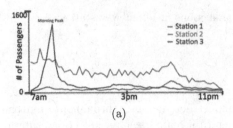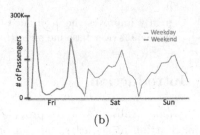

(a)   (b)

Fig. 1. (a) Time series of passenger arrivals at 3 stations during a Monday. (b) System wide traffic during three consecutive days.

features capture the time-invariant unique characteristics of stations, including each station's peak hour crowd flow, remoteness from the downtown area, and mean flow. (See Sect. 4)

- PULSE employs a novel stream feature selection algorithm and a model selection algorithm to select the most appropriate machine learning model for each target station and tune that model by choosing an optimal subset of stream traffic features from other stations. (See Sects. 5 and 6)
- We evaluate our PULSE framework using real transaction data of 11 million passengers from a subway system in Shenzhen, China. The results demonstrate that PULSE greatly improves the accuracy of predictions at all subway stations by up to 49 % over baseline algorithms. (See Sect. 7)

2   Related Work

In this section, we briefly discuss two research areas that are closely related to this work, namely, urban computing and traffic prediction.

Urban computing studies the impact and application of technology in urban areas, including the collection and usage of smart card transactions. Analyzing smart card records is an effective way of understanding human mobility patterns in urban areas [18,21]. Various studies [6,7,16,18] show that city wide mobility follows a common pattern that is consistent across cities and modes of transportation. These studies describe mobility patterns, but fall short of developing a framework for fine-grained predictions of human mobility. To our knowledge this study is the first to directly address the prediction of arrival crowd flow in a subway network.

Traffic prediction in road networks has been studied extensively [8,13,15,25, 26,28,31]. In this study, we compare and contrast the most commonly used machine learning models as baseline methods. One of these baselines (Multiple Linear Regression–MLR) is described in [26], where it is used to capture short term traffic trends. In another study [8] non-parametric models similar to K-Nearest Neighbours (KNN) are used for road traffic flow predictions. The concept of using ensembles of models is used in [25], where a state machine switches among different Auto-regressive Moving Average Models (ARIMA) [15].

In [13], Random Forest models are used for short term context aware predictions. All these traffic prediction methods are addressing vehicle traffic prediction problem and utilize a fixed (sometimes ensemble) model to conduct the traffic prediction. Thus, when applied to our crowd flow prediction problem at subway stations, these methods would fail to capture unique features and choose appropriate models for a subway system.

In summary, PULSE is the first framework that enables fine-grained arrival crowd flow predictions at subway stations, using smart card transaction data, weather data, and calendar data.

3   Overview

In this section, we define the subway traffic prediction problem and outline the framework of our methodology.

3.1   Preliminary and Problem Definition

We worked on transaction data generated from the subway system in Shenzhen, China. Similar to many other subway systems in different cities, such as Beijing Subway[1], and London Subway[2], a passenger needs to swipe his smart card at both the entering and leaving stations. Such paired transaction records capture the trip information of passengers. Below, we explicitly highlight the key terms used in the paper, and define the subway station traffic prediction problem.

Definition 1 (Trip). $tr = (p_{id}, s_d, t_d, s_a, t_a)$ *represents a trip made by a passenger with ID p_{id}, who departs from station s_d at time t_d and arrives at the station s_a at time t_a. **TR** represent the set of all trips, i.e., $tr \in$ **TR**.*

Definition 2 (Subway Trajectory). *A subway trajectory is a sequence of subway stations that a passenger enters and leaves in the subway system as a function of time. Each record thus consists of a passenger ID p_{id}, subway station ID s, and a time stamp t.*

Definition 3 (Subway Network). *A subway network consists of a set of subway stations connected by subway lines. We represent a subway network as a undirected graph $G = (V, E)$, where V represents the subway station set and E contains the edges between neighboring subway stations via subway lines.*

Problem Definition. Given a set of historical trips **TR**, the subway network G, and the current time t, we aim to predict the number of passengers arriving at a subway station $s \in V$ (from other stations) during the consecutive time intervals $[t + T * (k - 1), t + T * k]$, with $1 \le k \le K$. T is a time aggregation interval, which is usually 15 min. K denotes the number of future intervals to be predicted, and we use $K = 6$ in this paper.

---

[1] http://www.bmac.com.cn.
[2] https://oyster.tfl.gov.uk/oyster/entry.do.

3.2   The PULSE Framework

To tackle the above subway station traffic prediction problem, we introduce PULSE framework (**P**rediction Framework For **U**sage **L**oad on **S**ubway
Syst**E**ms) as shown in Fig. 2. PULSE takes the historical trip data, calendar
information, and weather data as input, to predict future traffic flows at each
subway station at fine-grained periodic intervals e.g., every 15 min. This task is
achieved in three core steps, namely, feature extraction, traffic prediction, and
model update, as outlined next.

Feature extraction module aggregates the time-varying data sources,
such as the transaction data,
weather data, calendar data, at certain time granularity, e.g., 15 min.
Then, we extract and model
both *streaming* and *profile* features. Streaming features are direct
aggregates of the time-varying
datasets, including aggregated traffic volumes entering and leaving a
subway station and weather statistics. Profile features describe relatively stable characteristics of each
station, including remoteness of a
station, peak-hour traffic, average
inflow at a station. See more details
in Sect. 4.2.

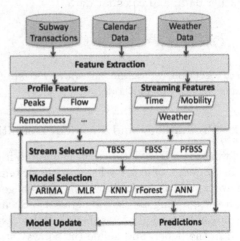

Fig. 2. The PULSE framework.

Traffic prediction. When predicting the entering and leaving traffic at a subway station s_i, the traffic prediction module employs an automatic feature and
model selection algorithm that achieves high prediction accuracy. A prediction
model is chosen and a subset of subway stations are selected to include their
streaming features as training data. The model and features selected are used
to perform predictions on the future entering and leaving traffic at each subway
station. Section 5 describes this process in more detail.

Model update module keeps track of the performance of the PULSE system
over time. It automatically re-selects features and rebuilds the models.

4   Feature Extraction for PULSE

The feature extraction module explores two sets of key features, namely streaming features and station profile features. The former capture the dynamics of
departing/arriving traffic at different stations and the meteorological features
over time; while the latter characterizes the time-invariant profiles of different
subway stations, including remoteness from the city center, the mean flow, peak-
hour traffic, etc.

4.1   Streaming Features

4.1.1   Time Features F^t

As discussed earlier, the departing and arriving transaction data
are aggregated at a certain time
granularity, e.g., $T = 15\,\mathrm{min}$. We
observe that the daily operation
time of a subway system, denoted
as T_0, is usually less than 24 h.
For example, in Shenzhen, the
subway system operates between
7 am and 11 pm every day, that is,

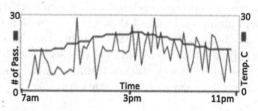

Fig. 3. Temperature and number of arrivals
during a Saturday.

a total of $T_0 = 16\,\mathrm{h}$ of operation time. Hence, given the time aggregation interval
T, the daily operation time T_0 is divided into a fixed number of time slots with
equal length of T minutes. For example, a total of 64 such intervals are obtained
given $T = 15\,\mathrm{min}$ and $T_0 = 16\,\mathrm{h}$. We then use the interval id $F_{int} \in [1, 64]$ to
represent the **time of day** as a feature. As observed in [6,16,18], this feature
is significant in urban human mobility predictions. Similarly, we introduce the
feature **day of the week**, that distinguishes between weekdays from Monday
to Sunday, which can be represented using the weekday id, namely, $F_{day} \in [1, 7]$.
As shown in Fig. 1b, The traffic patterns vary significantly during the different
days of the week as it is also observed in [16,18].

4.1.2   Traffic Stream Features F^s

Given an aggregation interval T, we can obtain the arrival and departure traffic
at each subway station during each time interval T. For one station s_i, we denote
the vector $F_i^{arr} = [a_1, a_2, \ldots, a_N]$ as the **arrival stream feature** of a station
s_i, where N is the total number of time intervals in the data. Given a starting
time t_0, each a_ℓ represents the number of passengers who arrived at the station
s_i, during the ℓ-th time interval, namely, $T_\ell = [t_0 + T * (\ell - 1), t_0 + T * \ell]$. Hence,
each a_ℓ can be obtained from the trip data as follows.

$$a_\ell = \sum_{tr \in \mathbf{TR}} I(tr.s_a = s_i, tr.t_a \in T_\ell), \tag{1}$$

where $I(\cdot)$ is the indicator function, which is 1 if the condition holds, and
0 otherwise. Similarly, we define the **departure stream feature** of a station s_i as a vector $F_i^{dep} = [d_1, d_2, \ldots, d_N]$. Each d_ℓ can be represented as
$d_\ell = \sum_{tr \in \mathbf{TR}} I(tr.s_d = s_i, tr.t_d \in T_\ell)$. When considering pair-wise flows between
station pairs, $F_{i,j}^{pair} = [p_1, p_2, \ldots, p_N]$ is the **pairwise flow feature**. p_ℓ representing the number of trips from station s_i to station s_j during the time inter-
val T_ℓ, namely, $p_\ell = \sum_{tr \in \mathbf{TR}} I(tr.s_d = s_i, tr.s_a = s_j, tr.t_d \in T_\ell, tr.t_a \in T_\ell)$.
We also take into account $F_{i,j}^{dur} = [\pi_1, \pi_2, \ldots, \pi_N]$ as the vector **average trip
duration feature** from station s_i to s_j during the time interval T_ℓ. Each
$\pi_\ell = \frac{1}{p_\ell} \sum_{tr \in \mathbf{TR}} (tr.s_a - tr.s_d) I(tr.s_d = s_i, tr.s_a = s_j, tr.t_d \in T_\ell, tr.t_a \in T_\ell)$.

4.1.3 Weather Features F^w

The traffic at subway stations is affected by meteorology. Hence, we identify two features that are correlated with the subway stations traffic, namely temperature and humidity. Figure 3 shows the correlation between the subway station traffic and the temperature feature, using the data we collected during 03/20/2014–03/31/2014 in Shenzhen. We can see that the temperature is positively correlated with subway station traffic, similarly our data indicates that humidity is correlated negatively with station traffic.

4.2 Station Profile Features

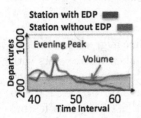

In this section, we present the time-invariant profile features extracted from each subway station. These features capture the unique profile of each subway station from different aspects, such as peak-hour traffic, mean flow, and remoteness from the city center.

4.2.1 Peak Traffic F^P

Fig. 4. Equivalent traffic volumes, but different peak patterns for stations with (green) and without (orange) an Evening Departure Peak (EDP). (Color figure online)

Crowd movement during commute hours shows unique and characteristic peak patterns that vary between stations, but are relatively stable over time. In our study, we choose the peak hours as 7–11 am and 5–11 pm. A naive way of characterizing the peak-hour behavior is to use total traffic volume. This approach may miss important information of the underlying traffic dynamics. For example, as shown in Fig. 4, two stations have the same peak-hour traffic volume, namely, the total area between the traffic curve and the x-axis. However, we observe that station 1 shows a flat traffic pattern during the peak-hour, while station 2 has one significant spike. To capture such spike, we employ the Tukey [27] outlier detection method to identify the outliers in the peak-hour, and count the number of outliers as the **peak-hour traffic feature**. In Fig. 5, we use the morning arrival peak-hour traffic as an example. Similarly, we can obtain the peak-hour traffic for evening arrival, evening departure, and morning departure, respectively.

4.2.2 Flow Related Features F^F

We introduce two types of flow related features, including attrition rate and mean flow of a station. **Attrition Rate.** For a station s_i, we define the attrition rate Att_i as the relative difference between departures and arrivals at s_i. As is observed in [18], most departure trips from a station s_i have a matching arrival trip. However, attrition rates in Shenzhen subway data vary considerably as illustrated in Fig. 6. $Att_i = (|F_i^{dep}| - |F_i^{arr}|)/|F_i^{arr}|$.

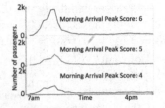

Fig. 5. Arrival streams with different morning peaks

Mean Flow of a station s_i (denoted by F_i^{flow}) is the average number of arrivals per interval, which can be calculated as $F_i^{flow} = |F_i^{arr}|/N$. Figure 7b illustrates the flow at each subway station. As expected, downtown areas and commercial centers show high concentrations of passenger arrivals.

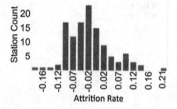

Fig. 6. Distribution of attrition rate.

4.2.3 Remoteness F^R

From the subway transaction data, we observe that in general stations located farther away from the downtown area tend to have similar traffic patterns and overall fewer traffic. This motivates us to extract the remoteness of station s_i as a feature, i.e., F_i^R. F_i^R is the average duration of the historical trips arriving at s_i, namely, $F_i^R = \sum_{tr \in \mathbf{TR}}(tr.t_a - tr.t_d).I(tr.s_a = s_i)$. Figure 7a illustrates the geographic distribution of remoteness.

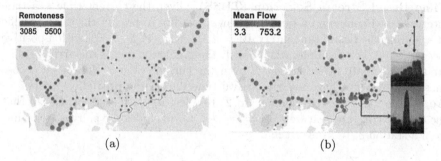

(a)   (b)

Fig. 7. Geographic distribution of (a) remoteness and (b) mean station flow.

5   Station Stream Selection

Our focus in this work is *arrival traffic prediction* at subway stations. Given a target station s_i, its historical traffic data as a time-series can be used to predict its future arrival traffic, e.g., [15]. In general, subway stations are interconnected, and the arrival traffic at one particular subway station s_i is affected and generated by the traffic from all other stations (in V/s_i). However, given s_i, it is computationally efficient in practice to include a subset of stations (instead of all stations), which contribute significantly to the arrival traffic at s_i, i.e., they are geographically close by, or they originate a significant amount of traffic flow to the target station. In this section, we present our stream selection algorithm, that can identify the subset of stations, whose departure traffic (as a key feature) contributes the most to the traffic at the target station. Our selection algorithm combines three criteria, including Time Based Stream Selection (TBSS),

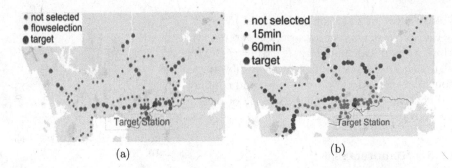

Fig. 8. Selecting streams based on (a) pairwise flow and (b) temporal distance.

Flow Based Stream Selection (FBSS), and Profile Based Stream Selection (PBSS). Below, we elaborate on each selection criterion and the overall stream selection algorithm.

Time Based Stream Selection (TBSS). Given the current time t, a time interval $T = 15\,\text{min}$, and a target station s_i, we aim to predict the arrival traffic at s_i during the future time interval $\phi = [t + T * (k-1), t + T * k]$ with a positive integer $k > 0$. For example, when $k = 1$, the prediction yields the arrival traffic for the immediate time interval T from the current time t. Hence, we choose those stations that have average arrival time during the prediction interval ϕ. We use the following criterion (in Eq. 2) to select θ_L such stations. Recall that the average trip time feature $F_{j,i}^{dur} = [\pi_1, \cdots, \pi_N]$ includes the pairwise trip time from a station s_i to s_j over time.

$$L_{i,\phi}(\theta_L) = \operatorname*{argmin}_{B^{\theta_L} \subset V/s_i} \sum_{s_j \in B^{\theta_L}} \left(\sum_{\pi \in F_{j,i}^{dur}} \left| T\left(k - \frac{1}{2}\right) - \pi \right| \right). \tag{2}$$

$L_{i,\phi}(\theta_L)$ is the set of θ_L selected stations. The value of θ_L is selected by the model selection module (See Sect. 6) to achieve high prediction accuracy.

Figure 8b illustrates the set of stations selected by TBSS for with $\theta_L = 20$, $T = 15\,\text{min}$, and two values of k (orange, $k = 1$ and green, $k = 4$).

Flow Based Stream Selection (FBSS). FBSS is based on the intuition that future traffic at station s_i will come from (departures of) stations with most historical trips to s_i. Recall that the pairwise flow feature $F_{j,i}^{pair} = [p_1, \cdots, p_N]$ includes the numbers of pairwise trips from a station s_i to s_j over time. $M_{i,\phi}(\theta_M)$ is the set containing θ_M stations with the highest number of trips to s_i, as illustrated in Eq. 3 where $|F_{j,i}^{pair}|$ indicates the total number of trips from station s_j to s_i and θ_M is again chosen by the model selection module. An example of stations selected by FBSS is given in Fig. 8a.

$$M_{i,\phi}(\theta_M) = \underset{B^{\theta_M} \subset V/s_i}{\text{argmax}} \sum_{s_j \in B^{\theta_M}} |F_{j,i}^{pair}|. \tag{3}$$

Profile Based Stream Selection (PBSS). Profile features characterize the overall traffic patterns of subway stations. Stations with similar profile features tend to have similar traffic patterns over time. Given a target station s_i, its profile feature vector is $PF_i = [F_i^P, F_i^F, F_i^R]$, where F^P, F^F and F^R represent the peak traffic features, flow related features, and remoteness features, respectively. PF_i is compared to PF_j for each $s_j \in V$ and a set $K_{i,\phi}(\theta_K)$ of the θ_K nearest (in terms of profile features) stations is selected as illustrated in Eq. 4. The optimal value for θ_K is determined during model selection.

$$K_{i,\phi}(\theta_K) = \underset{B^{\theta_K} \subset V/s_i}{\text{argmin}} \sum_{s_j \in B^{\theta_K}} \left(\sqrt{\sum_{n=1}^{|PF|} \left(PF_i^n - PF_j^n \right)^2} \right). \tag{4}$$

Stream selection. The final set of stations is simply the union set of the results from three criteria, i.e., $L_{i,\phi}(\theta_L) \cup M_{i,\phi}(\theta_M) \cup K_{i,\phi}(\theta_K)$.

The pseudocode for the stream selection is given in Algorithm 1. In Lines 2–6, the procedure iterates through all stations $s_j \in V/s_i$ and calculates the time distances, pairwise flows, and profile feature Euclidean distances between stations s_i and s_j. In lines 7–12, these distances are sorted, and the first θ_L, θ_M, and θ_K, streams are selected. Line 13 returns the union of the three stream sets.

Algorithm 1. Stream selection for station s_i

1 **function** StreamSelection $(s_i; \phi; F_{i,j}^{dur}; F_{i,j}^{pair}; PF; \theta_L; \theta_M; \theta_K)$;

Input : Station s_i. Prediction interval ϕ. Sets $F_{i,j}^{dur}, F_{i,j}^{pair}$, and PF. Number of streams to be selected defined by θ_L, θ_M, and θ_K.

Output: $L_{i,\phi}^{\theta} \cup M_{i,\phi}^{\theta} \cup K_{i,\phi}^{\theta}$

2 **for** $s_j \in V/s_i$ **do**

3      $timedistance[j] = |average(F_{i,j}^{dur}) - T * (k - 1/2)|$;

4      $flow[j] = |F_{i,j}^{pair}|$;

5      $pfdistances[j] = euclidiandistance(PF_i, PF_j)$;

6 **end**

7 $timedistances = sort(timedistances)$;

8 $flow = sort(flow)$;

9 $pfdistances = sort(pfdistances)$;

10 $L_{i,\phi}^{\theta} = getKeys(timedistances[1..\theta_L])$;

11 $M_{i,\phi}^{\theta} = getKeys(flow[1..\theta_M])$;

12 $K_{i,\phi}^{\theta} = getKeys(pfdistances[1..\theta_K])$;

13 **return** $L_{i,\phi}^{\theta} \cup M_{i,\phi}^{\theta} \cup K_{i,\phi}^{\theta}$;

6 Model Selection

To accurately predict the arrival traffic for a prediction interval ϕ at a target station s_i, we need to choose the right prediction model and the right set of stream features from other stations, namely, θ_L, θ_M, θ_K. We consider five candidate prediction models used in the literature for time-series data prediction, including Autoregressive integrated moving average (ARIMA) [15,25], Artificial Neural Networks (ANN) [19,28,30,31], K-Nearest Neighbours (KNN) [8,9,11], Random Forest (RF) [13,14,17], and Multiple Linear Regression (MLR) [26]. The system also needs to choose the optimal number of streams to be included using the methods described in Sect. 5. In our study, the Shenzhen subway system has five subway lines with 118 subway stations. Thus each parameter θ_L, θ_M, and θ_K can vary from 1 to 118, leading to a search space of 118^3. Each model configuration setup requires training and testing using historical data.

To find the optimal configuration of model and stream set for a station s_i and prediction interval ϕ requires examining all configurations with different model and stream combinations. **A naive method** is to brute force all such configurations, and choose the one with the highest prediction accuracy. However, this is too costly to be implemented in practice. To be precise, we have five prediction models and 118^3 possibilities of stream set sizes. Let's consider 6 future prediction intervals and different temporal partitions, which in this set of experiments is two (weekdays and weekends). In total, there are about 79 million different models. We ran our experiments in a server with 30 Intel(R) Xeon(R) CPU E5-4627 v2 @ 3.30 GHz Cores. Each model training and testing would take about 1 to 15 s, which leads to a total of 14 years to compare all configurations using our 30 core system. Thus, we are motivated to employ the profile features to conduct **Gradient-based optimization of hyper-parameters** [4,5] to optimize this process. Initially this method uses a pure gradient search approach to discover parameters. As more station profiles are matched to models, PULSE can initiate subsequent searches with model parameters from stations with similar profiles as described by Eq. 5. Henceforth, we refer to this method as Model Select (MSELECT). After a large number of stations have been assigned with prediction models, the process only takes a few seconds. Therefore this method is suitable as an online process for model updates based on changes in the profile features. Our gradient based model search takes approximately 2 h to find the optimal prediction configuration for all 118 stations in this study.

$$Model_i = \underset{Model_j \in Models}{\arg\min} \left[\sqrt{\sum_{n=1}^{|PF|} \left(PF_i^n - PF_j^n \right)^2} \right]. \tag{5}$$

Model update. PULSE monitors the prediction performance over time. It automatically re-selects features and rebuilds the models when the average prediction accuracy goes below a certain threshold value.

7 Evaluation of PULSE Model

To evaluate the performance of our PULSE framework on arrival traffic prediction, we conducted comprehensive experiments using a real subway transaction dataset collected from Shenzhen subway system for 21 days in March 2014. By comparing with baseline algorithms, the experimental results demonstrate that PULSE can achieve a 26 %–94 % relative prediction accuracy, which is on average 20 % higher than baseline algorithms. Below, we present the datasets, baseline algorithms, experiment settings and results.

7.1 Dataset Description

For this work, we used 60 million smart card transactions from the subway system in the city of Shenzhen, China between March 10th and March 31st, 2014. The dataset contains 11 million unique passengers (identified by their smart card ids). Each transaction contains a timestamp, location coordinates, and whether the transaction is a departure from or an arrival at a station. During data preprocessing we matched entry and exit transactions for each passenger in order to generate a trip record $tr = (p_{id}, s_d, t_d, s_a, t_a)$ containing a passenger identifier p_{id}, a starting station s_d, a destination s_a and respective departure and arrival times t_d and t_a.

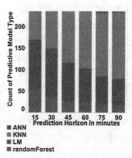

Fig. 9. Distribution of the best performing models over the prediction horizon.

7.2 Evaluation Settings

PULSE predicts the number of arrivals at a station s_i at future time intervals in $[t + T * (k − 1), t + T * k]$ with $1 \leq k \leq K$. In our evaluation of PULSE, we used a variable $k \in [1, \cdots, 6]$.

Prediction models for both PULSE and the baseline methods are trained using a sliding window containing a week of historical data to predict the arrival traffic of a future interval specified by k. The accuracy of the predictions is defined as $accuracy = 1 − \frac{\sum |\hat{y}_i − y_i|}{\sum y_i}$. Again, we consider five prediction models used in the literature for time-series data prediction, including Autoregressive integrated moving average (ARIMA) [15,25], Artificial Neural Networks (ANN) [19,28,30,31], K-Nearest Neighbours (KNN) [8,9,11], Random Forest (RF) [13,14,17], and Multiple Linear Regression (MLR) [26]. All these methods can be setup as both single stream (only using the features of the target station) or multi-stream models (using features from both the target station and other selected stations)[3]. In our experiments, we evaluate the PULSE framework in two stages.

---

[3] Note that ARIMA can only be setup as a single stream model by its design in nature.

In the *first stage*, we run all prediction models in a single-stream fashion using the arrival stream feature F_i^{arr} of the target station s_i, with vs without other streaming features, such as time feature F^T and weather features F^W.

In the *second stage*, we evaluate the stream feature selection and model selection algorithms introduced in Sects. 5 and 6 in a multi-stream scenario. We compare our PULSE framework with each individual model under the single-stream mode. The evaluation results are summarized in the next subsection.

7.3   Evaluation Results

Stage 1: Single-stream models. In Table 1, the column *BaseL No SF* lists the baseline results of single stream models, that only use the arrival stream feature of the target station. The column *BaseL SF* lists the results of single stream models, that include both the arrival stream feature of the target station, and also other streaming features introduced in Sect. 4.1, such as the weather and time features. The results show that by introducing time and weather features, the prediction accuracy for the single-stream models is improved on average 13.4 % and up to 21.7 %, namely, from 60 %–75.8 % to 76.9 %–81.7 %, respectively.

When we look at the different prediction horizons from 15 min to 60 min ahead of time, the accuracy of all models (except ARIMA) decreases as the prediction horizon increases. This is reasonable since it is in general harder to predict the arrival traffic in a long term future interval than an immediate future interval.

Table 1. Overall performance evaluation at 118 stations.

	H.	KNN	MLR	RF	ANN	ARIMA	KNN	MLR	RF	ANN	MSEL
		BaseL No SF					BaseL SF				
W	*15*	0.738	0.735	0.735	0.750	0.746	0.872	0.848	0.860	0.836	0.884
D	*30*	0.658	0.647	0.657	0.672	0.745	0.872	0.846	0.855	0.840	0.883
a	*45*	0.575	0.560	0.574	0.595	0.745	0.870	0.837	0.850	0.840	0.882
y	*60*	0.526	0.509	0.525	0.548	0.745	0.868	0.831	0.848	0.834	0.881
	75	0.498	0.477	0.498	0.524	0.745	0.865	0.824	0.845	0.832	0.880
	90	0.488	0.462	0.489	0.516	0.744	0.862	0.818	0.842	0.825	0.879
W	*15*	0.752	0.784	0.749	0.780	0.772	0.770	0.726	0.801	0.724	0.845
E	*30*	0.712	0.760	0.707	0.755	0.772	0.768	0.667	0.791	0.718	0.841
n	*45*	0.639	0.702	0.631	0.698	0.771	0.761	0.603	0.763	0.705	0.833
d	*60*	0.585	0.662	0.578	0.649	0.771	0.760	0.573	0.745	0.693	0.827
	75	0.540	0.623	0.535	0.610	0.769	0.762	0.572	0.731	0.687	0.820
	90	0.518	0.601	0.516	0.590	0.771	0.770	0.590	0.728	0.699	0.813
Av.		0.602	0.627	0.600	0.641	**0.758**	**0.817**	0.728	0.805	0.769	**0.856**

Stage 2: Multi-stream models. In Table 1, the last column *MSEL* lists the results of multi-stream models, when stream feature selection and model selection algorithms are applied to include departure stream features from other

Table 2. Stations with top improvement in prediction accuracy.

	Rank	Station ID	ML	H	TBSS	FBSS	PBSS	KNN	M.Select	Diff
Week days	1	260011	LM	90	0	0	0	0.709	0.769	0.060
	2	260024	RF	30	30	10	20	0.465	0.523	0.058
	3	260024	RF	45	30	40	0	0.465	0.521	0.056
	4	268028	RF	15	40	40	20	0.469	0.522	0.053
	5	268023	KNN	90	40	40	40	0.871	0.921	0.050
Week ends	1	261006	RF	45	0	0	10	0.264	0.755	0.491
	2	268023	KNN	60	30	0	40	0.334	0.814	0.481
	3	268012	KNN	60	20	20	30	0.618	0.854	0.236
	4	261006	KNN	90	0	20	10	0.481	0.716	0.234
	5	263013	KNN	15	30	0	10	0.512	0.739	0.228

stations than the target station. We observed that the average prediction accuracy is further improved to 85.6 % over single-stream models, with an average of 7.6 % improvement over *BaseL SF* and 21 % improvement over *BaseL No SF*.

Table 2 lists the evaluation results of the stations with the top five improvement on the prediction accuracy for weekdays and weekends, respectively. During weekends, the first ranked station (in terms of model improvement) has a prediction accuracy as low as 26.4 % at 45 min prediction horizon when using KNN (the best performing single-stream baseline) with all streaming features. By applying stream feature selection and model selection algorithms, PULSE increases the prediction accuracy of this model to 75.5 % with a total of 49.1 % improvement. This was achieved by using a Random Forest model with 10 streams that were selected using profile based stream selection (PBSS). Overall, the stream feature selection and model selection algorithms improve the prediction accuracy more during the weekends (up to 49.1 % improvement) than the weekdays (up to 6 %). This happens primarily because the arrival traffic in weekends is less stable than during weekdays, and single-stream models have low prediction accuracy, providing more room to improve the performance when stream feature selection and model selection algorithm are used.

Summary and Observations. The above results with single-stream models demonstrate that by introducing time and weather features, the prediction accuracy is improved on average 13.4 %. For multi-stream models, our PULSE framework further improves the prediction accuracy by an average of 7.6 %. To better understand the evaluation results, Fig. 10(a, b) presents the prediction accuracy distribution at all stations as a function of their mean arrival flow for single stream model (KNN) in Fig. 10(a) vs multi-stream models in Fig. 10(b). We observed that stations with lower mean arrival traffic had the most improvement. When we looked at the best models being selected by our model selection algorithm over different prediction horizons, we noticed that there is a clear shift in the machine learning models with increasing prediction horizons (Fig. 9). For example, linear model (LM) and Random forest (RF) are used more for smaller

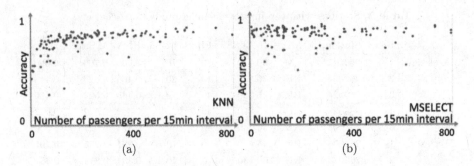

Fig. 10. (a) KNN vs (b) MSELECT weekend prediction accuracy at 60 min horizon, for stations with different mean passenger flow.

prediction horizons (i.e., predicting the near future), while k-nearest neighbors (KNN) in general performs better for larger prediction horizons (i.e., predicting the long term future intervals). These observations shed light on the performances of different models in subway station traffic predictions.

8   Conclusion

In this study we present PULSE, a real-time system to predict arrival crowd flow at metropolitan subway stations. The system extracts streaming features and station profile features from heterogeneous urban data, including subway transaction data, weather data, and calendar data. PULSE employs novel stream feature selection and model selection algorithms to improve the prediction accuracy and running time. Experimental results on real subway transaction data from 11 million passengers in Shenzhen, China demonstrated that PULSE can increase the prediction accuracy by up to 49 % over baseline algorithms.

References

1. Statistic Brief: World Metro Figures, 1st ed. UITP, Brussels (2014)
2. United Nations: World Urbanization Prospects 2014: Highlights. United Nations Publications (2014)
3. Annez, P.C., Buckley, R.M.: Urbanization and growth: setting the context. Urbanization Growth **1**, 1–45 (2009)
4. Bengio, Y.: Gradient-based optimization of hyperparameters. Neural Comput. **12**(8), 1889–1900 (2000)
5. Bergstra, J.S., Bardenet, R., Bengio, Y., Kégl, B.: Algorithms for hyperparameter optimization. In: Advances in Neural Information Processing Systems, pp. 2546–2554 (2011)
6. Chakirov, A., Erath, A.: Use of public transport smart card fare payment data for travel behaviour analysis in Singapore. Eidgenössische Technische Hochschule Zürich, IVT - Institut für Verkehrsplanung und Transportsysteme (2011)

7. Cheng, Y.Y., Lee, R.K.W., Lim, E.P., Zhu, F.: Measuring centralities for transportation networks beyond structures. In: Kazienko, P., Chawla, N. (eds.) Applications of Social Media and Social Network Analysis, pp. 23–39. Springer, Switzerland (2015)

8. Clark, S.: Traffic prediction using multivariate nonparametric regression. J. Transp. Eng. **129**(2), 161–168 (2003)

9. Cover, T., Hart, P.: Nearest neighbor pattern classification. IEEE Trans. Inf. Theory **13**(1), 21–27 (1967)

10. Fu, L., Liu, Q., Calamai, P.: Real-time optimization model for dynamic scheduling of transit operations. J. Transp. Res. Board **1857**, 48–55 (2003)

11. Fukunaga, K., Narendra, P.M.: A branch and bound algorithm for computing k-nearest neighbors. IEEE Trans. Comput. **100**(7), 750–753 (1975)

12. Furth, P., Rahbee, A.: Optimal bus stop spacing through dynamic programming and geographic modeling. J. Transp. Res. Board **1731**, 15–22 (2000)

13. Hamner, B.: Predicting travel times with context-dependent random forests by modeling local and aggregate traffic flow. In: 2010 IEEE International Conference on Data Mining Workshops, pp. 1357–1359. IEEE (2010)

14. Ho, T.K.: Random decision forests. In Proceedings of the Third International Conference on Document Analysis and Recognition, vol. 1, pp. 278–282. IEEE (1995)

15. Jenkins, G.M., Reinsel, G.C.: Time Series Analysis: Forecasting and Control. Wiley, Hoboken (1976)

16. Li, Y., Zheng, Y., Zhang, H., Chen, L.: Traffic prediction in a bike-sharing system. In: Proceedings of the 23rd SIGSPATIAL International Conference on Advances in Geographic Information Systems, p. 33. ACM (2015)

17. Liaw, A., Wiener, M.: Classification and regression by randomforest. R News **2**(3), 18–22 (2002)

18. Liu, L., Hou, A., Biderman, A., Ratti, C., Chen, J.: Understanding individual and collective mobility patterns from smart card records: a case study in Shenzhen. In: 12th International IEEE Conference on Intelligent Transportation Systems, ITSC 2009, pp. 1–6. IEEE (2009)

19. McCulloch, W.S., Pitts, W.: A logical calculus of the ideas immanent in nervous activity. Bull. Math. Biophys. **5**(4), 115–133 (1943)

20. Moreira-Matias, L., Gama, J., Ferreira, M., Mendes-Moreira, J., Damas, L.: Predicting taxi-passenger demand using streaming data. IEEE Trans. Intell. Transp. Syst. **14**(3), 1393–1402 (2013)

21. Pelletier, M.P., Trépanier, M., Morency, C.: Smart card data use in public transit: a literature review. Transp. Res. Part C: Emerg. Technol. **19**(4), 557–568 (2011)

22. Qian, S., Cao, J., Mouël, F.L., Sahel, I., Li, M.: SCRAM: a sharing considered route assignment mechanism for fair taxi route recommendations. In: Proceedings of the 21th ACM SIGKDD International Conference on Knowledge Discovery and Data Mining, pp. 955–964. ACM (2015)

23. Rabiner, L.R., Gold, B.: Theory and Application of Digital Signal Processing, 777 p., 1. Prentice-Hall Inc., Englewood Cliffs (1975)

24. Salnikov, V., Lambiotte, R., Noulas, A., Mascolo, C.: Openstreetcab: exploiting taxi mobility patterns in new york city to reduce commuter costs. arXiv preprint arXiv:1503.03021 (2015)

25. Stathopoulos, A., Karlaftis, M.G.: A multivariate state space approach for urban traffic flow modeling and prediction. Transp. Res. Part C: Emerg. Technol. **11**(2), 121–135 (2003)

26. Sun, H., Liu, H.X., Xiao, H., He, R.R., Ran, B.: Short term traffic forecasting using the local linear regression model. In: 82nd Annual Meeting of the Transportation Research Board, Washington, DC (2003)

27. Tukey, J.W.: Exploratory Data Analysis, 1st edn. Addison-Wesley, Reading (1977)

28. Vlahogianni, E.I., Karlaftis, M.G., Golias, J.C.: Optimized and meta-optimized neural networks for short-term traffic flow prediction: a genetic approach. Transp. Res. Part C: Emerg. Technol. **13**(3), 211–234 (2005)

29. Weisstein, E.W.: Fast fourier transform. From MathWorld-A Wolfram Web Resource (2015). http://mathworld.wolfram.com/FastFourierTransform.html

30. Yegnanarayana, B.: Artificial Neural Networks. PHI Learning Pvt. Ltd., New Delhi (2009)

31. Zheng, W., Lee, D.H., Shi, Q.: Short-term freeway traffic flow prediction: Bayesian combined neural network approach. J. Transp. Eng. **132**(2), 114–121 (2006)

Finding Dynamic Co-evolving Zones in Spatial-Temporal Time Series Data

Yun Cheng$^{(\boxtimes)}$, Xiucheng Li, and Yan Li

Air Scientific, Beijing, China
chengyun.hit@gmail.com, xiucheng90@gmail.com, yan.li@coilabs.com

Abstract. Co-evolving patterns exist in many Spatial-temporal time series Data, which shows invaluable information about evolving patterns of the data. However, due to the sensor readings' spatial and temporal heterogeneity, how to find the stable and dynamic co-evolving zones remains an unsolved issue. In this paper, we proposed a novel divide-and-conquer strategy to find the dynamic co-evolving zones that systematically leverages the heterogeneity challenges. The precision of spatial inference and temporal prediction improved by 7 % and 8 % respectively by using the found patterns, which shows the effectiveness of the found patterns. The system has also been deployed with the Haidian Ministry of Environmental Protection, Beijing, China, providing accurate spatial-temporal predictions and help the government make more scientific strategies for environment treatment.

Keywords: Air quality · Time series clustering · Co-evolving

1 Introduction

Spatio-temporal time series data has become ubiquitous thanks to affordable sensors and storage. Those invaluable data shows a potential to extract and understand complex spatio-temporal phenomena and their dynamics. Additionally, the ubiquitous sensor stations continuously measure several geophysical fields over large zones and long (potentially unbounded) periods of time, which highlights the importance of unsupervised methods in monitoring spatio-temporal dynamics with little or no human supervision.

Time series clustering are rapidly becoming popular data mining techniques. Lots of methods have been proposed to solve the problem [18]. Different dissimilarity measures for time series have been tested for various purposes. Yet, the ubiquitous sensor monitoring data is always spatio-temporal heterogenous, which means that different clustering structure may exist during the whole period. Furthermore, in the geo-sensory applications wherein a bundle of sensors are deployed at different locations to cooperatively monitor the target condition, groups of sensors are spatially correlated and co-evolve frequently in their readings and how to find those spatial co-evolving patterns is of great importance to various real-world applications [21]. When dealing with dense and continuous

© Springer International Publishing AG 2016
B. Berendt et al. (Eds.): ECML PKDD 2016, Part III, LNAI 9853, pp. 129–144, 2016.
DOI: 10.1007/978-3-319-46131-1_20

spatio-temporal data, the co-evolving sensors (zones) may change their sizes, shape and statistical properties over time (see Fig. 1). The goal is to find those dynamic co-evolving zones and try to establish linkages between those found zones and give reasonable explanations.

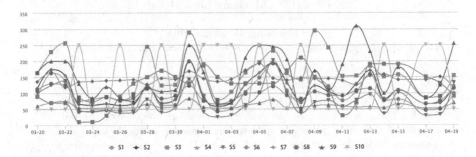

Fig. 1. The spatio-temporal air quality monitoring data (10 spatially adjacent sensor readings during one month).

In this paper, we propose a novel dynamic co-evolving zones discovery paradigm to identify co-evolving zones in continuous spatio-temporal field and establish linkages where the co-evolving zones may change their size, shape from time to time. Our paradigm first detects the overall breakout and divides the time series into uptrend and downtrend intervals. Then, we cluster the spatio-temporal time series data in each interval by using the specific dissimilarity measures. A hierarchical clustering method is used to deal with the found dynamic co-evolving zones in the previous step to give the final co-evolving structure. We evaluate our paradigm on a real world application of monitoring air quality which uses ubiquitous sensor stations on a regional scale (see our previous work [4] for details). The paradigm produced more stable and meaningful co-evolving zones and automatically found the segmentation intervals which helped better understand the evolving patterns of the pollution. We then use the found patterns to make spatial-temporal predictions and find an obvious improvement on the performance, which shows a potential usage area of the found dynamic co-evolving zones. Overall, our contribution has three parts:

- We proposed a novel paradigm to find the dynamic co-evolving zones and structures in the spatio-temporal time series data. The model uses three general key steps to deal with the spatio and temporal heterogenous to find co-evolving structures and patterns for future use.
- We use the patterns found in the co-evolving structures to increase the accuracy of spatial-temporal predictions and find a significant improvement compared with the original method.
- We use the proposed approach and result in a real world application, which has been used in the daily work of a environmental protection agency to help them make accurate predictions, do pollution causal analysis and make decisions or strategies.

2   Related Work

2.1   Problem Formulation

The goal of this work is to autonomously extract dynamic co-evolving zones from a continuous spatio-temporal field and give reasonable explanations. The dense deployment air quality monitoring data is an example of continuous spatio-temporal field, where each location has unique spatial coordinates and has co-evolving patterns with other sensors, which is changing over time. In the following subsection, we will first describe the existing approaches on the related topics, then gives our challenges.

2.2   Existing Approaches

Generally, our work is related to the following topics.

Time series change point detection. Sliding window, top-down, and bottom-up approaches [10] are popular methods to partition a time series into line segments. Wang et al. [17] proposed the pattern-based hidden Markov model that can segment a time series as well as learn the relationships between segments. Methods have also been proposed [9] to obtain piecewise polynomial approximations and/or perform on-line segmentation.

Change detection aims to find the time points where the statistical property of the time series changes significantly. It is closely related to time series segmentation as such points can be considered as the boundaries of different segments. Yamanishi et al. [20] unified the problems of change detection and outlier detection based on the on-line learning of an autoregressive model. Sharifzadeh et al. [15] used wavelet footprints to find the points where the polynomial curve fitting coefficients show discontinuities. Kawahara et al. [8] judged whether a point is a change by computing the probability density ratio of the reference and test intervals.

Our work uses the bottom-up segmentation approach due to its simplicity and practical effectiveness, it can be easily adapted to other segmentation algorithms. It is also worth mentioning that, the segmentation of this work is performed on short evolving intervals instead of the original long time series, which renders the segmentation process really fast.

Time series clustering. A crucial question in time series cluster analysis is establishing what we mean by similar data objects, i.e., determining a suitable similarity/dissimilarity measure between two time series objects. There exist a broad range of measures to compare time series and the choice of the proper dissimilarity measure depends largely on the nature of the clustering, i.e., on determining what the purpose of the grouping is. Current dissimilarity measures are grouped into four categories: model-free measures, model-based measures, complexity-based measures and prediction-based measures [13]. Considering the unsupervised feature of the problem and temporal heterogenous properties, we choose the model-free approaches.

The *Minkowski distance* is typically used to measure the proximity of two time series. This metric is very sensitive to signal transformations as shifting or time scaling. *Frechet distance* was introduced by Frechet [7] to measure the proximity between continuous curves, but it has been extensively used on the discrete case (see [6]) and in the time series framework. The dynamic time warping (DTW) distance was studied in depth by [14] and proposed to find patterns in time series by [2]. [5] introduce a dissimilarity measure addressed to cover both conventional measures for the proximity on observations and temporal correlation for the behavior proximity estimation, which includes both behavior and values proximity estimation.

In our scenario, we need to cluster the time series with both behavior and values similarity, we use an extension of the adaptive dissimilarity index covering both proximity on values and on behavior.

Co-evolving Zones. [16] studied the problem of finding regions that show similar deviations in population density using mobile phone data. They assume that the condition has periodicity, i.e., the daily population densities in a region are similar in different days. While this assumption is reasonable for population density, it does not hold in many geo-sensory applications like air quality monitoring. Moreover, they extract vertical changes in population density by comparing the same hour of different days. In contrast, we extract the horizontal changes, i.e., comparing the condition in current time interval with the previous time interval.

[21] studied problem of mining spatial co-evolving patterns from geo-sensory data, due to the sparse data they used, the paper only mines the spatial coevolving patterns (SCPs), i.e., groups of sensors that are spatially correlated and co-evolve frequently in their readings. In our situation, we first find the co-evolving zones, then give the causal explanations of the phenomenon, which can be used to further improve the accuracy of spatial inference and temporal prediction.

2.3 Challenges

In addition to the technical limitations, finding the co-evolving zones faces significant challenges in many real world applications. One significant challenge is the heterogeneity in space and time (see Fig. 1). Space heterogeneity refers to the case where data belonging to different clusters may have the same feature values. While heterogeneity in time refers to the instance where the sensor cluster membership may change over time, which all lead to one much debated question [12]: How long should the time series be? If too short, the clusters found can be spurious; if too long, dynamics can be smoothed out. Those heterogeneity challenges caused us to propose a novel paradigm to eliminate the limitations.

Another challenge is how to find the physical meaning of the found co-evolving zones, i.e., how to give the causal explanation, and find associate relationship between those zones to improve the performance of other application domains, e.g., space inference and temporal prediction.

3 Overview

To address the above-mentioned challenges, we propose a novel dynamic co-evolving zones data mining paradigm that systematically leverages the very challenges. Our paradigm consists of two main steps: finding the co-evolving zones under the spatial and temporal heterogeneity constraint; mining the association between the found co-evolving zones and give reasonable explanations. Figure 2-A outlines three key steps. The first step is to do the changepoint detection, which acts on the average value of the monitoring region and gets the uptrend or downtrend change intervals for the use of next step. The second step is to cluster the time series data in every change interval, the key here is to choose an appropriate dissimilarity measure under the space constraint. The final step is to mine the relationship between the previous found zones, which in the best case will give us the inner relationship between those co-evolving zones and causal explanations of the phenomenon, which also gives us the appropriate time series segementation length for clustering analysis.

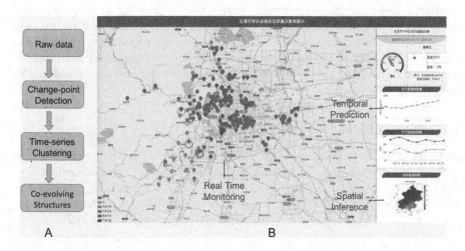

Fig. 2. A: dynamic co-evolving structure mining paradigm. B: web user interface of the deployed system.

The first step is essential. If we cluster the time series using the whole period, we will get bad and meaningless result for the space and temporal heterogeneity, which will be illustrated in the following experiment section. We cluster the segmented time series using an extension version of the adaptive dissimilarity index covering both proximity on values and on behavior in the second step. In the final step, we define a dynamic co-evolving zones' dissimilarity measure index and use the hierarchical clustering method to get the final co-evolving structure and give the dynamic segementation length used in clustering analysis.

Figure 2-B shows the real deployed web user interface in Haidian Ministry of Environmental Protection, where we can see the real time monitoring station readings and accurate spatial-temporal prediction results. The above proposed paradigm helps us improve the prediction precision significantly and makes the scientific environment treatment possible.

4   Proposed Model

The proposed model takes an divide-and-conquer strategy to find the dynamic co-evolving structures and give final causal explanations. It first breaks down the problem into multiple sub-problems of the same (or related) type (divide), until these become simple enough (uptrend/downtrend intervals) to be solved directly (conquer). The solutions to the sub-problems are then combined to give a solution or explanation to the original problem. The approach eliminates the effect of space and temporal heterogeneity on the original problem and help produce more reliable and reasonable result for future use in related domains.

In this section, we will first describe our change point detection algorithm, which is simple and efficient, then follows the co-evolving time series clustering algorithm under the space constraint. Lastly, the co-evolving structure learning framework is proposed to build the relationship between the found co-evolving zones. The above steps belong to the divide-and-conquer approach, which divide the whole time series clustering problem into sub-problems and then cluster segmented co-evolving zones respectively to produce more stable and meaningful co-evolving zones.

4.1   Change Point Detection

Definition 1. *Uptrend/Downtrend Interval. Given a sensor reading s, an uptrend (downtrend) interval is a consecutive subsequence of measurement $\mathcal{I} = \langle s[t_i], s[t_{i+1}], \ldots, s[t_{i+m-1}] \rangle$ and $\forall j \in \{i, i+1, \ldots, i+m-2\}, s[t_{j+1}] - s[t_j] > 0 (< 0,$ for downtrend interval), where m denotes the length of the subsequence and $t_i, t_{i+1}, \ldots, t_{i+m-1}$ are the timestamps of every measurement in $\mathcal{I}$.*

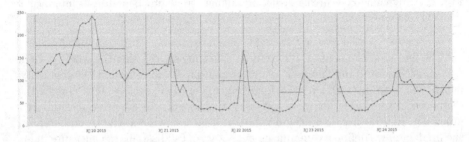

Fig. 3. The figure is the mean value of one monitoring region with almost 200 sensors; Blue lines is the segemented uptrend/downtrend intervals. (Color figure online)

Since the geo-sensory data is typically overwhelmed by various trivial fluctuations, we apply the wavelet transform to capture the multi-resolution evolving intervals by following the previous work [21]. Recall that we aim to discover the co-evolving sensor reading patterns, especially during the pollutant propagation period, which correspond to the uptrend or downtrend intervals of the geo-sensory data. Consequently, we adopt their method as well as the break and segment strategy [10] to extract the uptrend and downtrend intervals. Note that the uptrend and downtrend intervals we extract do not exactly follow Definition 1. Instead, we allow small fluctuations. Figure 3 shows the extracted uptrend and downtrend intervals in blue lines.

4.2 Time Series Clustering

From the previous step, we can get mounts of time series uptrend/downtrend intervals. For each uptrend (downtrend) interval, we adopt the selected time series clustering method to get the co-evolving zones. We will first describe the dissimilarity measures used for time series clustering, then follows the description of the clustering method.

Dissimilarity Measures. The key in the dissimilarity measures, namely, how to define the similarity of two time series object. In our scenario, the spatial constraint also need to be considered to define the spatio-temporal distance function. Two time series objects are similar if they are spatially adjacent and have similar temporal characteristic. It is a function as below.

$$d_{st}(x, y) = \begin{cases} d_t(x, y) & \text{if } x \text{ and } y \text{ are spatial neighbors} \\ 0 & \text{otherwise} \end{cases} \tag{1}$$

where, $d_t(x, y)$, is a time-series distance function.

The choice of time series distance function is related to the application. Commonly used time-series distance functions include proximity on value, proximity on behavior or both in the view of what the purpose of the grouping is. The conventional measures ignore the interdependence relationship between measurements, characterizing the time series behavior. The proximity is only based on the closeness of the values, while the proximity on behavior measure the growth behavior of the time series without considering the closeness of the values.

Previous work [5] introduced an adaptive dissimilarity index covering both proximity on values and on behavior, which is able to cover both conventional measures for the proximity on observations and temporal correlation for the behavior proximity estimation. These characteristics make it an ideal dissimilarity measures in our scenario.

First of all, temporal correlation for the behavior proximity estimation has been given. The proximity between the dynamic behaviors of the series is evaluated by means of the first order temporal correlation coefficient, which is defined by

$$CORT(\mathbf{X}_T, \mathbf{Y}_T) = \frac{\sum_{t=1}^{T-1}(X_{t+1}-X_t)(Y_{t+1}-Y_t)}{\sqrt{\sum_{t=1}^{T-1}(X_{t+1}-X_t)^2}\sqrt{\sum_{t=1}^{T-1}(Y_{t+1}-Y_t)^2}} \tag{2}$$

In the above equation, $CORT(\mathbf{X}_T, \mathbf{Y}_T)$ belongs to the interval $[-1, 1]$, The value $CORT(\mathbf{X}_T, \mathbf{Y}_T) = 1$ means that both series show a similar dynamic behavior, i.e., their growths (positive or negative) at any instant of time are similar in direction and rate, while $CORT(\mathbf{X}_T, \mathbf{Y}_T) = -1$ implies a similar growth in rate but opposite in direction (opposite behavior). Finally, $CORT(\mathbf{X}_T, \mathbf{Y}_T) = 0$ expresses that there is no monotonicity between X_T and Y_T, and the growth rates are stochastically linearly independent (different behaviors). In all, $CORT(\mathbf{X}_T, \mathbf{Y}_T)$ gives the similarity measures of time series.

The dissimilarity index proposed by [5] modulates the proximity between the raw-values of two time-series X_T and Y_T using the coefficient $CORT(X_T, Y_T)$. Specifically, it is defined as follows.

$$d_{CORT}(\mathbf{X}_T, \mathbf{Y}_T) = \phi_k[CORT(\mathbf{X}_T, \mathbf{Y}_T)] \cdot d(\mathbf{X}_T, \mathbf{Y}_T) \tag{3}$$

where $\phi_k(\cdot)$ is an adaptive tuning function to automatically modulate a conventional raw-data distance $d(X_T, Y_T)$ according to the temporal correlation. The modulating function should work increasing (decreasing) the weight of the dissimilarity between observations as the temporal correlation decrease from 0 to -1 (increase from 0 to $+1$). In addition, $d_{CORT}(\mathbf{X}_T, \mathbf{Y}_T)$ should approach the raw-data discrepancy as the temporal correlation is zero. In our scenario, we choose an exponential adaptive function given by

$$\phi_k(u) = \frac{2}{1 + exp(ku)}, k \geqslant 0. \tag{4}$$

The above exponential tuning function will cover both proximity on values and behavior, which is an appropriate choice in our situation.

Hierarchical Clustering Groups. Partitioning clustering methods meet the basic clustering requirement of organizing a set of objects into a number of exclusive groups [19], while in our situations we want to partition our data into groups at different levels such as in a hierarchy, which works by grouping data objects into a hierarchy or tree of clusters.

We use the agglomerative hierarchical clustering method based on the bottom-up strategy. It typically starts by letting each object form its own cluster and iteratively merges clusters into larger and larger clusters, until all the objects are in a single cluster or certain termination conditions are satisfied. The single cluster becomes the hierarchys root. For the merging step, it finds the two clusters that are closest to each other (according to some similarity measure), and combines the two to form one cluster. Because two clusters are merged per iteration, where each cluster contains at least one object, an agglomerative method requires at most n iterations.

In our scenario, using the method in [11], we can divide the sensor readings, which have similar proximity on values and behavior, in each change interval into k different co-evolving groups.

4.3   Co-evolving Structures

From the previous step, we get k co-evolving zones in each evolving interval, as shown in the left of Fig. 4. Our problem is to find the relationship between the found co-evolving zones and build the tree structure to show the inner causal associations among the co-evolving zones of different time period. In our situation, we also use hierarchical clustering method to build the co-evolving tree, which is illustrated in the right of Fig. 4.

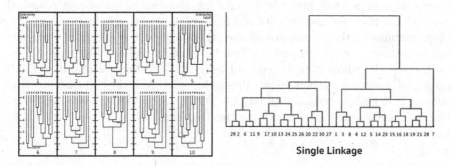

Fig. 4. The left figure is the co-evolving zones found in each co-evolving intervals, while the right figure shows how we restruct the relationship between those zones.

The key here is to define the similarity measures of the co-evolving zones in different time period. Suppose $A = \{A_1, ..., A_k\}$ and $B = \{B_1, ..., B_k\}$ are two co-evolving zones at two time interval of k clusters, the similarity measures of the two co-evolving zones is defined as follows:

Definition 2. *Cluster Similarity Measures. The Cluster Similarity Measures* $Sim(A, B)$ *of two co-evolving zones,* $A = \{A_1, ..., A_k\}$ *and* $B = \{B_1, ..., B_k\}$, *is defined by:*

$$Sim(A, B) = \frac{1}{k} \sum_{i=1}^{k} \max_{1 \leqslant j \leqslant k} Sim(A_i, B_j) \tag{5}$$

where

$$Sim(A_i, B_j) = \frac{|A_i \cap B_j|}{|A_i| + |B_j|} \tag{6}$$

in which $|\cdot|$ denoting the cardinality of the elements in the set.

In the merging step of the Hierarchical Clustering, we use the above similarity measure as the closeness index of two clusters (in our situation, we use single similarity linkage of two clusters) to combine the two to form one cluster.

The Hierarchical Clustering method works by grouping data objects (in our case, the co-evolving zones of different time intervals) into a hierarchy or tree of clusters, which reflects the relationship and inner association between those co-evolving zones. The groups at different levels of the hierarchy can give us more valuable information about the co-evolving zones, such as the relationship, causal association etc. This would help us have a better understanding of the evolving of the co-evolving patterns and casual association, which can help us make better spatial inference and temporal predictions.

5   Experiment

In this section, we will give the experiment and evaluation of our proposed approach. At first, we give the data and features used in the evaluation Sect. 5.1, which contains all the feature description, then follows the evaluation using real world data Sect. 5.2, we compare the found co-evolving zones with the clustering result using the whole data to test and verify the effectiveness of our method. The found co-evolving structure provides a clear picture of the pollution evolving patterns and has the potential to improve the accuracy of spatial inference and temporal prediction result, even give the recommendation of new air quality stations' locations, which is illustrated in Sect. 5.3.

5.1   Data and Features

We utilize real air quality monitoring datasets collected from Haidian district of Beijing, China. The datasets consist of three parts, as elaborated in the following.

- **Air Quality Records.** The data contains the real-valued AQI of two kinds of pollutions, $PM_{2.5}$ and PM_{10}, measured by almost 200 air quality monitoring stations every 30 min. This dataset is collected over 11 months (from March 1, 2015 to February 1, 2016).
- **Meteorological Data.** Previous study has shown that the concentration of air pollutants is influenced by meteorology. Especially, wind speed, wind power, humidity and barometer pressure all have a big influence on the concentration of the air quality. We choose the four aspects and the weather condition as the five features to evaluate the co-evolving structure result. The fine-grained meteorological data is collected hourly from a public website [1].
- **Point-Of-Interests (POIs).** In the urban area, the land use and the function of the region is well reflected by the category and density of POIs in the area, which is valuable in making accurate spatial inference. In our setting, we extract 8 POI features by using a POI database of Baidu Maps of Beijing (see Table 1).

5.2   Evaluation Using Real World Data

To illustrate the effectiveness of the proposed approprach, we use almost 11 months $PM_{2.5}$ sensor data to evaluate the algoritm. Figure 5 shows the result of almost 2 months data. Figure 5-A shows the mean value of the time series data,

Table 1. Category of POIs

C1: Culture & education	C5: Shopping malls and Supermarkets
C2: Parks	C6: Entertainment
C3: Sports	C7: Decoration and furniture markets
C4: Hotels	C8: Vehicle Services (gas station, repair)

the blue line is the segment result of the uptrend/downtrend detection algorithm, the algorithm get 89 intervals in total. For each change interval, we use the dissimilarity measures defined above to cluster all the sensor readings and divide them into 10 different classes. Then, using the Cluster Similarity Measures, we get the final co-evolving structures, as shown in Fig. 5-B. The co-evolving structure has four obvious sub-clusters: 1, 2, 3, 4. When mapping the sub-clusters into the time dimension, we found a clear temporal correlation, which can be seen in Fig. 5-A. This result shows the heterogeneity of the dynamic co-evolving zones and may provide a novel way to get the appropriate segementation length for future clustering analysis.

Using the above found co-evolving structure and the new segment interval: 1, 2, 3, 4. We get the co-evolving zones for each of the time interval, which is shown in Fig. 5-D, E, F, G. Compared with the clustering result using the whole two months data, as shown in Fig. 5-C, Fig. 5-D, E, F, G show more meaningful and stable results. In this scenario, two month data is too long for the clustering algorithm and the dynamics is smoothed out. While in Fig. 5-D, E, F, G, we can see clear co-evolving zones (the sensors in the same zone show similar patterns in both behavior and value) and the zones are dynamic between two different time intervals, which shows the necessity and effectiveness of the proposed paradigm. In the following section, we will use the found patterns to help improve the accuracy of spatial inference and temporal prediction result and show the effectiveness of the found patterns.

In our experiment settings, we set significant_delta, significant_length in change point algorithm is 35 and 3, and get 508 uptrend/downtrend time intervals. For each time interval, the distance between x, y is below 2 km if they are spatial neighbours, and k is set to 10, which means that there are 10 different sub-clusters for each co-evolving time interval. Using the co-evolving structure clustering algorithm, we get 28 different co-evolving intervals, which all shows an obvious co-evolving zones structure, in the following section, we will use the above found results to evaluate the effect on spatial-temporal prediction result.

5.3   Effect on Prediction and Inference

Spatial Inference. In the previous work [3], we compared the spatial inference accuracy using linear, cubic spline and gaussian process regression method, which shows the effectiveness of Gaussian Process regression in spatial $PM_{2.5}$ concentration inference. However, one big disadvantage of the GP method is the

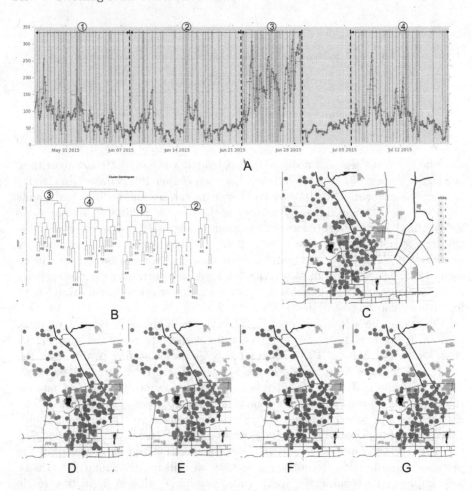

Fig. 5. Evaluation using the real world data (2 months). A: mean value of almost 200 sensors and segementation result; B: final co-evolving structure; C: co-evolving zones clustering result using whole data; D, E, F, G: co-evolving zones clustering result using segmentation intervals 1, 2, 3, 4. (Color figure online)

time complexity. Since an exact inference in Gaussian Process involves computing K^{-1}, the computation cost is $O(n^3)$ (n is the number of the training cases), when the deployment is large (in our situation, almost 200 devices), the compution cost is a big challenge in real time online systems. In this section, we try to decrease the number of devices (n) used for Gaussian Process algorithm with the help of the found co-evolving zones (C-zones).

In experiment, the real deployment dataset of more than 11 month was used to evaluate the performances of the algorithm. There are totally 200 monitor stations deployed in an area with the size of $30\,km \times 30\,km$ and each station reports its measurements every 30 min, the deployment map is shown in

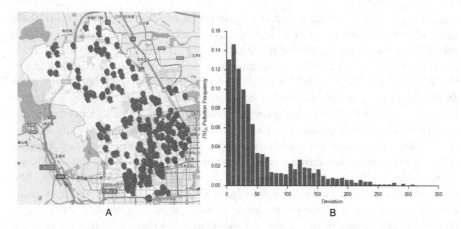

Fig. 6. (**A**) The deployment map of the monitor stations; (**B**) The distribution of the deviation between station S_1 and S_2 over one month

Fig. 6-(**A**). We deliberately remove one station as ground truth and infer its value using the remaining stations' reading at each timestamp. The Fig. 6-(**B**) also shows the distribution of deviation between our two monitor stations, S_1 and S_2. The geospatial distance of the two stations is about 6 km shown in Fig. 6-(**A**), over 21 % cases have a deviation greater than 100, which also shows the need for an efficient and accurate spatial inference algorithm.

Table 2. Inference errors

Measure Method	$\|x\|_1$	$\frac{1}{n}\|x\|_1$	$\|x\|_2$	RMSE	$\|x\|_\infty$
Gaussian Process	593429.56	25.12	4322.19	26.74	161.23
Gaussian Process + C-zones	569328.64	17.44	4011.73	20.14	145.08

Table 2 lists the inference errors of the two methods measured via different rules (assume that x is the absolute error vector). Gaussian Process uses all of the devices for training, while Gaussian Process + C-zones only uses the devices in the same co-evolving zones for training process, which only use almost 38 devices in average. From the comparison result, we can see that the inference accuracy has a significant increase by using the co-evolving zones, specially the Chebyshev norm $\|x\|_\infty$ achieved by Gaussian Process is 161.23 while the Gaussian Process + C-zones obtains a smaller value 145.08, which proves that the Gaussian Process + C-zones is more stable in the inference of $PM_{2.5}$ concentrations. The result also shows the efficiency of the found dynamic co-evolving zones.

Temporal Prediction. Over the past decades, some statistic models, like linear regression, regression tree and neural networks, have been employed in atmospheric science to do a real-time prediction of air quality. However, these methods simply feed a variety of features about a location into a single model to predict the future air quality of the location [22]. In work [22], they use a *Temporal Predictor* to predict the air quality of a station in terms of the data about the station. Instead, the *Spatial Predictor* considers spatial neighbor data, such as the AQIs and the wind speed at other stations, to predict a station's future air quality, The two predictors generate their own predictions independently for a station, which are combined by the *Prediction Aggregator* dynamically according to the current weather conditions of the station. In this way, they improve the prediction accuracy significantly. However, the meteorological data is almost same for devices in dense deployment scenario, using the spatial partition method in work [22] equals to feed all the data from a station's neighbors into a machine learning model. In this way, there are too many inputs for an ANN, leading to too many parameters in the model. Consequently, we cannot learn a set of accurate parameters for the ANN based on the limited training data, which may lead to some problems and can not be used directly in practice (see details in [22]).

In this experiment, we use the devices in the same co-evolving zones as the selected "spatial partition devices" to evaluate the accuracy of the algorithm. Long period prediction may need more data in large scale, so we only evaluate the next 6 h $PM_{2.5}$ concentrations in this experiment, which can be extend to next 48 h prediction in the similar method.

For the next 1–6 h, we measure the prediction of each hour $\widehat{y_i}$ against its ground truth y_i, calculating the accuracy according to Eq. 7, We also calculate the absolute error of each time interval according to Eq. 8, where n is the number of instances measured for a time interval. We random select 30 devices for this evaluation for almost 5 months.

$$p = 1 - \frac{\sum_i |\widehat{y_i} - y_i|}{\sum_i y_i} \tag{7}$$

$$e = \frac{\sum_i |\widehat{y_i} - y_i|}{n} \tag{8}$$

Table 3 shows the prediction result using different methods, LR and ANN only use the local monitor station readings as the data source and make predictions. In general, LR has a similar performance in predicting normal instances but less effective than ANN in dealing with sudden drops. Also, the results presented in Table 3 justify the advantages of the *ANN + C-zones* which use local and devices in same co-evolving zones for prediction which acquires a big improvement in the performance of overall accuracy, especially in the sudden drops scenario.

6   Conclusion

In this paper, we propose a novel divide-and-conquer strategy to find the dynamic co-evolving zones that systematically leverages the sensor readings'

Table 3. Prediction Result of different methods.

Methods	All instances		Sudden drops	
	p	e	p	e
LR (Linear Regression)	0.684	27.5	0.298	103.2
ANN (Artifical Neural Network)	0.646	29.9	0.221	73.7
ANN + C-zones	**0.725**	**20.1**	**0.302**	**51.4**

spatial and temporal heterogeneity challenges. The paradigm produced more stable and meaningful co-evolving zones and automatically found the segmentation intervals which shows the inner pollution change paterns. We use the found result to evaluate the performance on spatial-temporal prediction result and found a significant improvement, which proves the effectiveness of the found patterns. What's more, the found zones and dynamic patterns may provide recommendation for new planned public monitoring stations and future city planning. The system has also been deployed with the Haidian Ministry of Environmental Protection (in Haidian district of Beijing, China) to make accurate spatial-temporal predictions and help the government better understand the pollution evolving patterns to make more scientific strategies for environment treatment. The current implementation still needs manual parameter tuning and has some limitations, for future work, we plan to eliminate those disadvantageous and make the algorithm more scaleable to use in the real production environment.

References

1. aqi.cn: Beijing air pollution. http://aqicn.org/city/beijing/
2. Berndt, D.J., Clifford, J.: Using dynamic time warping to find patterns in time series. In: KDD Workshop, Seattle, WA, vol. 10, pp. 359–370 (1994)
3. Cheng, Y., Li, X., Li, Z., Jiang, S., Jiang, X.: Fine-grained air quality monitoring based on gaussian process regression. In: Loo, C.K., Yap, K.S., Wong, K.W., Teoh, A., Huang, K. (eds.) ICONIP 2014, Part II. LNCS, vol. 8835, pp. 126–134. Springer, Heidelberg (2014)
4. Cheng, Y., Li, X., Li, Z., Jiang, S., Li, Y., Jia, J., Jiang, X.: Aircloud: a cloud-based air-quality monitoring system for everyone. In: Proceedings of the 12th ACM Conference on Embedded Network Sensor Systems, pp. 251–265. ACM (2014)
5. Chouakria, A.D., Nagabhushan, P.N.: Adaptive dissimilarity index for measuring time series proximity. Adv. Data Anal. Classif. **1**(1), 5–21 (2007)
6. Eiter, T., Mannila, H.: Computing discrete fréchet distance. Technical report, Citeseer (1994)
7. Fréchet, M.M.: Sur quelques points du calcul fonctionnel. Rendiconti del Circolo Matematico di Palermo (1884–1940) **22**(1), 1–72 (1906)
8. Kawahara, Y., Sugiyama, M.: Change-point detection in time-series data by direct density-ratio estimation. In: SDM, vol. 9, pp. 389–400. SIAM (2009)
9. Keogh, E., Chu, S., Hart, D., Pazzani, M.: An online algorithm for segmenting time series. In: Proceedings IEEE International Conference on Data Mining, 2001, ICDM 2001, pp. 289–296. IEEE (2001)

10. Keogh, E., Chu, S., Hart, D., Pazzani, M.: Segmenting time series: a survey and novel approach. Data Min. Time Ser. Databases **57**, 1–22 (2004)
11. Langfelder, P., Zhang, B., Horvath, S.: Defining clusters from a hierarchical cluster tree: the dynamic tree cut package for r. Bioinformatics **24**(5), 719–720 (2008)
12. Marti, G., Andler, S., Nielsen, F., Donnat, P.: Clustering financial time series: How long is enough? arXiv preprint arXiv:1603.04017 (2016)
13. Montero, P., Vilar, J.A.: Tsclust: an r package for time series clustering. J. Stat. Softw. **62**(1), 1–43 (2014)
14. Sankoff, D., Kruskal, J.B.: Time warps, string edits, and macromolecules: the theory and practice of sequence comparison. Addison-Wesley Publication, Reading (1983). Edited by Sankoff, D., Kruskal, J.B
15. Sharifzadeh, M., Azmoodeh, F., Shahabi, C.: Change detection in time series data using wavelet footprints. In: Medeiros, C.B., Egenhofer, M., Bertino, E. (eds.) SSTD 2005. LNCS, vol. 3633, pp. 127–144. Springer, Heidelberg (2005)
16. Trasarti, R., Olteanu-Raimond, A.M., Nanni, M., Couronné, T., Furletti, B., Giannotti, F., Smoreda, Z., Ziemlicki, C.: Discovering urban and country dynamics from mobile phone data with spatial correlation patterns. Telecommun. Policy **39**(3), 347–362 (2015)
17. Wang, P., Wang, H., Wang, W.: Finding semantics in time series. In: Proceedings of the 2011 ACM SIGMOD International Conference on Management of Data, pp. 385–396. ACM (2011)
18. Warren Liao, T.: Clustering of time series data-a survey. Pattern Recogn. **38**(11), 1857–1874 (2005). http://dx.doi.org/10.1016/j.patcog.2005.01.025
19. Xu, R., Wunsch, D., et al.: Survey of clustering algorithms. IEEE Trans. Neural Netw. **16**(3), 645–678 (2005)
20. Yamanishi, K., Takeuchi, J.i.: A unifying framework for detecting outliers and change points from non-stationary time series data. In: Proceedings of the Eighth ACM SIGKDD International Conference on Knowledge Discovery and Data Mining, pp. 676–681. ACM (2002)
21. Zhang, C., Zheng, Y., Ma, X., Han, J.: Assembler: efficient discovery of spatial co-evolving patterns in massive geo-sensory data. In: Proceedings of the 21th ACM SIGKDD International Conference on Knowledge Discovery and Data Mining, pp. 1415–1424. ACM (2015)
22. Zheng, Y., Yi, X., Li, M., Li, R., Shan, Z., Chang, E., Li, T.: Forecasting fine-grained air quality based on big data. In: Proceedings of the 21th ACM SIGKDD International Conference on Knowledge Discovery and Data Mining, pp. 2267–2276. ACM (2015)

ECG Monitoring in Wearable Devices by Sparse Models

Diego Carrera[1]([⊠]), Beatrice Rossi[2], Daniele Zambon[2],
Pasqualina Fragneto[2], and Giacomo Boracchi[1]

[1] Dipartimento di Elettronica, Informazione e Bioingegneria,
Politecnico di Milano, via Ponzio 34/5, 20100 Milano, Italy
{diego.carrera,giacomo.boracchi}@polimi.it
[2] STMicroelectronics, via Olivetti 2, 20864 Agrate Brianza, Italy
{beatrice.rossi,daniele.zambon-ext,pasqualina.fragneto}@st.com

Abstract. Because of user movements and activities, heartbeats recorded
from wearable devices typically feature a large degree of variability in
their morphology. Learning problems, which in ECG monitoring often
involve learning a user-specific model to describe the heartbeat morphology, become more challenging.

Our study, conducted on ECG tracings acquired from the Pulse Sensor
– a wearable device from our industrial partner – shows that dictionaries
yielding sparse representations can successfully model heartbeats acquired
in typical wearable-device settings. In particular, we show that sparse representations allow to effectively detect heartbeats having an anomalous
morphology. Remarkably, the whole ECG monitoring can be executed
online on the device, and the dictionary can be conveniently reconfigured
at each device positioning, possibly relying on an external host.

1 Introduction

In this paper we deal with the problem of monitoring electrocardiogram (ECG)
tracings through wearable devices like the Pulse Sensor [1], which is shown in
Fig. 1 and developed in a joint collaboration between MR&D and STMictroelec-
tronics. Wearable devices have a huge potential in health and fitness scenarios,
and in particular in the transitioning from hospital to home/mobile health monitoring. However, to make these devices operational in real-world applications,
it is necessary to address relevant machine-learning and data-science challenges.
In particular, to provide prompt interaction with the user and prevent massive data-transfer which can spoil their battery life, wearable devices have to
autonomously process the sensed data.

In the case of ECG tracings, this processing typically consists in classifying or
detecting anomalies in the heartbeats. These tasks are traditionally performed
by computing expert-based features like those in [2–6], which tend to mimic
the criteria clinicians use to interpret ECG tracings. Examples of these features
are the ECG values in specific locations of the heartbeat, interval features (e.g.,
the duration of the QRS, ST-T or QT complex, or the distance between two

B. Berendt et al. (Eds.): ECML PKDD 2016, Part III, LNAI 9853, pp. 145–160, 2016.
DOI: 10.1007/978-3-319-46131-1_21

consecutive peaks, namely the RR distance), and the average ECG energy over these intervals.

Often, expert-based features are combined with data-driven ones, that do not tend to reproduce some clinical evidence but they are directly learned from data [7,8], possibly by clustering heartbeats [9,10]. In practice, learning data-driven features boils down to learning a model to represent heartbeats. Since heartbeats of each user are characterized by their own morphology [11] (see the examples of Fig. 2), global models are not able to properly describe heartbeats of different users, and lead to poor classification [12] or anomaly detection performance even when trained on large datasets. Therefore, it is convenient to make these models user-specific or at least user-adaptable [2,13].

Fig. 1. Pulse Sensor. The two external electrodes inject the current, while those in the middle read the difference in electric potential. The Pulse Sensor can either analyze onboard, store or transmit the ECG tracings.

Here we focus on data-driven models for learning the morphology that characterize each user heartbeats, and to this purpose, we consider dictionaries yielding sparse representations of heartbeats. Sparse representations are nowadays one of the leading models in image and signal processing [14,15], and dictionary learning has been successfully used for modeling ECG tracings for anomaly-detection [16,17] and person-identification [18] purposes. Intuitively, learning a dictionary yielding sparse representations corresponds to learning a union of low-dimensional subspaces where user heartbeats live.

ECG tracings acquired by wearable devices are different from those typically considered in the literature, like the MIT-BIH Arrhythmia Database [19], which contains relatively short segments of good-quality Holter recordings. In the Pulse Sensor, for instance, the electrodes are closer than in an Holter device and these could be mispositioned since they are typically placed by users themselves rather than by clinicians. Moreover, during long-term monitoring, user movements might also cause device displacements. These issues might affect the morphology of heartbeats [11,20] and have implications on the model used to describe the heartbeats of each user, which are better discussed in Sect. 2.2.

We here show that dictionaries yielding sparse representations are the right choice for modeling ECG recordings in wearable devices, and that they allow to detect anomalies directly on the device. To this purpose, we consider an anomaly-detection algorithm similar to [21], and study its applicability on the

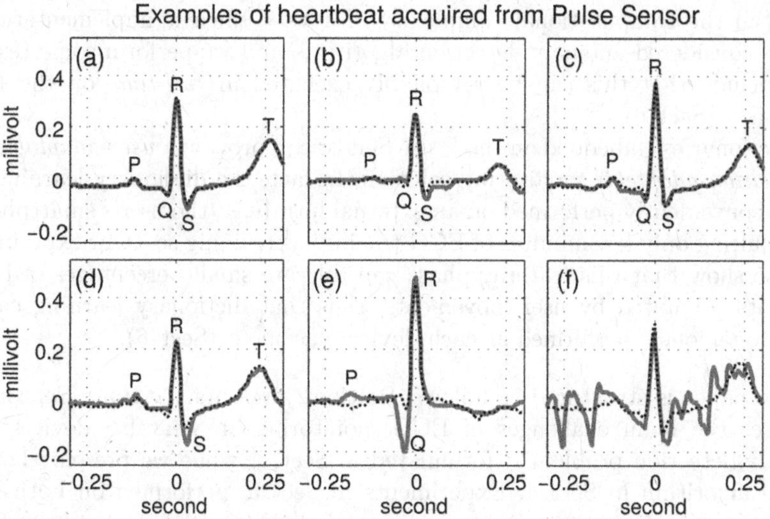

Examples of heartbeat acquired from Pulse Sensor

Fig. 2. Examples of heartbeats morphology. The top row (**a**, **b** and **c**) contains heartbeats acquired from user 1 with the Pulse Sensor placed in position 1. In all these heartbeats we depict also the P-waves, the QRS-complexes and the T-waves. The small variations in the morphology of these heartbeats are also due to different heart rates (72 bpm in **a**, 93 bpm in **b** and 77 bpm in **c**). Note also that the morphology remains unaltered over time, since **c** was acquired more than 100 min after **a** and **b**. The bottom row (**d**, **e** and **f**) contains heartbeats featuring a different morphology. In particular, **d** reports an heartbeat from the user 1 acquired in position 2 (heart rate of 81 bpm), **e** reports an heartbeat of user 2 (84 bpm) and **c** reports an example of artifact due to movements of user 1. We also report heartbeats reconstructed by the sparse coding with respect to the dictionary learned from user 1 position 1 (dotted lines). Heartbeats in the top row are properly reconstructed (reconstruction errors $r_a = 0.07$, $r_b = 0.15$ and $r_c = 0.10$) since these are from the same user and position. In contrast, the heartbeats in the bottom row show a poor reconstruction quality ($r_d = 0.18$, $r_e = 0.50$ and $r_f = 0.63$). The reconstruction error can be thus used to detect anomalous heartbeats, namely heartbeats that do not feature the morphology characterizing a specific user and electrodes placement.

Pulse Sensor. This algorithm is tested over a large dataset of ECG tracings from healthy users, where every heartbeat featuring morphology different from the training ones is considered anomalous. Our experiments show that:

(1) Dictionaries yielding sparse representations can successfully describe the *overall variability* in the morphology of heartbeats acquired by wearable devices like the Pulse Sensor. These models do not seem likewise necessary in more controlled situations, as for example in the MIT-BIH Arrhythmia Database, where there is less variability in the normal heartbeats and anomalies are easier to detect (Sect. 5).

(2) It is possible to detect heartbeats that do not conform the user morphology (i.e., anomalous heartbeats) directly on the device. Indeed, we analyze in

detail the computational complexity of a very efficient implementation of the considered anomaly-detection algorithm, and we perform some tests to conclude that this can be reasonably executed in *real-time* on the Pulse Sensor (Sect. 6).

(3) Dictionaries embedded on the Pulse Sensor can provide a *user-adaptable* and *position-adaptable* monitoring solution. In fact, the dictionary learning can be conveniently performed on an external host (e.g., the user's smartphone), requiring only few minutes of ECG tracings as training set. Our experiments also show that this learning phase can tolerate small percentages of heartbeats corrupted by user movements, thus that dictionary learning can be autonomously performed at each device placement (Sect. 6).

The paper is structured as follows. Section 2 presents the Pulse Sensor and discusses the main challenges of ECG monitoring on wearable devices. The anomaly-detection problem is formulated in Sect. 3, while we present the considered algorithm in Sect. 4. Experiments in Sect. 5, performed on both ECG tracings acquired from the Pulse Sensor and the MIT-BIH Arrhythmia Database, show the that dictionaries yielding sparse representation can effectively model heartbeats and detect those having a different morphology. In Sect. 6 we study the overall feasibility of this monitoring solution on the Pulse Sensor, while in Sect. 7 we draw conclusions along with future works.

2   The Pulse Sensor

2.1   Device Description

The Pulse Sensor [1] is a wearable device developed by MR&D in collaboration with STMicroelectronics. It is a battery-powered device, designed for monitoring ECG tracings correlated to other physiological information. In particular, this device continuously acquires, stores and periodically transmits: ECG tracings, measurements of heart rate and breathing rate.

The sensor suite of the Pulse Sensor is made up of one microelectromechanical systems (MEMS) accelerometer, dedicated to estimate both the physical activity and the body position, and four electrodes embedded in a patch (see Fig. 1). The outer ones inject AC current with intensity 100 μA and frequency 50 kHz, while the central ones – placed at a distance of 8 cm – read a single-lead ECG (thus a single univariate signal) and a bioimpedance signal.

The main block of electronic components comprises a signal amplifier, three light-emitting diodes (LEDs), a Bluetooth module and a battery. The Bluetooth module connects the Pulse Sensor with a host device (e.g. a smartphone, tablet or a computer) in order to periodically transmit all the acquired signals. The internal battery is a rechargeable Lithium-ion one (3.7 VDC with 350 mAh capacity). The LEDs provide information on the battery charge-status, on the current operational mode of the device (engage, streaming and monitoring) and warnings on the incoming signals. The adopted microcontroller is the STM32F103 which incorporates the

ARM® Cortex™-M3 32-bit RISC core operating at a 72 MHz and embeds up to 32 Kbytes of flash memory and up to 10 Kbytes of SRAM.

A peculiarity of this device is modularity, which allows to tailor the sensor suite around specific application requirements. In fact, it is very easy to add new types of electrodes, scaling the software or replacing the microcontroller with a more powerful one as far as this is compatible with the firmware and pinout.

2.2   Issues of ECG Monitoring on Wearable Devices

We here discuss the main issues that makes real-time monitoring of physiological signals particularly challenging in wearable devices like the Pulse Sensor.

Variety of ECG morphology. First of all, during long-time monitoring, the heart-beat morphology might be subject to variations due to changes in the heart rate. This makes ECG tracings acquired by the Pulse Sensor more heterogeneous than ECG tracings acquired in more controlled situations, as for example those of MIT-BIH Arrhythmia Database, which refer to relatively short time intervals. Moreover, the sensing capabilities of the Pulse Sensor are lower than those of devices typically used in clinical trials, since a single ECG tracing is acquired from two electrodes placed at a relatively close distance. The overall variability in the morphology of heartbeats acquired by wearable devices is thus quite large, and difficult to describe.

Computational Constraints. In wearable devices meant for real-time monitoring of physiological signals, sensors continuously acquire data, producing a massive amount of information to be analyzed and possibly stored or transmitted. Needless to say, if the device were periodically transmitting the whole ECG tracings to an host, its battery would be spoiled soon. Data transmission between the wearable and the host can be reduced by enabling the device to autonomously process the sensed data, thus transmitting only the most relevant information, like heartbeats having an anomalous morphology. As such, algorithms used to analyze heartbeats should be compliant with the device computing-capabilities.

Changes in user and device position. ECG tracings do not only depend on the specific user, but also on the specific placement of the ECG electrodes [11,20]. While this is not an issue when electrodes are placed by clinicians (that at the meantime analyze the ECG tracings) this represents a serious problem in the typical application scenario of wearable devices. In fact, the Pulse Sensor is meant to be positioned by users themselves and, as such, electrodes could be mispositioned, making the model used for automatic analysis unreliable since the heartbeats morphology has changed (see Fig. 2.d). The same problem happens during long-term monitoring, when user movements might cause device displacements. Therefore, the model learned on the device has to be easily re-trainable every time the device is positioned, without requiring any supervision by an expert clinician. Also, the device configuration should tolerate at least a small fraction of heartbeats affected by user movements.

3   Problem Formulation

We denote by $s \colon \mathbb{N} \to \mathbb{R}$ the ECG tracing which has been uniformly sampled in time, and we assume that the heartbeats have been already segmented e.g., by [22]. We define the i-th heartbeat $\mathbf{s}_i \in \mathbb{R}^p$ as

$$\mathbf{s}_i = \{s(t_i + u) \ : \ u \in \mathcal{U}\}, \tag{1}$$

where $\mathcal{U}$ is a neighborhood of the origin containing p samples, and t_i denotes the sample in the ECG tracing corresponding to the i-th R peak of the ECG tracing. We assume that the normal heartbeats of each wearable-device user are generated by a stochastic process $\mathcal{P}_N$, which characterizes the heartbeats' morphology. Our goal is to learn a model representing the heartbeats morphology; to quantitatively assess the effectiveness of the learned model, we consider the anomaly-detection problem, which is itself of primary concern in ECG monitoring. More precisely, *anomalous* heartbeats are generated by a process $\mathcal{P}_A \neq \mathcal{P}_N$ and exhibits different morphology than heartbeats generated by $\mathcal{P}_N$. Anomalies might be due, for instance, to arrhythmias (as those in the MIT-BIH Arrhythmia Database), movements (as it typically happens in long-term monitoring, e.g. see Fig. 2), acquisition errors (which might occur in consumer devices), or simply because these have been acquired from a different user or by changing the electrodes placement (see Fig. 2). Anomalies are detected by analyzing each heartbeat $\mathbf{s}_i$ and determining whether it conforms or not the morphology characterizing $\mathcal{P}_N$. When this is not the case, we consider the beat $\mathbf{s}_i$ as anomalous. Since we analyze each beat independently we ignore anomalies that affect, for instance, the heart-rate or that require inspecting multiple heartbeats. We assume only that a training set TR of normal heartbeats is provided, as this allows us to learn a model approximating $\mathcal{P}_N$. We do not require any example of anomalous heartbeats, thus $\mathcal{P}_A$ remains completely unknown. This is a reasonable assumption since normal heartbeats are quite easy to collect and, at least in healthy users, it is enough to record few minutes after having placed the device; in contrast, anomalies are rare and difficult to gather thus the wide range of signals covered by $\mathcal{P}_A$ cannot be properly characterized.

4   The Considered Anomaly-Detection Algorithm

We consider a simple, yet effective, anomaly-detection algorithm that leverages a dictionary yielding sparse representations of the normal heartbeats. In practice, this follows the approach in [21], where a change-detection algorithm was used to monitor rock faces and detect structural changes in fixed-length signals acquired by triaxial MEMS accelerometer. While we use the same model for describing normal data and we analyze the reconstruction error as in [21], we adopt an outlier-detection technique rather than a sequential change-point method for monitoring ECG. This choice better conforms the considered scenario, since the ECG tracings are typically affected by sporadic anomalies rather than permanent changes. In what follows we describe the two main steps of the considered algorithm.

4.1 Modeling Normal Heartbeats

Our modeling assumption is that the normal heartbeats $\mathbf{s}_i \in \mathbb{R}^p$ of a user are generated from the process $\mathcal{P}_N$ and can be well approximated by the following linear model

$$\mathbf{s}_i \approx D\mathbf{x}_i, \tag{2}$$

where $D \in \mathbb{R}^{p \times n}$ is a matrix called *dictionary* and the coefficient vector $\mathbf{x}_i \in \mathbb{R}^n$ is *sparse* [23]. Sparsity means that $\mathbf{x}_i \in \mathbb{R}^n$ has few of nonzero components, thus in practice that the ℓ^0 "norm" of $\mathbf{x}_i$ is bounded, i.e., $\|\mathbf{x}_i\|_0 \leq \kappa$, where $\kappa > 0$ is the maximum number of nonzero coefficients allowed in these representations.

Example of learned dictionary

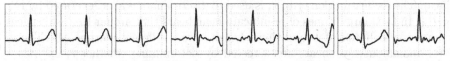

Fig. 3. Atoms of the dictionary learned from the user yielding normal heartbeats in Fig. 2. The parameters adopted for the training are $n = 3$, $\kappa = 8$ and $m = 500$.

The dictionary D is learned from a training set containing normal heartbeats of a single user. We stack the m heartbeats provided for training in the columns of a matrix $S \in \mathbb{R}^{p \times m}$. *Dictionary learning* consists in solving:

$$[D, X] = \underset{\widetilde{D} \in \mathbb{R}^{p \times n}, \widetilde{X} \in \mathbb{R}^{n \times m}}{\arg \min} \|\widetilde{D}\widetilde{X} - S\|_2, \quad \text{such that } \|\widetilde{\mathbf{x}}_i\|_0 \leq \kappa, \quad i = 1, \dots, n \tag{3}$$

where the sparsity constraint applies to each column of the matrix $X \in \mathbb{R}^{n \times m}$, which stacks the coefficient vectors of all the heartbeats in S. In practice, (3) can be solved by the KSVD algorithm [24], which alternates the calculation of the dictionary D and the sparse representations of the training heartbeats X.

Thus, the dictionary D is user-specific: its columns, which are referred to as *dictionary atoms*, depict the most relevant morphologies characterizing user heartbeats, as shown in Fig. 3. Equation (2) implies that each heartbeat $\mathbf{s}_i$ is approximated by a linear combination of at most κ dictionary atoms.

4.2 Detecting Anomalous Heartbeats

Learning D such that (2) holds for normal heartbeats corresponds to learning a union of low-dimensional subspaces of $\mathbb{R}^p$ where normal heartbeats live. In particular, since the κ atoms can be arbitrarily chosen among the n columns of D, these subspaces can be at most κ-dimensional. The sparse representation $\mathbf{x}_i$ of an heartbeat $\mathbf{s}_i$ can be computed by projecting $\mathbf{s}_i$ on the closest of such subspaces. This problem is referred to as *sparse coding* and it is formulated as

$$\mathbf{x}_i = \underset{\widetilde{\mathbf{x}} \in \mathbb{R}^n}{\arg \min} \|D\widetilde{\mathbf{x}} - \mathbf{s}_i\|_2 \text{ such that } \|\widetilde{\mathbf{x}}\|_0 \leq \kappa. \tag{4}$$

The problem (4) is NP-Hard, and it is typically addressed by greedy algorithms. In particular we here adopt the Orthogonal Matching Pursuit [25], an iterative algorithm which selects the best column of D at each iteration. The OMP can be well implemented in the Pulse Sensor, as discussed in Sect. 6.

We detect anomalies by assessing whether each heartbeat s_i to be tested falls in the union of low-dimensional subspaces that characterizes normal heartbeats for a specific user. In particular, we solve (4) and obtain x_i, the coefficients of the closest projection over subspaces of D. Then, we measure the reconstruction error as

$$r_i = \|D x_i - s_i\|_2, \tag{5}$$

where $D x_i$ denotes the linear combination of dictionary atoms that best reconstruct s_i (the reconstruction of the examples in Fig. 2 is reported with dashed lines). The reconstruction error r_i is used to discriminate if s_i is generated by $\mathcal{P}_{\mathcal{N}}$ or $\mathcal{P}_{\mathcal{A}}$. In fact, large values of r_i indicate heartbeats that are far from subspaces spanned by columns of D and that as such have a different morphology. Therefore, anomalous heartbeats are detected by determining whether r_i exceeds a suitable threshold $\gamma > 0$, which has to be defined experimentally.

We remark that r_i is a data-driven and user-specific feature, as it is entirely defined from the dictionary D that is learned from the training set without any a-priori information about the heartbeat morphology. Finally, other dictionary-learning and sparse-coding algorithms have been proposed in the literature, and in particular, some of them replace the constrained problems (3) and (4) with their convex relaxation where sparsity is measured by the ℓ^1 norm of the coefficient vectors. These lead to basis pursuit denoising (BPDN) formulation [26]. In the considered settings (see Sect. 5) these are however more computationally demanding than the OMP, which can be reliably embedded on the Pulse Sensor. It is also worth commenting that, when changing the problems (3) and (4), monitoring the reconstruction error might not be the best option [27].

5   Experiments

In this section we consider two different datasets of ECG tracings: the former was acquired using the Pulse Sensor, the latter is the MIT-BIH Arrhythmia Database [19] that is commonly used in the literature. We consider the algorithm described in Sect. 4 in a few anomaly-detection scenarios, as a way to quantitatively assess the effectiveness of sparse representations in modeling heartbeats.

5.1   Datasets Description

The *Pulse dataset* contains 20 ECG tracings recorded from 10 healthy users[1] (two tracings per user). The two acquisitions from each user have been performed in different times, repositioning the Pulse Sensor such that the morphology of heartbeats changes. Each ECG tracing lasts from 40 min up to 2 h and is acquired

---

[1] The dataset can be made available upon request.

during normal-life activities, thus the heart rate can significantly vary along the same tracing. Due to motion artifacts or temporary device detachments, these tracings sometimes contain low-quality segments (depicting heartbeats as in Fig. 2.e), which have been discarded by an experienced cardiologist with the aid of a commercial software. While these heartbeats are not anomalous from a clinical point-of-view, we exclude them as they do not show the same morphology of others. Possibly, these heartbeats could be removed directly on the Pulse Sensor by monitoring the MEMS recordings. Each ECG tracing is preprocessed as in [3] in order to remove the baseline wander and unwanted power-line and to attenuate high-frequency noise.

The *MIT-BIH Arrhythmia Database* [19] contains 48 ECG tracings lasting around 30 min each, that have been extracted from long-term Holter recordings. These segments have been selected by expert cardiologists which discarded the low-quality parts of these traces. Each ECG tracing contains a few arrhythmias, and every heartbeat is provided with annotations by the cardiologists. Both the heart rate and the morphology of normal heartbeats in this dataset are characterized by less variability than in the Pulse dataset.

In all our experiments, we extract heartbeats using a temporal window $\mathcal{U} = [-0.3, 0.3]$ centered in each R-peak, which yield heartbeats having $p = 155$ and $p = 216$ samples in the Pulse Sensor and MIT-BIH dataset, respectively[2].

5.2   Figures of Merit

We consider figures of merit traditionally used to assess the anomaly-detection performance: *(i) False Positive Rate* (FPR), namely the percentage of normal heartbeats identified as anomalous and *(ii) True Positive Rate* (TPR), namely the percentage of heartbeats correctly identified as anomalous. Since both FPR and TPR depend on the threshold $\gamma > 0$ (see Sect. 4.2), we consider the Receiving Operating Characteristic (ROC) curve, which are obtained by varying γ and plotting the corresponding TPR against the FPR. An example of ROC curve is provided in Fig. 5: the closer the curve to the point (0,1), the better. To get a quantitative assessment of the anomaly-detection performance, we measure the area under the curve (AUC), which for the ideal detector (namely the one having no false positives and no false negatives) is 1.

5.3   Experiments on the Pulse Dataset

Even though ECG recordings from the Pulse dataset were acquired from healthy users and contain no clinical anomalies, we design two anomaly-detection experiments to show that the considered algorithm can effectively detect heartbeats having a different morphology. In particular, we consider as normal (i.e., generated from $\mathcal{P}_N$) heartbeats acquired form a specific user with a specific positioning of the Pulse Sensor. Anomalous heartbeats (i.e., generated from $\mathcal{P}_A$) are

---

[2] Pulse Sensor has a sampling frequency of 256 Hz, while the sampling frequency in the MIT-BIH Arrhythmia Database is 360 Hz.

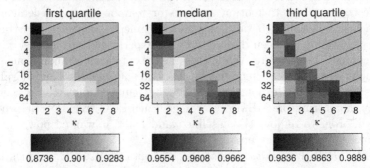

Fig. 4. Performance of several configurations of the considered algorithm in the inter-user anomaly detection. The three figures report the first quartile, the median and the third quartile of the AUC values computed on the Pulse dataset. The best configuration corresponds to $n = 8$, and $\kappa = 3$, as the performance degrades when considering simpler models (small n, κ) and more flexible ones (large n, κ). The intensity ranges in the three images are different for visualization sake.

acquired from a different user or from a different device position. We use the KSVD algorithm [24] to learn a dictionary D from each of these 20 ECG tracing, using 500 randomly selected heartbeats[3]. Thus, for each dictionary D we consider normal those heartbeats belonging to the same tracing used to learn D (namely the same pair user-position), and anomalous those heartbeats from any different tracing.

We test the following number of atoms $n \in \{1, 2, 4, 8, 16, 32, 64\}$ in D and levels of sparsity $\kappa \in \{1, 2, 3, \ldots, \lceil n^{1/2} \rceil\}$. These settings are quite different from those traditionally used in image and signal processing, where $n > p$, yielding redundant dictionaries. However, we experienced heartbeats can be properly described by fewer atoms.

Figure 4 shows the performance on *inter-user* anomalies, where the anomalous heartbeats come from different users. More precisely, we report the three quartiles of the AUC values computed over the $20 \cdot 18 = 360$ combinations of ECG tracings from different users. Overall, the AUC values are quite large and this indicates that the considered algorithm can effectively discriminate between users. The *best* performance are achieved when $n = 8$ and $\kappa = 3$. Observe that the *single-atom* configuration ($n = 1$ and $\kappa = 1$) which reconstructs heartbeat by scaling a single atom to match at best the heartbeats, achieves significantly lower performance, as confirmed by a Wilcoxon signed-rank test (p-value $\approx 10^{-16}$).

Figure 5(a) shows the ROC curves on *intra-user* anomalies, where we consider as anomalous heartbeats acquired from the same user but with the device in a different position. These curves are averaged over all the possible 20 combinations of the ECG tracings, and we report only the *best* and *single-atom*

---

[3] We have observed that larger training sets do not lead to an improvement in the anomaly-detection performance.

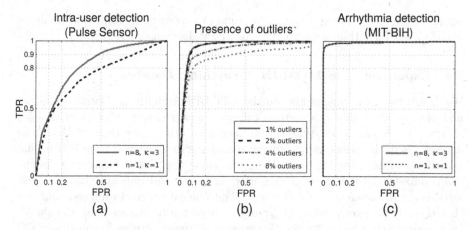

Fig. 5. ROC curves computed on Pulse dataset (a), (b), and on MIT-BIH Arrhythmia Database (c). In (a) two different configurations of parameters n and κ are considered in the intra-user anomaly detection. The best configuration clearly outperforms the single-atom one, confirming that a too simple model can not properly represent the structure of normal heartbeats. In (b) we consider the inter-user anomaly detection problem when the training set is corrupted by different percentages of outliers. This algorithm can tolerate small percentages of outliers, as its performance clearly degrades when the outliers reach 8 % of training data. In (c) we compare the *best* and *single-atom* configuration in the arrhythmia detection problem. The Wilcoxon signed-rank test reveals no statistical evidence between the performance of the two configurations (p-value $= 0.13$), and both achieve very high performance.

configurations. Still, changes in the device positioning can be better detected when using multiple atoms than a single one. The AUC values are typically lower than in the inter-user case (the median AUC here 0.81 and 0.77 in the best and single-atom settings, respectively), and this indicates that in this dataset, intra-user differences are more subtle than inter-user differences.

These experiments confirm that it is necessary to use a quite flexible model to properly characterize the variety of normal heartbeats acquired by the Pulse Sensor and that dictionaries yielding sparse representations can successfully learn the heartbeat morphology of each user.

Finally, we remark that in ECG tracings acquired from wearable devices, user's movements can introduce low quality heartbeats, i.e., outliers, that might impair dictionary learning. Thus, we repeat the inter-user anomaly-detection experiment to assess whether the considered algorithm can tolerate small percentage of outliers in the training data. In particular, we consider the best configuration and introduce in the training sets of 500 heartbeats, 1 %, 2 %, 4 %, 8 % of outliers, which are selected among those heartbeats that were initially discarded. This experiment is repeated 15 times, and the average ROC curves are reported in Fig. 5(b). It can be seen that the performance of the anomaly detection are stable when including only 1 % and 2 % of outliers, but dramatically decreases when outliers are 8 %. This suggests that it is necessary to reduce the number of outliers from the training set, e.g., by some prescreening method that analyzes MEMS recordings that are embedded on the Pulse Sensor.

5.4 Experiments on MIT-BIH Arrhythmia Database

We design two experiments also on the MIT-BIH Arrhythmia Database. In the first one, we show that our method can successfully detect *inter-users* changes also in this dataset, and that the performance are higher than in the Pulse dataset. As in the previous experiment we learn a dictionary D from 500 normal heartbeats of each tracing, considering the same range of parameters as in the Pulse dataset. AUC values are reported in Fig. 6 and indicate that the best settings are the same ($\kappa = 3$ and $n = 8$). The Wilcoxon signed-rank test confirms that these parameters yield significantly superior performance than the single-atom settings (p-value $\approx 10^{-16}$). However, in all these settings, the median AUC is very close to 1, indicating very good detection performance independently of the parameters adopted. This suggests that the ECG tracings in the Pulse dataset are more difficult to model than in the MIT-BIH Arrhythmia Database, and we speculate that this is due to the fact that the heartbeats from MIT-BIH Arrhythmia Database present a low variability than in the Pulse dataset.

Finally, we assess the performance of the considered algorithm in an arrhythmia-detection task, using the annotations provided in the MIT-BIH Arrhythmia Database. In particular, we consider as anomalous the arrhythmias from the same patient used for dictionary learning. The ROC curves averaged over the entire dataset, for the *best* and *single-atom* configuration are very similar, and are reported in Fig. 5(c). The Wilcoxon signed-rank test on the corresponding AUC values confirms that there is not a clear statistical evidence to claim that one configuration is better than the other (p-value $= 0.13$). This

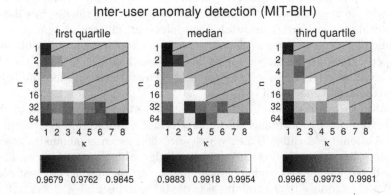

Fig. 6. Performance of several configurations of the considered algorithm for inter-user anomaly detection. The three figures reports the first quartile, the median and the third quartile of the AUC values computed on the MIT-BIH Arrhythmia Database. The best configuration corresponds to $n = 8$, $\kappa = 3$, as for the experiment in Fig. 4. The intensity ranges in the three images are different for visualization sake.

results can be explained by the fact that the arrhythmias show a very different morphology with respect to normal heartbeats, which allows these two methods to perform equally good.

6   Feasibility on the Pulse Sensor

We now investigate the overall feasibility of the considered anomaly-detection solution on the Pulse Sensor. In particular, we study both the requirements of dictionary learning, which is conveniently performed at each device positioning on an external host, and the computational complexity of the sparse coding, which has to executed in real time on the Pulse Sensor.

6.1   Dictoinary Learning and Device Configuration

Figures 4 and 5 confirm the need of learning the dictionary D every time the device is positioned, and at the same time indicate that 500 heartbeats are enough for this purpose. At an average heart rate, 500 heartbeats correspond to 7 min of ECG tracings, which can be conveniently transmitted via Bluetooth to an external host, e.g., the user smartphone, where the KSVD algorithm [24] can be executed[4] to learn the dictionary D, which is then sent back to the Pulse Sensor.

In the Pulse dataset these 7 min for training were acquired from users that were typically working in their office, thus performing normal actions and movements, while not in a rest state. In particular, experiments with outliers in the training set indicate that the dictionary learning in our specific settings (i.e., $n = 8$, $\kappa = 3$), can well tolerate a small percentage of heartbeats affected by user movements. Whenever higher robustness is requested, it is possible to leverage robust dictionary-learning algorithms that adopt an ℓ^1 norm for the data-fidelity term in (3), as in [28]. Alternatively, some form of pre-screening of the training set could be performed analyzing the MEMS recordings.

Let us finally remark that even if the device would provide sufficient computing power and memory for running the KSVD algorithm, it is nevertheless convenient to keep track of the training sets and learned dictionaries on an external host. It is in fact desirable to assess the quality of the recent acquisitions, for instance, by testing them with dictionaries previously learned.

6.2   Anomaly Detection on the Pulse Sensor

The ECG preprocessing we performed is the same as in [3], which consists in two median and a low-pass, convolutional, filter. Heartbeats are then segmented by locating the R-peaks using the Pan-Tompkins algorithm [22] and extracting a suitable temporal window centered in the R-peaks as in Sect. 5.1. All these operations are definitively compliant with computational capabilities of the Pulse Sensor.

---

[4] In the considered settings, the KSVD algorithm takes only few seconds on an ordinary laptop.

Anomalous heartbeats are then detected by solving the sparse coding problem (4), which represents the most time-demanding operation to be executed on the device. For this task, we adopt the OMP algorithm [25]: other anomaly-detection solutions based on sparse representations, like those in [16,27], are way more computationally demanding and cannot be implemented on the Pulse Sensor. The OMP is a greedy algorithm that solves (4) iteratively. In what follows we briefly illustrate the main steps of the OMP in its efficient implementation described in [29] (the same we used in our experiments), and we describe its computational complexity in terms of floating point operations (flop). At the very beginning $\mathbf{z} = D^T\mathbf{s}$ is computed at the cost of $\leq 2pn$ flop and the residual vector is defined as $\mathbf{r}^{(0)} = \mathbf{s}$. Then, the OMP iterates at most κ times the following steps, where l is used as an iteration index:

Correlation compute the inner product of the residual with each atom, i.e., $\mathbf{d}_k^T\mathbf{r}^{(l-1)}$, $k = 1,\ldots,n$ with an overall cost of $2pn$ flop.

Maximum select the atoms that is most correlated with $\mathbf{r}^{(l-1)}$, thus maximizing $|\mathbf{d}_k^T\mathbf{r}^{(l-1)}|$, $k = 1,\ldots,n$ which costs $\leq 2n$ flop.

Projection compute the coefficients $\mathbf{x}^{(l)}$ by orthogonal projection of $\mathbf{s}$ on the subspace spanned by the l atoms selected so far. This involves solving the linear system $\mathbf{z} = D_l^T D_l\mathbf{x}$, where D_l denotes the matrix containing all the selected atoms. Exploiting Cholesky factorization of $D_l^T D_l$, $\mathbf{x}^{(l)}$ can be computed at a cost $\leq 2pl + 3l^2$ flop.

Update update the residual $\mathbf{r}^{(l)} = \mathbf{s} - D\mathbf{x}^{(l)}$ at a cost of $\leq 2lp + p$ flop.

Considering the best parameters identified in our experiments (i.e., $p = 155, n = 8$, and $\kappa = 3$), a full execution of the OMP algorithm requires approximately 16 K flop, which seems to be compliant with real-time operations on the Pulse Sensor. However, to make sure that the overall Pulse Sensor computing capabilities can guarantee real-time operation, we have performed some tests (with the same parameter values) directly on the device. In particular, thanks to the sensor suite modularity, we measured the execution times on the STM32F401 processor embedding a Cortex^{TM}-M4F CPU with floating point unit (FPU). Tests were conducted using two versions of the CMSIS DSP library[5]: disabling/enabling the FPU optimization. The execution times when disabling the FPU optimization are a good estimate of the execution times on the STM32F103 processor that is actually used on the Pulse Sensor (which embeds a Cortex^{TM}-M3 CPU without FPU). In this case, the OMP algorithm took 58.13 ms allowing 17 executions per second at the maximum frequency of 72MHz, confirming the concrete possibility of executing the algorithm on this device within the period of an heartbeat. When enabling the optimization for FPU the OMP algorithm took only 6.58 ms allowing 152 executions per second. These results are particularly encouraging since, realistically, a STM32F401 processor with FPU is going to be adopted in future embodiments of the Pulse Sensor.

---

[5] CMSIS DSP Software Library, https://www.keil.com/pack/doc/CMSIS/DSP/html/index.html.

7   Conclusions

In this paper we investigate the problem of learning models to represent heartbeat morphology, in particular for monitoring ECG tracings acquired from wearable devices. Our study, conducted on ECG tracings form the Pulse Sensor, shows that dictionaries yielding sparse representations can effectively model the heterogeneous morphology of these heartbeats. In particular, we show that dictionaries can be successfully used to detect heartbeats having a morphology that is different from the training ones, and that this model can be effectively used in online monitoring schemes, implemented directly on the Pulse Sensor. Dictionary learning instead can be conveniently performed on an external host as it requires a limited amount of data to be transferred. Ongoing work concerns techniques to make the device configuration robust to user movements during the acquisition of the training set, which can be reasonably performed by pre-screen outliers in the MEMS recordings.

References

1. The Pulse Sensor Project. http://www.mrd-institute.com/#!projects/c1iwz
2. Wiens, J., Guttag, J. V.: Active learning applied to patient-adaptive heartbeat classification. In: Advances in Neural Information Processing Systems, pp. 2442–2450 (2010)
3. De Chazal, P., Dwyer, M.O., Reilly, R.B.: Automatic classification of heartbeats using ECG morphology and heartbeat interval features. IEEE Trans. Biomed. Eng. **51**(7), 1196–1206 (2004)
4. Hughes, N.P., Tarassenko, L., Roberts, S.J.: Markov models for automated ECG interval analysis. In: Advances in Neural Information Processing Systems, pp. 611–618 (2003)
5. Melgani, F., Bazi, Y.: Classification of electrocardiogram signals with support vector machines and particle swarm optimization. IEEE Trans. Inf. Technol. Biomed. **12**(5), 667–677 (2008)
6. Osowski, S., Hoai, L.T., Markiewicz, T.: Support vector machine-based expert system for reliable heartbeat recognition. IEEE Trans. Biomed. Eng. **51**(4), 582–589 (2004)
7. Castells, F., Laguna, P., Sörnmo, L., Bollmann, A., Roig, J.M.: Principal component analysis in ECG signal processing. EURASIP J. Appl. Sig. Process. **2007**(1), 98–98 (2007)
8. Kiranyaz, S., Ince, T., Gabbouj, M.: Real-Time patient-specific ECG classification by 1D convolutional neural networks. IEEE Trans. Biomed. Eng. **63**(3), 664–675 (2015)
9. Syed, Z., Guttag, J., Stultz, C.: Clustering and symbolic analysis of cardiovascular signals: discovery and visualization of medically relevant patterns in long-term data using limited prior knowledge. EURASIP J. Appl. Signal Process. **2007**(1), 97–97 (2007)
10. Aidos, H., Lourenço, A., Batista, D., Bulò, S.R., Fred, A.: Semi-supervised consensus clustering for ECG pathology classification. In: Bifet, A., May, M., Zadrozny, B., Gavalda, R., Pedreschi, D., Bonchi, F., Cardoso, J., Spiliopoulou, M. (eds.) ECML PKDD 2015. LNCS, vol. 9286, pp. 150–164. Springer, Heidelberg (2015)

11. Hoekema, R., Uijen, G.J., Van Oosterom, A.: Geometrical aspects of the interindividual variability of multilead ECG recordings. IEEE Trans. Biomed. Eng. **48**(5), 551–559 (2001)

12. Hu, Y.H., Palreddy, S., Tompkins, W.J.: A patient-adaptable ecg beat classifier using a mixture of experts approach. IEEE Trans. Biomed. Eng. **44**(9), 891–900 (1997)

13. de Chazal, P., Reilly, R.B.: A patient-adapting heartbeat classifier using ECG morphology and heartbeat interval features. IEEE Trans. Biomed. Eng. **53**(12), 2535–2543 (2006)

14. Bruckstein, A.M., Donoho, D.L., Elad, M.: From sparse solutions of systems of equations to sparse modeling of signals and images. SIAM Rev. **51**(1), 34–81 (2009)

15. Mairal, J., Bach, F., Ponce, J.: Sparse modeling for image, vision processing, arXiv preprint arXiv:1411.3230 (2014)

16. Adler, A., Elad, M., Hel-Or, Y., Rivlin, E.: Sparse coding with anomaly detection. J. Signal Process. Syst. **79**(2), 179–188 (2015)

17. Mailhé, B., Gribonval, R., Bimbot, F., Lemay, M., Vandergheynst, P., Vesin, J.-M.: Dictionary learning for the sparse modelling of atrial fibrillation in ECG signals. In: IEEE International Conference on Acoustics, Speech and Signal Processing, pp. 465–468. IEEE (2009)

18. Wang, J., She, M., Nahavandi, S., Kouzani, A.: Human identification from ECG signals via sparse representation of local segments. IEEE Signal Process. Lett. **20**, 937–940 (2013)

19. Moody, G.B.: MIT-BIH Arrhythmia database directory, May 1997. https://www.physionet.org/physiobank/database/html/mitdbdir/mitdbdir.htm

20. Willems, J.L., Poblete, P.F., Pipberger, H.V.: Day-to-day variation of the normal orthogonal electrocardiogram and vectorcardiogram. Circulation **45**(5), 1057–1064 (1972)

21. Alippi, C., Boracchi, G., Wohlberg, B.: Change detection in streams of signals with sparse representations. In: IEEE International Conference on Acoustics, Speech and Signal Processing, pp. 5252–5256. IEEE (2014)

22. Pan, J., Tompkins, W.J.: A real-time QRS detection algorithm. IEEE Trans. Biomed. Eng. **3**, 230–236 (1985)

23. Elad, M.: Sparse, Redundant Representations: From Theory to Applications in Signal and Image Processing. Springer Science & Business Media, New York (2010)

24. Aharon, M., Elad, M., Bruckstein, A.: K-SVD: an algorithm for designing overcomplete dictionaries for sparse representation. IEEE Trans. Signal Process. **54**(11), 4311–4322 (2006)

25. Tropp, J.A.: Greed is good: algorithmic results for sparse approximation. IEEE Trans. Inf. Theory **50**, 2231–2242 (2004)

26. Chen, S.S., Donoho, D.L., Saunders, M.A.: Atomic decomposition by basis pursuit. SIAM **20**(1), 33–61 (1998)

27. Boracchi, G., Carrera, D., Wohlberg, B.: Novelty detection in images by sparse representations. In: IEEE Symposium Series on Computational Intelligence, pp. 47–54. IEEE (2014)

28. Lu, C., Shi, J., Jia, J.: Online robust dictionary learning. In: IEEE International Conference on Computer Vision and Pattern Recognition, pp. 415–422 (2013)

29. Rubinstein, R., Zibulevsky, M., Elad, M.: Efficient implementation of the k-svd algorithm using batch orthogonal matching pursuit. CS Technion **40**(8), 1–15 (2008)

Do Street Fairs Boost Local Businesses? A Quasi-Experimental Analysis Using Social Network Data

Ke Zhang[(✉)] and Konstantinos Pelechrinis

School of Information Sciences, University of Pittsburgh, Pittsburgh, USA
{kez11,kpele}@pitt.edu

Abstract. Local businesses and retail stores are a crucial part of local economy. Local governments design policies for facilitating the growth of these businesses that can consequently have positive externalities on the local community. However, many times these policies have completely opposite from the expected results (e.g., free curb parking instead of helping businesses has been illustrated to actually hurt them due to the small turnover per spot). Hence, it is important to evaluate the outcome of such policies in order to provide educated decisions for the future. In the era of social and ubiquitous computing, mobile social media, such as Foursquare, form a platform that can help towards this goal. Data from these platforms capture semantic information of human mobility from which we can distill the potential economic activities taking place. In this paper we focus on street fairs (e.g., arts festivals) and evaluate their ability to boost economic activities in their vicinity. In particular, we collected data from Foursquare for the three month period between June 2015 and August 2015 from the city of Pittsburgh. During this period several street fairs took place. Using these events as our case study we analyzed the data utilizing propensity score matching and a quasi-experimental technique inspired by the difference-in-differences method. Our results indicate that street fairs provide positive externalities to nearby businesses. We further analyzed the spatial reach of this impact and we find that it can extend up to 0.6 miles from the epicenter of the event.

Keywords: Quasi-experimental design · Difference-in-Differences · Social media · Urban informatics · Local businesses

1 Introduction

A healthy local business sector is important for the prosperity of the surrounding community. City governments design policies and community organizations take actions that aim in boosting the growth of such businesses. This growth can have rippling positive externalities, such as, reducing local unemployment rates, keeping the local economy alive[1] and facilitating regional resilience to name just a few.

---

[1] As per the New Economics Foundation "local purchases are twice as efficient in terms of keeping the local economy alive".

© Springer International Publishing AG 2016
B. Berendt et al. (Eds.): ECML PKDD 2016, Part III, LNAI 9853, pp. 161–176, 2016.
DOI: 10.1007/978-3-319-46131-1_22

These are even more important during periods of economic crises and recession, similar to the recent one in 2008 that US is just getting itself out of.

However, these efforts might not have the results expected. For example, many local governments during the "Small Business Saturday" (last Saturday of November) offer free curb parking. The rationale behind this policy is to give incentives to city dwellers (i.e., reduced trip cost to the business) to shop locally. However, the outcome is in many cases radically different. The underpricing of curb parking creates latent incentives for drivers to keep their cars parked for longer than normal periods of times. This leads to low turnover per parking spot and hence, ultimately to fewer number of customers in the local stores [22]. Therefore it is crucial to evaluate the efficiency of similar *interventions*. Knowing what boosts the local economy and what not, can allow the involved parties to make educated decisions for their future actions and ultimately lead to *urban intelligence* through data-driven decisions and policy making. In this study we are interested in a specific question and in particular, we are studying a research hypothesis related with the **impact of street fairs on neighboring local businesses**.

The golden standard for evaluating public policies is randomized experiments. However, in many cases designing and running the experiment is impossible from a practical point of view. Hence, quasi-experimental techniques [21] have been developed to analyze observational data in such a way that resembles a field experiment. To complicate things more with respect to our specific research hypothesis, evaluating the economic impact of street fairs requires access to the appropriate revenue data. While a city government office can obtain access to information such as sales tax revenue, local business advocates and citizens organizations will certainly face obstacles in obtaining such kind of data. This type of information is not part of the Open Data released by local governments and are accessible (if at all) in a very limited form through pay-per-request APIs (e.g., http://zip-tax.com/pricing). This lack of transparency can be compensated to a certain extend by utilizing information from social networks and social media. While similar types of data can potentially suffer from well-documented biases (e.g., demographic biases), they form an open platform that can be easily accessed and analyzed by citizens themselves to facilitate further investigation of issues, leading to a grassroots approach to urban governance.

In our case, given that we do not have actual revenue data for the businesses in the area of Pittsburgh as aforementioned, we collect Foursquare *check-ins* from the city of Pittsburgh over a three-month period (June-August 2015) and evaluate the effect of summer street fairs on local economy. The check-in information can serve as a proxy - even though not perfect - for the revenue ρ generated [24]. We would like to emphasize here that, our study aims in evaluating the impact of street fairs on the brick-and-mortar stores that are adjacent to the event location and not that on the participating entities – which is expected to be positive in order for them to participate.

In order to analyze our data we rely on two quasi-experimental techniques. First, an increase in the check-ins for the venues near the street fair does not necessarily mean that this was due to the event. One or more control areas need to be used for comparison. However, our data are not generated through a randomized experiment but they are purely observational. For our analysis, this essentially means that we cannot assume that the area hosting a street fair event is chosen at random. Consequently, we cannot assume that the areas that do not host street fairs exhibit the same characteristics with respect to unobserved confounding features and hence, we cannot compare the revenue in the treated area with any untreated area. For overcoming this problem, we rely on quasi-experimental design techniques that identify appropriate control areas. In particular, we rely on propensity score matching [20], adopted in our setting by utilizing expert domain knowledge, in order to pick a set of *matched* areas $\mathcal{A}_m$ with the treated area α that will serve as our control subjects. Second, once the matched areas for comparison are chosen, we adopt the difference-in-differences method [3] in our setting in order to quantify the impact of the street fairs on local businesses. In a nutshell, the difference-in-differences is a regression model that examines the average change of the treatment group once the treatment has been applied and compares it with the control group. The implicit assumption is that this difference would be zero if the treatment had not been applied. We elaborate further on these two methods in the following section.

The main contributions of our work can be summarized as follows:

- We provide quantifiable evidence that support the positive impact of street fairs on local businesses.
- We show how social media data - despite their potential biases - can be useful to public policy makers and local governments since they are transparent, accessible and are able to provide good evidence when analyzed properly.

Scope of our work: While in the current study we are focusing on the effect of street fairs on local businesses the method can be applied in a variety of scenarios that include an external event/stimulant. For example, one can use our framework to quantify the effect of short-term road closures and/or constructions on the local economy. This is especially important during the bidding phase of a construction project since these effects should be included in the calculation of liquidated damages [9]. However, they are not currently included since there is not a framework to estimate this effect.

Roadmap: In the following section we present our method. We then describe our experimental setup and results, while we further discuss the limitations of our study. Finally, we discuss relevant to our work studies and conclude our study.

2 Analytical Methods

Let us denote the total volume of revenue within area α at day t with $\rho_{t,\alpha}$. Furthermore, $\mathcal{T}_\alpha$ is the set of days that a street fair took place within area α. The trending of $\rho_{t,\alpha}$ by itself cannot reveal anything with respect to the contribution of the street fair at the revenue generated in area α. Hence, in order to account for various confounding factors and other externalities we will need to get a "baseline" for comparison. When experimental design and implementation is possible this happens with random assignment of the treatment (in our case the street fair) to the experimental subjects. However, in our case this is not possible and hence, we rely on matching techniques and more specifically we use propensity score matching. Matching techniques provide us with the ability to analyze observational data in a way that mimics some of the particular characteristics of a randomized trial. In particular, we choose a matched, with area α, neighborhood, say, α_m, to analyze and compare the corresponding revenues generated.

Our analysis is inspired by the difference-in-differences method [3]. In brief, we compare the daily revenue differences between the area with the street fair and the corresponding matched area(s) both during the period of the street fairs as well as during the period without any street fair. The comparison with the matched area(s) - that are exposed to the same externalities - accounts for various confounding factors that can affect revenues, and hence, any observed difference can be attributed to the treatment, i.e., the street fairs in our case. In what follows, we describe in detail the building blocks of our analysis, i.e., propensity score matching and difference-in-differences.

2.1 Propensity Score Matching

Propensity score matching can be used to reduce (or even eliminate) the effect of confounding variables on the analysis of observational data. To reiterate propensity score matching allows an analysis in a way that mimics a randomized trial. In our own context, the **treatment** of interest is whether or not there is a street fair in neighborhood i. The propensity score of each (untreated) instance (i.e., every untreated neighborhood) represents the probability of this instance to be treated, conditional on a set of confounding variables. In a real randomized experiment, the instances are randomly assigned to the treatment and control groups. This ensures (given sufficiently large number of instances) that on average the two groups will only differ with respect to the reception of the treatment. In the case of observational data, the treatment is not randomly assigned but usually the "treated" instances are chosen due to some specific characteristics (i.e., the confounding factors). Therefore, in order to identify an appropriate control group we need to calculate the probability of the untreated instances obtaining the treatment.

In order to calculate the propensity scores, i.e., the conditional probabilities of the instances receiving the treatment, we employ a logistic regression model similar to [1]. In particular, given a feature vector $\boldsymbol{Z}$ that is formed by a set of neighborhood characteristics (i.e., the confounding factors) we estimate the following conditional probability:

$$\Pr(b_i = 1 | \boldsymbol{Z}_i) = \frac{\exp(\boldsymbol{w}_i^T \cdot \boldsymbol{Z}_i)}{1 + \exp(\boldsymbol{w}_i^T \cdot \boldsymbol{Z}_i)} \tag{1}$$

where b_i is a binary indicator variable, which takes the value 1 if area i is treated and 0 otherwise. In our case, Z_i includes **three types of features** for every type of establishment T that exists in neighborhood i that captures **(a)** the fraction of type T venues in i, as well as, **(b)** the fraction of the revenue (check-ins in our case) within α that was generated by venues of type T. Finally, for every business venue type, we use **(c)** the "stickiness" of the users in this type as an additional feature. The "stickiness" is defined as the ratio between the total number of check-ins in the corresponding category over the number of unique users that generated these check-ins.

After training the aforementioned logistic regression model, we estimate the probability from Eq. (1) for all neighborhood instances $i \in \mathcal{N}$ (both treated and untreated), where $\mathcal{N}$ is the set of areas/neighborhoods. Then we match the treated neighborhood α, with:

$$\alpha_m = \min_{i \in \mathcal{N} \setminus \{\alpha\}} |\Pr(b_i = 1|Z_i) - \Pr(b_\alpha = 1|Z_\alpha)| \qquad (2)$$

Essentially, this means that area α_m is the one that has the closest probability of hosting a street fair to that of area α, under the assumption that the only features that affect the decision are the ones captured by the observable confounding variable vector Z.

In many scenarios (such as in our case study) we might only have one treated area α, i.e., only one area has hosted a street fair. In this case, evaluating Eq. (2) is *trivial*, since, the minimum is observed for the area i for which the vector distance $d(Z_i, Z_\alpha)$ is minimized. Simply put, the matched area α_m is the one whose feature vector Z_{α_m} is closer to that of the treated area Z_α. We would like to emphasize here that, there might be other, unobserved, factors that lead to the choice of an area for a street fair. This is a limitation of the quasi-experimental techniques in general and propensity score matching can only account for observable confounders Z.

One way we propose to use in order to alleviate some of the potential problems associated with the aforementioned limitation is to initialize the matching process with expert knowledge. In particular, the matched area α_m can be chosen using expert knowledge (e.g., urban planners in our case). The benefit of this approach is that the domain expert is - implicitly or explicitly - considering various (potentially unobserved) confounders simultaneously. We can then use the expert matching as a "seed" for matching more than one neighborhoods to α using the propensity scores.

In particular, with $\pi_{m,e}$ being the propensity score of the (domain expert) matched area $\alpha_{m,e}$, we can pick the following set of matched areas:

$$\mathcal{A}_m = \{\alpha_{m_j} : |\pi_{m_j} - \pi_\alpha| < |\pi_{m,e} - \pi_\alpha| + \epsilon\} \qquad (3)$$

Essentially, as per Eq. (3), the set $\mathcal{A}_m$ includes neighborhoods that have propensity scores that are closer to the score of the treated area (within a tolerance factor ϵ) as compared to the expert matched area. Once set $\mathcal{A}_m$ is obtained we can analyze the corresponding revenues generated using the difference-in-differences method described in what follows.

2.2   Difference-in-Differences

The difference in differences (DD) method [3] is a quasi-experimental technique that aims in identifying the effect of an intervention using observational data. DD requires observations obtained in different points in time, e.g., t_1 and t_2 ($t_1 < t_2$), for both the control (e.g., $y_{m,1}$ and $y_{m,2}$) and the treatment (e.g., $y_{\tau,1}$ and $y_{\tau,2}$) groups. The treatment group is exposed to the intervention only during t_2. The difference between $y_{\tau,2}$ and $y_{m,2}$ does not only include the effect of the intervention but it also includes other "intrinsic" differences between the two groups. The latter can be captured by their difference during time t_1, i.e., $y_{\tau,1} - y_{m,1}$, where the treatment group has not been exposed to the intervention. The DD estimate is then:

$$\delta_{\tau,m} = (y_{\tau,2} - y_{m,2}) - (y_{\tau,1} - y_{m,1}) \tag{4}$$

If $\delta_{\tau,m} > 0$ ($\delta_{\tau,m} < 0$), then the treatment has a positive (negative) impact on y, while if $\delta_{\tau,m} = 0$ there is not any impact from the intervention. Eq. (4) captures the impact of the intervention assuming that both the treatment and control follow a **parallel trend**. In particular, in order for the conclusions drawn from a difference-in-differences analysis to be reliable, the parallel trend assumption needs to hold. This assumption essentially states that the average change in the control group represents the counterfactual change expected in the treatment group if there was no treatment. Simply put, if there was not any treatment applied, we would have: $(y_{\tau,2} - y_{m,2}) = (y_{\tau,1} - y_{m,1})$, that is, the two groups would have a stable difference. This assumption is crucial for the conclusions from a difference-in-differences analysis to hold and is many times overlooked when the method is applied.

The exactly same estimate for the DD can be formally derived through a linear regression that models the dependent variable y. In particular, we have the following model:

$$y_{ilt} = \gamma_0 + \gamma_1 \cdot \alpha_l + \gamma_2 \cdot \beta_t + \delta \cdot D_{lt} + \epsilon_{ilt} \tag{5}$$

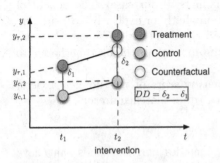

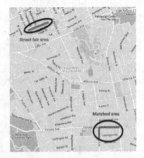

Fig. 1. The difference in differences method.

Fig. 2. Treated and domain expert matched neighborhood. (Color figure online)

where y_{ilt} is the dependent variable for instance i (at time t and location l), α_l and β_t are binary variables that capture the fixed effects of location and time respectively, D_{lt} is a dummy variable that represents the treatment status (i.e., $D_{lt} = \alpha_l \cdot \beta_t$) and ϵ_{ilt} is the associated error term. The coefficient δ captures the effect of the intervention on the dependent variable y. It is then straightforward to show that the DD estimate $\hat{\delta}$ is exactly Eq. (4). In particular, if $\overline{y}_{lt}$ is the sample mean of y_{ilt} and $\overline{\epsilon}_{lt}$ is the sample mean of ϵ_{ilt}, and using Eq. (5) we have:

$$(\overline{y}_{11} - \overline{y}_{01}) - (\overline{y}_{10} - \overline{y}_{00}) = \delta(D_{11} - D_{01}) - \delta(D_{10} - D_{00}) + \overline{\epsilon}_{11} - \overline{\epsilon}_{01} + \overline{\epsilon}_{00} - \overline{\epsilon}_{10}$$

Taking expectations and considering the i.i.d. assumptions for the errors for the ordinary least squares we further get:

$$E[(\overline{y}_{11} - \overline{y}_{01}) - (\overline{y}_{10} - \overline{y}_{00})] = \delta(D_{11} - D_{01}) - \delta(D_{10} - D_{00}) \qquad (6)$$

Given that the dummy variable D is equal to 1 only when $l = 1$ and $t = 1$ (i.e., for the treatment group after the intervention), we finally get for the DD estimator:

$$\hat{\delta} = (\overline{y}_{11} - \overline{y}_{01}) - (\overline{y}_{10} - \overline{y}_{00}) \qquad (7)$$

which is essentially the same as Eq. (4). Therefore, one can estimate the DD using either of the Eq. (4) or (5). Figure 1 further visualizes the estimation process. The control and treatment subjects in our setting are urban neighborhoods. Treated subjects includes neighborhoods that host street fairs.

2.3 Hypothesis Development

Having introduced our basic methodology we are ready to formally state the research hypotheses that are the focus of our study. In particular, we will examine the following two hypotheses.

Hypothesis 1 [Street fairs impact on local businesses]: *Street fair events lead to an increase in customer visitations for nearby business venues.*

Hypothesis 2 [Spatial impact of street fairs]: *The impact of street fairs on the customer visitations is geographically contained in a very small area.*

In order to support or reject Hypotheses 1 and 2 we will rely on data we collected from Foursquare described in the next section, utilizing the difference-in-differences method described in Sect. 2.2. We will further examine contextual dependencies, i.e., whether specific types of business venues benefit more than others.

3 Experimental Setup and Results

In this section we will present the dataset we collected, as well as, the setup for our analysis. We will then present our results and finally, we will discuss the implications and the limitations of our analysis.

3.1 Dataset

For the purposes of our study we collected time-series data using Foursquare's venue public API. We queried daily all Foursquare venues in Pittsburgh for the three-month period between **06/01/2015 - 08/30/2015**. This period includes six street fairs/events[2] that took place at a specific neighborhood in the city of Pittsburgh (see the street marked with red in Fig. 2).

Our time-series data include information with respect to the number of check-ins $c_v[t]$ that have been generated in venue v during day t. To reiterate, given the fact that we do not have actual revenue data for the businesses in Pittsburgh we rely on the check-in information as a proxy for the corresponding revenue of venue v, $\rho_v[t]$. This information will allow us to build the aggregate volume daily check-ins c_α within area α, i.e., $c_\alpha[t] = \sum_{v \in \alpha} c_v[t]$. Every area is defined as a circle of radius r centered at the centroid of the neighborhood under consideration. In our experiments, we examine various values for r in order to explore the spatial distribution of the impact.

We have also collected meta-data information. In particular, Foursquare associates each venue v with a type/category T (e.g., restaurant, school etc.). This classification is hierarchical and at the top level of the hierarchy there were 9 categories at the time of data collection. In order to obtain the feature vector Z, we use the top-level categories and hence Z includes 21 features (2 for each category and 3 for the stickiness of each type of business venue). Our final dataset includes 27,263 venues in the city of Pittsburgh, where 21.53 % (5,869) are business venues (i.e., *Nightlife Spots*, *Food* and *Shops & Services*). There are in total 32,501 check-ins in our dataset, among which 44.46 % were generated in business venues.

3.2 Experimental Setup

In our study we consider a single area α that has hosted street fairs during our data collection period. This area is a small business center, with a number of restaurants, cafes, retail stores (e.g., clothing stores, galleries etc.) and services (e.g., bank branches). The *treated* area is also accessible through public transportation, Pittsburgh's shared bike system as well as through private vehicle with parking facilities nearby. We (initially) perform the matching process based on the *expertise*[3] of local urban planners. Based on their recommendations we choose another small business area, with a similar urban form and accessibility patterns not very far from the treated area (approximately 2 miles away - green area in Fig. 2). We have further used Eq. (3) to build a set of matched areas. More specifically, we first pick 2,000 random points in the city of Pittsburgh and create a neighborhood of radius 0.3 miles around this point. We further eliminate areas with less than 60 venues. We consequently obtain the matched area set $\mathcal{A}_m$ using Eq. (3) with $\epsilon = 0$ and we filter out overlapping matched neighborhoods,

---

[2] http://thinkshadyside.com/events/.

[3] We have consulted with urban planners familiar with the city of Pittsburgh.

in order to remove possible dependencies in our datasets originating from the overlapping regions. In particular, when k matched areas overlap we only keep the final matched set the area with a propensity score matching closest to the treated area. We would like to emphasize here that we have examined different values for the radius of the control neighborhood area selection and the tolerance factor ϵ and the results obtained were very similar.

3.3 Results

The metric of interest for our analysis is the mean number of daily check-ins in area α, denoted with y_α. For every area α we compute the average number of daily check-ins during the treatment period, $y_{\alpha,\mathcal{T}_\alpha}$, as well as, during the days with no street fair, $y_{\alpha,\mathcal{T}_\alpha^c}$, where $\mathcal{T}_\alpha^c$, represents the complement of $\mathcal{T}_\alpha$, i.e., the set of days in our dataset where no street fair took place in α. With this setting the difference-in-differences coefficient is equal to **4.95 (p-value < 0.001)**. Simply put, there are 5 more check-ins every day with a fair in area α on average. This corresponds to an almost 100% increase in the check-ins in the area, since the average daily check-ins for the days with no event is 5.3.

As mentioned in Sect. 2.2 one of the crucial assumptions for the difference-in-differences to provide robust results is the parallel trend assumption. Typically the way that has been followed in the literature for verifying this assumption is to calculate the difference-in-differences coefficient for periods that the treatment has not been applied [16,17]. Hence, for the days that in reality no street fair occurred we randomly assign pseudo-treatments in order to calculate a null coefficient δ. Figure 3 depicts the distribution of the corresponding coefficients obtained from 100 randomizations. As we can see the mass of the distribution is concentrated around $\delta = 0$, while the 95% confidence interval is $[-0.42, 0.37]$. Hence, we cannot reject the hypothesis that the null coefficient δ is actually 0, hence, verifying the parallel trend assumption needed for the difference-in-differences method.

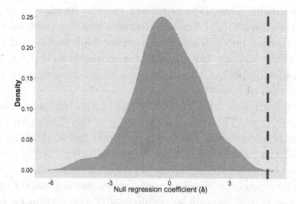

Fig. 3. The null difference-in-differences coefficient is practically equal to 0, hence, allowing us to apply the model with high confidence.

We also want to examine the spatial extent of this impact, i.e., how the impact decays with space. For this, we compute the difference-in-differences coefficient for zones of different radius around the treated area making sure that there is not any overlap with control areas. In particular, we examine zones of $[0, 0.1]$, $[0.1, 0.3]$, $[0.3, 0.6]$ miles. Our results are depicted in Fig. 4 where as we can see there is a clear decreasing trend of the impact. In fact, the coefficient for the range $[0.1, 0.3]$ miles is much smaller, and equal to 0.89 (p-value < 0.1), while going further away from the area of the event (i.e., $[0.3, 0.6]$ miles) the effect is practically eliminated ($\delta_{[0.3, 0.6]} = 0.33$, p-value = 0.61). These results indicate - as one might have expected - that the impact of a street fair event is highly localized within a very small area around the epicenter of the event.

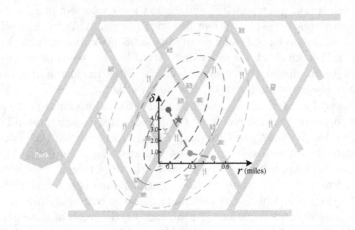

Fig. 4. The impact of street fairs on local businesses rapidly decays with the spatial distance from the event.

We further examine the impact of each event individually, i.e., we consider a single day treatment. Table 1 presents our results. As we can see every event contributes to the overall local business sector a positive increase to the check-ins, which can further be translated to increase foot traffic and revenue. The only exception is the Vintage GP Car show. Compared to the other events, this attracts a very specific part of the population - i.e., car-lovers - and this might have affected its overall impact.

Our analysis until now has considered all of the business venues together regardless of their type. This essentially captures the aggregate impact of the street fair in the neighborhood. However, we would like to decompose this effect in order to understand better what type of establishments benefit from the fairs. In particular, we compute the difference-in-differences regression coefficient for the three different types of business venues our dataset contains. Figure 5 depicts our results, where the 95 % confidence interval of the estimated coefficients is also presented. As we can see shopping venues are the ones that benefit the most from the street fairs, while nightlife and food establishment exhibit a much

Table 1. All events - except the Vintage GP Car Show - exhibit a statistically significant and positive coefficient δ. The reason why the car show does not impact the nearby businesses could potentially be attributed to the fact that compared to other events, it attracts a specific part of the population only. Significance codes: 0 '***' 0.01 '**' 0.05 '*' 0.1 '.' 1 ' '.

Event	Difference-in-differences coefficient δ
Jam On Walnut 1	9.7***
Vintage GP Car Show	-2.01***
Jam on Walnut 2	5.45***
Jam on Walnut 3	6.64***
Arts Festival on Walnut 1	4.45***
Arts Festival on Walnut 2	5.53***

(but significant and positive) lower coefficient δ. However, one crucial point here is that the coefficient provides the cumulative - additional to the counterfactual - check-ins recorded in all venues of the specific type. Hence, if a specific venue type is overrepresented in the area the estimated DD coefficient might be *inflated*[4]. In order to avoid similar issues, we can normalize the obtained coefficients from the regression model by the number of venues for every establishment type. In particular, the number of shop, nightlife and food venues in the treated area are 60, 13 and 25 respectively. Therefore, the normalized coefficients for the shop and nightlife are practically equal (0.066 and 0.061 respectively). However, the food venues still have a much smaller normalized coefficient, that is, 0.014.

Overall, we can say that our results support the two research hypotheses put forth in Sect. 2.3. In particular, street fairs have a positive impact on nearby businesses as captured by the check-ins on Foursquare and the difference-in-differences method. Furthermore, this impact is highly concentrated in the areas around the street fair (i.e., 0.1, 0.2 miles) and drops extremely fast as we move further away.

3.4   Discussion and Limitations

One of the main critics that studies relying on social media get is that of the potential demographic biases that the data include. This is certainly true and is one of our study's limitation as well. Nevertheless, location-based social media is a very good, and accessible, proxy for the economic activities in urban areas. Certainly there will be noise in the obtained signal, but this information is valuable for providing supporting (or not) evidence in a variety of research hypotheses

---

[4] Note here that, this is not an issue when we applied the difference-in-differences at the level of a neighborhood. In that case, we were interested in the total additional check-ins in the neighborhood as compared to the counterfactual. Hence, if a control area had a different number of venues this would not impact the results.

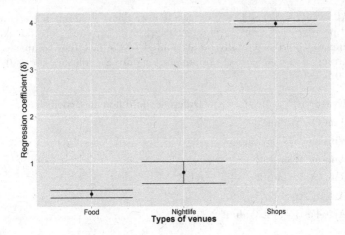

Fig. 5. The shopping businesses appear to have the largest benefit from the street fairs among the local establishments around the area.

similar to ours. For example, similar datasets have been used to study urban gentrification, deprivation, emotions in a city [7,11,23] etc.

In our difference-in-differences regression model we included fixed time and location effects. One might argue that we should also control for the day of the week. However, this is not necessary since the null regression model essentially shows us that the different days of the week will exhibit the same "trending" on average (of course the absolute values of the check-ins will be different). To verify this we run the regression model by adding an independent variable that captures the day of the week. Our results for the various zones around the treated neighborhood are presented in Table 2.

Table 2. Even when controlling for the day of the week, the impact of the street fair remains. The 95 % confidence intervals of the difference-in-differences coefficient for the ranges [0.1, 0.3] and [0.3, 0.6] overlap; hence, we cannot support with confidence the presence of a larger impact in the further zone. Significance codes: 0 '***' 0.01 '**' 0.05 '*' 0.1 '.' 1 ' '.

Radius r	Difference-in-differences coefficient δ
[0, 0.1]	[3.71, 4.1]
[0.1, 0.3]	[0.17, 0.85]
[0.3, 0.6]	[0.51, 1.61]

As we can see even when controlling for the day of the week the impact is strong and significant. In fact, when controlling for the day of the week the impact appears to be significant even for distances beyond the 0.1 miles.

Nevertheless, the impact itself is weak (i.e., the coefficient is small). Furthermore, even though it appears that the further zone has a stronger effect, the 95 % confidence intervals for the two coefficients overlap, and hence, we cannot confidently support the presence of a trend.

4 Related Work

In this section we briefly discuss related methodological literature as well as literature relevant to the specific application domain.

Quasi-experimental methodologies: The gold standard for evaluating the impact of a policy is a field experiment. However, when it comes to public policy many times this is not possible for a variety of reasons. In this case we need to rely on quasi-experimental techniques [21] in order to quantify the potential impact. Quasi-experimental designs allows to control the assignment to the treatment condition, but using some criterion different than random assignment as in field experiments.

There are various techniques that can be used depending on the type of observational data one has. For example, the difference-in-differences method [3] compares the average change over time in the outcome variable for the treatment group to the average change over time for the control group. One of the major problems when applying this method is the parallel trend assumption, that is, that the two groups exhibit the same temporal trend on their averages without the treatment. Regression discontinuity [12] is another technique that can be used to quantify the effects of treatments that are assigned by a threshold. The key idea is that observations lying very closely on either side of the threshold while differing in the reception of the treatment, they are *equal* for all practical purposes. Hence, their treatment assignment mimics that of a randomized control trial. It should be clear that not all quasi-experimental designs are applicable in all scenarios (for example regression discontinuity cannot be applied in our setting), while there can be settings were no method is applicable. A nice survey of various quasi-experimental techniques can be found in [10].

Local businesses and urban economy: Small shops and businesses are the backbone of local economy and quantifying the effect of external events and policies on their prosperity is of utmost importance. Given the absence of large scale data, most of the existing studies have been based on survey data. For instance, a survey research conducted by Lee *et al.* [14] during the 2002 World Cup identified that the event-related tourists yielded much higher expenditure as compared to *regular* tourists, indicating that such mega-events could have a positive economic impact for local businesses. As another example, a report from a Toronto-based think tank has identified the positive impact that bike lanes have on the revenue of local businesses despite the fact that business owners systematically underestimate it [2]. In a similar direction, based on merchant and pedestrian surveys in Toronto's Annex Neighborhood, the "Clean Air Partnership" [5] recommended reallocating a curb parking lane to bike lanes, since this is likely to increase commercial activity. A recent study further showed that the installation of shared bike system can lead to an increase of the housing property values [17]. Moreover, in a briefing paper DeShazo et al. [6] using a survey conducted over a small sample of businesses quantified the effect of CicLAvia on local businesses. CicLAvia[5] is a car-free event that happens once every year in various areas in Los Angeles. Furthermore, anecdotal hard evidence from Seattle [18] show that increasing the price of curb parking can be beneficial to restaurants and local businesses mainly due to the increased turnover of each parking spot [22].

During the last years, and driven by the proliferation and availability of geo-tagged social media data, there has been a surge of studies on business analytics. For instance, Qu and Zhang [19] proposed a framework that extends traditional trade area analysis and incorporates location data of mobile users. Their framework can answer crucial questions in retail management such as "where are the customers of a business coming from?". As another example, Karamshuk et al. [13] proposed a machine learning framework to predict the optimal placement for retail stores, where they extracted two types of features from a Foursquare check-in dataset. Furthermore, these platforms can serve as mobile "yellow pages" with business reviews that can influence customer choices and business revenue. For example, Luca [15] has identified a causal impact of Yelp ratings on restaurant demand using the regression discontinuity framework. Closer to our study, Georgiev et al. [8] using data collected from Foursquare study the impact of the 2012 Olympic Games on the businesses in London, while Zhang et al. [24] quantify the effectiveness of special deals offered through location-based services as an affordable advertisement for local businesses.

To the best of our knowledge no one has examined the impact of street fairs on the adjacent businesses, even though local authorities expect this policy to have a positive outcome for businesses[6]. Studies that examine the economic effects of special events/festivals exist (e.g., [4]) but their focus is slightly different, focusing on the participating entities/kiosks themselves. On the contrary, our study is focused on the "network" effects a street fair can have for the nearby businesses.

5   Conclusions and Future Work

In this study we have used social media data and quasi-experimental techniques to evaluate the effect of street fairs on the local business sector. In particular, we have adopted quasi-experimental techniques, i.e., difference-in-differences, and synthesized them with domain expert knowledge. We consequently applied our method on street fairs and outdoors arts festivals that took place on a specific neighborhood in the city of Pittsburgh as a case study. Our results indicate that similar street fairs can boost local businesses and stimulate and contribute to a healthy local economy. Similar approaches can be used to evaluate the impact of

---

[5] http://www.ciclavia.org.
[6] E.g., http://tinyurl.com/zdved39.

different interventions (e.g., installation of new transportation modes, alterations on the street network etc.).

Of course, our specific case study exhibits limitations with respect to the available data as we elaborated earlier. In particular, while check-in information is intuitively a good proxy for the underlying revenues, demographic biases can provide us with a skewed view of the exact magnitude of the impact. Nevertheless, similar analysis can provide advocate citizens' organizations with a case for further scrutiny of any public policy in place. Social media data are "readily" available and accessible (at least most of the times) and can provide the basis for grassroots innovation in the space of policy evaluation. In the future we plan in examining other potential sources (e.g., sales tax data) and analyze information from other cities as well, in order to obtain a cross-city comparison with respect to street fairs and their impact on local economy. Furthermore, even though we have verified the parallel trend assumption, the increase in the check-ins (revenues) in the treated area might be partly attributed to a decrease in the rest of the areas. This interaction between neighborhoods is extremely interesting and can potentially be captured and analyzed through a network between the urban areas. Finally, the long-term effect of these events is also important. In particular, even though the street fair can potentially increase the revenues during its lifetime, does it have the ability to create new clientele for the area? We will further explore these points in our future work.

Acknowledgments. We would like to thank Bob Gradeck from the University Center of Urban & Social Sciences, for his suggestions on matching the treated area in our case study.

References

1. Aral, S., Muchnik, L., Sundararajan, A.: Distinguishing influence-based contagion from homophily-driven diffusion in dynamic networks. Proc. Nat. Acad. Sci. **106**(51), 21544–21549 (2009)
2. Arancibia, D.: Cycling economies: economic impact of bike lanes. Report. Toronto Cycling, Think and Do Tank (2012)
3. Ashenfelter, O., Card, D.: Using the longitudinal structure of earnings to estimate the effect of training programs. Rev. Econ. Stat. **67**(4), 648–660 (1985)
4. Carter, R.D., Zieren, J.W.: Festivals that say cha-ching! measuring the economic impact of festivalst. In: Main Street Now (2012)
5. CleanAir-Partnership: Bike lanes, on-street parking and business. Report (2009)
6. DeShazo, J., Callahan, C., Brozen, M., Heimsath, B.: Economic impacts of ciclavia: study finds gains to local businesses. In: Briefing Paper - UCLA Luscin School of Public Fairs (2013)
7. Gallegos, L., Lerman, K., Huang, A., Garcia, D.: Geography of emotion: where in a city are people happier? In: WWW (2016)
8. Georgiev, P., Noulas, A., Mascolo, C.: Where businesses thrive: predicting the impact of the olympic games on local retailers through location-based services data. In: AAAI ICWSM (2014)

9. Goetz, C.J., Scott, R.E.: Liquidated damages, penalties and the just compensation principle: some notes on an enforcement model and a theory of efficient breach. Columbia Law Rev. **77**(4), 554–594 (1977)
10. Harris, A.D., McGregor, J.C., Perencevich, E.N., Furuno, J.P., Zhu, J., Peterson, D.E., Finkelstein, J.: The use and interpretation of quasi-experimental studies in medical informatics. J. Am. Med. Inform. Assoc. **13**(1), 16–23 (2006)
11. Hristova, D., Williams, M., Musolesi, M., Panzarasa, P., Mascolo, C.: Measuring urban social diversity using interconnected geo-social networks. In: ACM WWW (2016)
12. Imbens, G., Lemieux, T.: Regression discontinuity designs: a guide to practice. Working Paper 13039, National Bureau of Economic Research, April 2007
13. Karamshuk, D., Noulas, A., Scellato, S., Nicosia, V., Mascolo, C.: Geo-spotting: mining online location-based services for optimal retail store placement. In: ACM SIGKDD (2013)
14. Lee, C.K., Taylor, T.: Critical reflections on the economic impact assessment of a mega-event: the case of 2002 FIFA world cup. Tourism Manag. **26**(4), 595–603 (2005)
15. Luca, M.: Reviews, reputation, and revenue: The case of yelp. com. Technical report, Harvard Business School (2011)
16. Mora, R., Reggio, I.: Treatment effect identification using alternative parallel assumptions (2012)
17. Pelechrinis, K., Kokkodis, M., Lappas, T.: On the value of shared bike systems in urban environments: evidence from the real estate market. Available at SSRN (2015)
18. de Place, E.: Are parking meters boosting business? http://daily.sightline.org/2012/03/28/is-metered-parking-boosting-business/
19. Qu, Y., Zhang, J.: Trade area analysis using user generated mobile location data. In: ACM WWW (2013)
20. Rosenbaum, P.R., Rubin, D.B.: The central role of the propensity score in observational studies for causal effects. Biometrika **70**(1), 41–55 (1983)
21. Shadish, W.R., Cook, T.D., Campbell, D.T.: Experimental and Quasi-Experimental Designs for Generalized Causal Inference. Cengage Learning, Belmont (2001)
22. Shoup, D.: The High Cost of Free Parking. American Planning Association, Chicago (2011)
23. Venerandi, A., Quattrone, G., Capra, L., Quercia, D., Saez-Trumper, D.: Measuring urban deprivation from user generated content. In: ACM CSCW. pp. 254–264 (2015)
24. Zhang, K., Pelechrinis, K., Lappas, T.: Analyzing and modeling special offer campaigns in location-based social networks. In: AAAI ICWSM (2015)

Intelligent Urban Data Monitoring
for Smart Cities

Nikolaos Panagiotou[1], Nikolas Zygouras[1(✉)], Ioannis Katakis[1],
Dimitrios Gunopulos[1], Nikos Zacheilas[2], Ioannis Boutsis[2],
Vana Kalogeraki[2], Stephen Lynch[3], and Brendan O'Brien[3]

[1] National and Kapodistrian University of Athens, Athens, Greece
nzygouras@di.uoa.gr
[2] Athens University of Economics and Business, Athens, Greece
[3] Dublin City Council, Dublin, Ireland

Abstract. Urban data management is already an essential element of
modern cities. The authorities can build on the variety of automatically
generated information and develop intelligent services that improve citizens daily life, save environmental resources or aid in coping with emergencies. From a data mining perspective, urban data introduce a lot of
challenges. Data volume, velocity and veracity are some obvious obstacles. However, there are even more issues of equal importance like data
quality, resilience, privacy and security. In this paper we describe the
development of a set of techniques and frameworks that aim at effective
and efficient urban data management in real settings. To do this, we
collaborated with the city of Dublin and worked on real problems and
data. Our solutions were integrated in a system that was evaluated and
is currently utilized by the city.

1 Introduction

Technological advancement led to the generation of massive amounts of data
originating from a variety of urban sources. Smart cities equipped with the
appropriate infrastructure are producing many gigabytes of information on a
daily basis and data sources range from static to dynamic sensors. Examples
include GPS trajectory traces, aggregated logs of mobile phone activity as well
as user generated content from social media. Such data variety offers the potential for novel applications that support decision making in multiple situations.

Interestingly enough, the nature of smart city data brings a lot of challenges
to data mining researchers and practitioners. Data volume and velocity impose
great challenges in performing any type of analysis in real time. On top of that,
data veracity hinders many sophisticated learning algorithms. Data quality issues
demand extra attention in the case of smart cities and the challenges extend even
further to resilience issues, data privacy and security. More specifically, tools
established in smart cities need to address the following challenges:

(i) *Identify events in real-time:* Exploring and detecting events of interest from
complex and voluminous urban data streams is extremely challenging [41].

© Springer International Publishing AG 2016
B. Berendt et al. (Eds.): ECML PKDD 2016, Part III, LNAI 9853, pp. 177–192, 2016.
DOI: 10.1007/978-3-319-46131-1_23

Fig. 1. DCC's traffic control center, INSIGHT system shown at middle top screen.

Urban sensors are transmitting ambiguous and contradictory information. In such occasions exploiting the wisdom of the crowd through crowdsourcing is necessary. Furthermore the incoming stream may be massive requiring efficient online solutions. For example, in Twitter, thousands of tweets per second need to be processed in order to identify a few that are relevant to the task.

(ii) *Handle varying loads:* In the city setting, data loads deviate significantly over time. During rush hours many vehicles send out position information, while during the night the load is significantly lower. Twitter follows similar 'normality' patterns and significant events are correlated with large number of tweets.

(iii) *Noisy data and erroneous measurements:* Very often sensors report faulty measurements due to miss-calibration or hardware problems. For example we observed that bus data from Dublin had many inaccuracies: buses reported erroneous locations, extreme vehicle speed, irrationally high delays, and more. All these issues hamper data exploration and analysis. Moreover, Twitter due to informal and short text hinders language processing algorithms.

In this work we discuss the above challenges inherent in urban data. We then describe the provided solutions in the context of the city of Dublin, where we have collaborated on real problems and data. The application we targeted is the intelligent urban data management for event detection and emergency response. Dublin City Council operates a Traffic Management Center analyzing information from multiple sources like buses moving around the city or measurements of traffic flow in junctions. However, their work-flow is hindered by the data volume and the raw data that it are difficult to interpret.

We provide with a number of novel modules specifically designed to detect multiple types of incidents. These components use state of the art event detection techniques and are built with the focus on scalability and efficiency. They are generic and can be utilized in multiple settings analyzing in real-time GPS trajectories, data coming from sensors installed in junctions, or social media textual information. Specifically, we present:

(a) a dynamically scalable pipeline for streaming data (Sect. 3.1),

(b) an adaptive monitoring framework for social media such as twitter, which optimises recall of traffic related information (Sect. 3.2),
(c) an efficient crowdsourcing system for collecting urban data and for minimizing uncertainty (Sect. 3.3).

Through multiple feedback loops of the development process, expert knowledge was integrated in order to improve functionality, tuning appropriately the parameters. We report on the integration of the above techniques into a system that addresses information comprehensibility and it is useful for the end user. The components were evaluated within the work flow of Dublin City's Traffic Management team where the system INSIGHT [18] is established (see Fig. 1).

2   Related Work

The first architectures and technological innovations on the area of smart cities were built in the early 90s. Early framework examples include the AOL cities, a virtual simulation environment, and the first digital cities Kyoto and Amsterdam [33]. Years later many projects that aim at urban data analysis and solutions were developed by utilizing distributed sensors. The Ubiquitous Sensor Network proposed in [16] is an architecture where decentralized and geographically diverse sensors across the city are aggregated in a central database (IoT). Similarly the SOFIA architecture [13] was built with the purpose of an ecosystem of heterogeneous sensors, devices and appliances.

Nowadays, many cities use real-time analysis mechanisms to measure city functionality. IBM in partnership with the government of Brazil built a system for the city of Rio De Janeiro that aggregates multiple streams and combines them in a control center where algorithms analyze the data in real time, describe the state of the city and inform operators about disastrous events such as floods. The authors in [3] suggest a similar real-time architecture for managing city-wide critical equipment detecting faults. For example a fault could be an electricity distribution failure. According to [39] urban management can be also used for a greener environment. Their proposed system aims to ensure that the environmental policies, set up by the city, are satisfied (e.g. levels of CO_2). The system is used in many cities including Barcelona and Edinburgh.

In recent years many algorithms have been proposed that aim to analyze and extract information from urban data. The authors in [29] proposed an algorithm that detects anomalous traffic behaviour analyzing GPS data from taxis in Beijing. A traffic event detection approach using k-means algorithm was presented in [26]. The authors in [9] used time varying scalar functions in order to detect events from urban data. The detection of users' transportation mode, using their GPS trajectories along with information regarding the road network was presented in [34]. In [22] the authors describe a holistic technique that uses a hierarchical Markov model with multiple abstraction levels is able to infer the user's destination or their mode of transportation. [27] discovers traffic congestion on the road network examining co-occurring congestion locations.

Distributed systems have been widely used for traffic monitoring in smart cities environments [4,19]. In [4], the authors exploit IBM-Streams, [17] a scalable stream processing platform, to perform traffic monitoring in the city of Stockholm. Their system is able to continuously derive current traffic statistics from vehicle reports and also can provide useful information such shortest-time routes from real-time and estimated traffic conditions. While in [19] the authors combine Apache's Hadoop [1] and Apache's Spark [2] to detect traffic congestion in the Greater Toronto Area. Due to the varying volume of data that needs to be processed in such traffic monitoring applications (e.g. during peak hours more input data will be received), it is common practice to exploit elasticity techniques [15,25], for automatically adjusting the amount of processing nodes used for the data processing.

Crowdsourcing in smart cities is the process of soliciting contributions from citizens that actively participate by contributing real-time information from their mobile devices about city events. Crowdsourcing has been widely used recently both from applications driven by city authorities, such as the JRA Find and Fix app [14] where users report road defects for the Johannesburg's road network and from applications driven from organizations such as in Waze, [36] where users are asked to report traffic events from their current location. Crowdsourcing in smart cities introduces a number of challenges like unpredictable user response delays [5], human characteristics that are difficult to be estimated such as reliability [7], expertise [24] and availability [30], as well as dealing with privacy issues [6].

3 Methods and Techniques

To meet the requirements of Smart Cities applications we aid users to identify events in (near) real-time or flag emergencies and anomaly events so that authorities can quickly allocate assets to address these problems. For example, in a real setting we may have data coming from different streaming sources including static or mobile sensors, social media, citizen reports, etc. Each of these sources requires a comprehensive set of techniques to analyze them. We identify generic problems that come up in these settings. In this section we describe how each problem is addressed in a real context in the city of Dublin. The streams that are used are presented in Table 1.

Table 1. Real time data sources from the city of Dublin

Source	Attributes
BUS	Timestamp, GPS location and route information (delay, closest bus stop, is bus stopped at a bus stop)
SCATS	Traffic flow and degree of saturation measurements based on vehicles that cross a specific road segment
Twitter	Timestamp, text, location information, user information

3.1 Flexible and Dynamic Pipeline for Complex Event Processing

One challenge was to process and analyze massive data streams and detect events in near-real time, using scalable techniques. To deal with the above issue we used a data pipeline that consists of distributed stream and batch processing components. This pipeline is able to perform Complex Event Processing (CEP) in the streaming data. Finally we instantiated the pipeline using the Dublin data.

3.1.1 Distributed Stream and Batch Processing Pipeline

In order to identify complex events in a scalable way, we adapted the architecture that is illustrated in Fig. 2 and presented in [41]. This module, exploits the Lambda architecture [20] combining a well-known stream processing framework, Apache Storm, with a highly expressive CEP system, Esper [12]. Our approach is modular as different stream processing frameworks could be used instead of Storm, for example Spark [38]. We decided to use Apache Storm as it supports very low per tuple latency by processing each tuple separately and not in the form of mini-batches as Spark [38]. This feature is extremely useful in our case as we want to identify events as soon as possible. Initially Storm preprocesses the raw incoming tuples and extracts meaningful information. Then the tuples are forwarded to multiple concurrently running Esper engines. Each engine runs at different cluster nodes to exploit the cluster's parallelism. These engines are responsible to invoke several rules on the incoming data and trigger events when they are satisfied. Additionally a batch processing framework is used, Apache's Hadoop, to compute several statistics regarding the rules. These map-reduce jobs run periodically and their output is used in order to update the rules' thresholds or models.

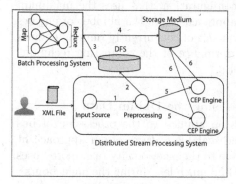

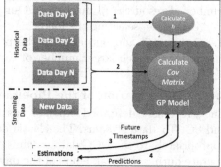

Fig. 2. Distributed complex event processing [41]

Fig. 3. Predict future input rate and tuples' latency [37]

We selected to enable the dynamic adjustment of the number of running Esper engines (*elasticity*) to exploit fully the parallelism offered by Storm during peak hours and avoid wasting resources when the input load is reduced. To

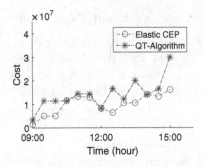
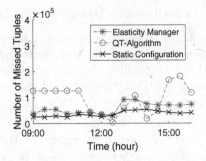

Fig. 4. Comparison of the cost overtime using different elasticity techniques [37]

Fig. 5. Comparison of the missed tuples overtime for the different techniques [37]

achieve this we used the technique presented in [37], where the expected input load for the upcoming time windows is estimated using Gaussian Processes (Fig. 3) and exploit it to automatically adjust the number of CEP engines in order to avoid information loss and without over-utilizing the system's resources. In Fig. 4, we illustrate the benefits of the Elastic CEP approach, compared to another commonly applied technique (QT-Algorithm [25]) which models the problem using queueing theory and assumes that the input rate follows Poisson distribution. The experiments run in our local 8-VMs cluster. As you can see in Fig. 4 our proposed technique, Elastic CEP, is able to minimize a cost function that considers both the information loss and the amount of resources (i.e. engines) that we bind. Finally, the scalability of our approach is presented in Fig. 5, where the amount of tuples that failed to be processed within a specified time window is illustrated over time. Our approach is able to process a similar number of data as the static configuration, that uses the maximum number of resources (8 machines). Our approach varies the number of engines used over time, selecting the following sequence of concurrently running engines $[1, 3, 7, 4, 5, 6, 7]$ and outperforms the performance of the comparing technique that misses a much larger number of tuples.

3.1.2 Instantiation of the Pipeline Using the Dublin Data

The previously described pipeline has been applied to the processing of bus and SCATS data streams. The elastic and scalable features of our approach fit appropriately to the DCC data streams due to their periodicity (e.g. more buses operate during the peak hours of the day and much less during the night hours). Using the pipeline to the real data had two main problems that we needed to overcome in order to identify meaningful events. The main issues were: (1) the *noisy measurements* for both bus and SCATS sensors that need specialized solutions and (2) the *end user requirements* that needed to be discussed with DCC traffic operators in order to set up the CEP rules appropriately.

(1) *Noisy measurements:* The first problem that we had to deal with was to clean the data and extract meaningful information from them. In our initial

analysis we identified that the reported raw data were noisy. More specifically, in some occasions buses reported as closest bus stop a stop that was many miles away from the actual closest bus stop. Also, some buses due to faulty sensors reported that they were stopped at a particular bus stop while they were actually moving. We clean the data solving the issues described above using the list of bus stops for each route, checking the spatial distance between the bus and the stops of the route. Also for the moving buses we set the at stop field to be *False*. In order to extract information from the raw data we calculated the time needed for a bus to go from one bus stop to the next. Finally, we calculated the approximate speed of each bus, using its previously reported coordinates. For the **SCATS Data** we identified that several sensors were faulty reporting extreme or unreasonably high measurements. To solve the above issue we used the technique presented in [40] that checks using a multivariate ARIMA model, whether the sensor's reports deviate significantly and unexpectedly from the measurements reported by neighbouring sensors.

(2) *End user requirements:* After discussing the problem with DCC traffic operators we set up several rules that when they are triggered a potential traffic anomaly may occur in the city. The rules that we used for the **Bus Data** are described bellow:

- Report a traffic anomaly when the time needed to travel from one stop to the next exceeds the expected time by some orders of magnitude.
- Report a noisy sensor when the bus moves while it is at stop.
- Report a noisy sensor when the bus seems to move with extreme speed.

The rules that we used for the **SCATS Data** are simple, but as described later in Sect. 4 are able to identify accurately traffic congestion. More specifically, we defined the following rules:

- Raise a traffic alarm when the moving average of the streaming values of the degree of saturation exceeds a predefined threshold.
- Report a faulty sensor if one is identified to variate significantly from its neighbors.

3.2   Building a Pipeline for Twitter Monitoring

Twitter was successfully used to detect meaningful events exploiting users as human sensors. Examples include the detection of earthquakes [31], floods [32], or even crimes [21]. We build a monitoring pipeline that utilizes Twitter to detect events of interest.

Usually when researchers perform topic-specific event detection on Twitter stream they query tweets containing particular keywords, generated by a specific user or located at a particular area, respecting at the same time the Twitter API constraints [35]. A great challenge that arises is how to use the Twitter's query filters efficiently in order to acquire more topic related tweets. Thus, we develop a *Twitter Fetcher* that is responsible to gather tweets relevant to a given topic by

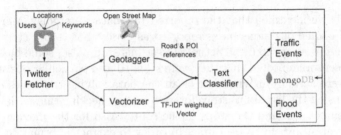

Fig. 6. The Twitter analysis component architecture.

tracking a dynamic set of keywords, users and locations and then forward them to the pipeline. The set of keywords and users is dynamically updated aiming to maximize the topic coverage. Another challenge is the fact that many messages do not contain location information. That said, a *Geotagger* able to assign exact GPS coordinates exploiting a tweet's text is the next pipeline component. Following the approach described in [8] and using the OpenStreetMap API [28] and Lucene, [23] the *Geotagger* assigns GPS coordinates according to road and POI references in tweets. The resulting set of geotagged tweets should be filtered and only the relevant tweets should be kept. In order to accomplish that a Text Classifier is used as the last unit of our pipeline. We used SVM to classify the incoming tweets, using as features the tweet's TF-IDF weighted vector, and the existence or absence of roads or POIs references. We tuned the classifier using an annotated dataset and under 10-fold cross validation we achieved an F-Measure of 88.3 %, a Precision of 93 % and a Recall of 84.2 %. The output of this pipeline is stored to a MongoDB instance for further usage and post-processing (Fig. 6).

The above framework is modular enough to be applied in any city given that citizens are active Twitter users. It was instantiated for the Dublin city. The Twitter Fetcher was set to return a Twitter stream from the Dublin city relevant to the topics of traffic and flood. The Geotagger was loaded with information about the Dublin road network from Open Street Maps. Finally, the Text Classifier was able to identify Traffic and Flood related messages. The training of the classifier was done using traffic and flood tweets originating from Dublin services such as 'AARoadwatch' and 'Livedrive'.

Dublin citizens proved to be very active Twitter users reporting on daily basis many events observed in the city. This fact was reflected in the evaluation where the Twitter monitoring pipeline was found very useful. A useful suggestion by the traffic operators during the feedback loops was to use a set of blacklisted words in order to avoid receiving tweets from towns nearby Dublin. Some example tweets detected from the Twitter monitoring pipeline are presented in Table 2.

3.3   Crowdsourcing Component

Crowdsourcing is a key part of our system as it allows us to extract information from the ubiquitous citizens, complementing the information extracted from the rest of the sensors and solving disagreements. Our experience from employing

Table 2. Traffic tweets coming from Dublin identified by our approach.

▶ M50 North: Jammed from J12 Firhouse due to drivers rubbernecking a collision in the south bound lanes

▶ @LiveDrive the M50 collision has the N4 inbound backed up to hermitage golf club ...

▶ Stuck in traffic on the M50 and my plc interview is at half six ... I'm gonna cry

▶ My bus has taken 40 min to get from Harolds Cross Road to Aungier Street

---

crowdsourcing in Smart City environments has shown that there are several challenges that need to be addressed to use crowdsourcing effectively: (i) users have different characteristics when processing crowdsourcing tasks in terms of response delays, user reliability and biases in their responses, (ii) user privacy is an important aspect that needs to be considered so that users will not be averted to send feedback, and (iii) scalability issues need to be taken into account when deploying crowdsourcing in such large scale environments. Our implementation addresses these challenges as explained in the remainder of the section.

3.3.1 How Crowdsourcing Is Invoked

The crowdsourcing is invoked whenever the different modules report contradictory types of anomalies. In order to identify such disagreements we developed an engine responsible to identify joint spatiotemporal anomalies. This component combines information from different components by grouping close (in space and time) reported anomalies together. When there is a disagreement in the type of anomalies reported from different modules, and hence uncertainty, it issues a query to the Crowdsourcing component to obtain direct information about this anomaly. It receives anomalies that contain the spatial area, where an event occurred, the timestamp of the event and the type of the event. In addition, in order to identify events it uses a R-tree and a queue data structure to store both the received anomalies and the candidate events.

The R-Tree data structure is used in order to efficiently detect spatial intersections. If a spatial intersection between two reported anomalies is found an event candidate is created and stored to the R-Tree. If a new anomaly intersects with an existing event candidate, the latter is updated. The event candidate affected area is set to the spatial intersection with the new anomaly. An event candidate may update to an event when a set of empirical rules is satisfied. For example, if multiple sensors (e.g. Twitter and SCATS) contribute to an event candidate then this candidate will evolve to an event.

In order to have guarantees that memory and time requirements will not grow unbounded over time the received anomalies and the candidate events are stored in a FIFO queue. These data are removed from both the queue and the R-Tree when they temporally differ from the latest data received more than a time threshold. The dataflow of the described engine is depicted on Fig. 7.

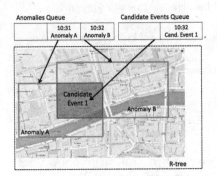

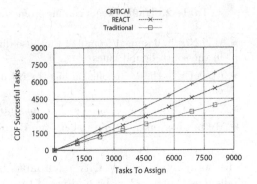

Fig. 7. The data structures used by the crowdsoucing invoker engine

Fig. 8. Task assignment in crowdsourcing

3.3.2   Making Crowdsourcing Efficient in a City-Wide Environment

Our crowdsourcing component comprises two main entities: (i) The Crowdsourcing server, that is responsible to act as a middleware between the streaming spatiotemporal event identification engine and the human users, by extracting information from the citizens and propagating this information to the event identification engine, and (ii) The CrowdAlert app, that was implemented aiming to allow users interact with our system and provide real-time information and observe ongoing events.

The Crowdsourcing server exploits the Misco framework [10], which is based on the MapReduce paradigm and tailored for mobile devices (Fig. 10) to assign crowdsourcing tasks to the citizens dynamically and aggregate the extracted information in a scalable manner. Assigning task to the human crowd is performed using techniques developed from our group [5,7] that consider the individual characteristics of the users. We have investigated the benefit of using different task assignment approaches by presenting the amount of tasks that have been accomplished correctly and within a predefined time interval, as shown in Fig. 8. Our approaches REACT [5], that considers the real-time constraints of the individual users, and CRITICAl [7], that considers both reliability and real-time constraints for groups of users, improve significantly the number of tasks processed successfully compared to traditional approaches that assign tasks randomly. Moreover CRITICAl performs better than REACT although there is a trade-off with the execution time needed to execute the algorithm, since REACT is faster. Hence, we use different task assignment strategies depending on the requirements (*e.g.*, critical tasks require fast responses even if the responses are unreliable). After we retrieve responses, we try to eliminate user bias to further improve the accuracy of our results. Finally, we note, that, the Crowdsourcing server is also responsible to receive user reports regarding ongoing events and propagate them to the spatiotemporal event identification engine as well as to inform the citizens for traffic and unusual events in their area.

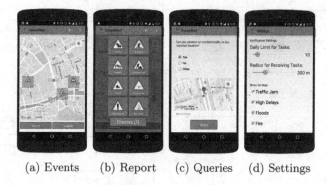

(a) Events   (b) Report   (c) Queries   (d) Settings

Fig. 9. CrowdAlert App

The CrowdAlert app allows both citizens and DCC operators to observe the ongoing events, identified from the system, in real-time, as shown in Fig. 9a and to provide valuable feedback. Thus, the users can participate at the central part of CrowdAlert which is to report events that take place near their location, including Accidents, Hazards, Constructions, etc., as shown in Fig. 9b. Additionally the users can also classify the traffic events that appear on the map to provide more accurate information regarding the events. All these reports are forwarded to the spatiotemporal event identification engine through the Crowdsourcing server to be processed along with data arriving from different components. We also note that the users of the CrowdAlert app report periodically their approximate location. Such information is used by the system to be able to ask user feedback dynamically when the users are near an ongoing event. An example of such a query is shown in Fig. 9c where the users need to respond in the question: "Can you observe an incident/traffic in the reported location?". The approximate location is provided by the Android API that exploits the cell network and WiFi. We chose to use the approximate location instead of the accurate GPS location for energy efficiency and privacy reasons. In addition, we intégrated our privacy preserving approach [6] to prevent privacy exposure of the user mobility when participating in CrowdAlert. Finally, the users can tune the CrowdAlert settings (Fig. 9d), such as the amount of Crowdsourcing tasks that they wish to receive per day, or the maximum distance from their current location for which they wish to answer to tasks.

4   User Evaluation and Lessons Learned

The Dublin City Council personnel evaluated our system in terms of effectiveness and usability. The evaluation of the previously described techniques was performed in the context of INSIGHT system presented in Fig. 10 and was performed using the INSIGHT Web Interface that allows the quick visualization and exploration of the real-time analysis output, as shown in Fig. 11. This interface offers a layered visualization of the identified events in order to help the operator to filter out the unnecessary information.

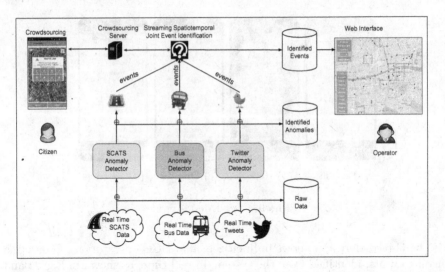

Fig. 10. Overview of system's layers: (i) the first layer receives the streaming data (ii) the second layer preprocess the raw data and detects anomalies, using scalable techniques (iii) the third layer groups the anomalies to identify events of interest or to send tasks to crowdsourcing users to resolve uncertainty

4.1 Evaluation Protocol

Relevant DCC personnel were invited to participate in the evaluation of the system. Using the Traffic Management Centre (TMC) and the adjoining Local Incident Room as a central hub allowed staff to evaluate the system all together, seeking clarification by asking questions and attending to concerns that may have arisen.

The evaluation was operated in two hour time windows and in two different days. The time windows were: (a) Day 1; 08.00–10.00, 16.00–18.00 and (b) Day2; 09.00–11.00, 15.00–17.00. These are typically periods of high volume of traffic in Dublin. In those time windows people involved in the evaluation were invited to utilize the system and complete a number of tasks. In each time window, participants were organized in groups based on their role in the department and completed only a part of the evaluation. For example people from the Traffic Management Centre Team worked on congestion related events. Participants from the Bus Priority Team evaluated the events identified from the Bus analysis component and Live-Bus layer of the system while the radio station team monitored social media. The following personnel were invited to participate in the evaluation of the system in the TMC: Traffic Management Personnel (12 people), LiveDrive Radio Station (5 people), Traffic ITS Officers (10 people). The personell confirmed or rejected the reported events based on manual CCTV cameras investigation. The cameras are able to capture a large portion of the city.

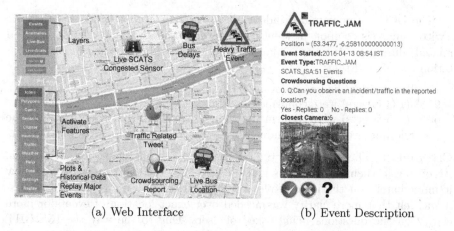

(a) Web Interface   (b) Event Description

Fig. 11. (a) The Web interface with different icons representing the locations of buses, locations with congestion and events from Twitter. The user can select which icons to display using the layers menu. (b) Details about a detected event.

4.2   Results

The evaluation team compiled an overall report for the whole system and more detailed reports for each component itself.

Bus Analysis Component. The bus priority team mentioned that the Bus anomaly detection mechanism proved useful in giving users information related to bus congestion events. They state that it proved more difficult to confirm these events using CCTV as by the time they tried to confirm the congestion event, the time may have lapsed in some cases, so verifying these anomalies was more difficult that the other events. They noted that the information given with each bus event was very useful in trying to isolate the location of the event. In particular they liked the 'view the bus stops' link, as this allows the user to know the direction of the bus and the current and next stop due.

SCATS Analysis Component. The SCATS anomaly detection proved very useful in allowing users to detect singular anomalies related to a junction. The anomaly detection triggers alerts about lanes with high degree of saturation, something that would not be possible using the existing SCATS system. Being able to use the map along with SCATS and CCTV proved useful to diagnose and confirm anomalies.

Twitter Analysis Component. According to the Livedrive radio station [11] team, Twitter event detection has been an excellent feature to the INSIGHT system. One issue they mentioned was the fact that tweets that refer to nearby regions of Dublin were mistakenly geolocated at Dublin. They suggested that with some additional negative keyword lists, such as counties outside of the Greater Dublin Area (e.g. Cork, Limerick, Galway), the location relevancy of the alerts will be more accurate. However, part of the issue is the fact that nearby countries share road names with Dublin confusing the system's geotagger.

The DCC team examined independently a complete list of Tweets that the Twitter analysis component identified as event tweets (Traffic related or Flood related) during the two day evaluation period. According to their manual annotation they found that:

- 179 Tweets identified as Relevant from the system
- 91 % (163 tweets) were confirmed as true positives
- 63 % (113 tweets) were confirmed as true positives in the city of Dublin. The rest relevant tweets were from nearby towns mistakenly identified at Dublin.

Crowdalert. The users found the crowdsourcing application very easy to use. According to them, the buttons make it very easy to report an event and view it immediately on the map display. With relation to users replying to alerts, it was felt that more clarity was needed over exact location of events or more clarity on the questions being asked of users. Due to the way the INSIGHT system is designed, it aggregates several events in adjacent geographic locations to issue crowdsourcing tasks and as a result CrowdAlert receives the approximate location of the area where the events take place rather than their exact location.

4.3   Lessons Learned

Component specific lessons learned could be summarized in the following points:

(i) The scalable and elastic framework enhanced with CEP engines helped to easily create or update the event detection rules. Finding the appropriate parameters for these rules is not an easy process, however, if the right rules are available, the processing can be simple and efficient.

(ii) Social media such as Twitter provide a valuable source of real time information about incidents in a city.

(iii) During the development of CrowdAlert we interacted with alpha and beta testers from DCC to improve it. Our conclusion is that such applications should provide a simple and easy to use interface, so that the user can interact immediately (*e.g.*, Yes/No answers), rather than providing too many options to the users. That way we modified the app so that the citizens will be willing to provide feedback and use the app.

5   Conclusions

In this work we describe three techniques able to cope with the challenges that arise in urban data analysis: the dynamic nature, the requirement for handling complex high velocity data streams, and information uncertainty. The proposed solutions address these challenges providing accurate, scalable real-time event detection. User oriented evaluation provided with evidence not only of the efficiency of the provided tools but also of their usability and positive impact in the user's work-flow.

Acknowledgments. This research has been financed by the European Union through the FP7 ERC IDEAS 308019 NGHCS project, the Horizon2020 688380 VaVeL project and a Yahoo Faculty award.

References

1. Apache-Hadoop: http://hadoop.apache.org
2. Apache-Spark: https://spark.apache.org
3. Attwood, A., Merabti, M., Fergus, P., Abuelmaatti, O.: Sccir: smart cities critical infrastructure response framework. In: Developments in E-systems Engineering (DeSE) 2011, pp. 460–464. IEEE (2011)
4. Biem, A., Bouillet, E., Feng, H., Ranganathan, A., Riabov, A., Verscheure, O., Koutsopoulos, H., Moran, C.: Ibm infosphere streams for scalable, real-time, intelligent transportation services. In: Proceedings of the 2010 ACM SIGMOD International Conference on Management of data, pp. 1093–1104. ACM (2010)
5. Boutsis, I., Kalogeraki, V.: Crowdsourcing under real-time constraints. In: IPDPS, Boston, MA, pp. 753–764, May 2013
6. Boutsis, I., Kalogeraki, V.: Privacy preservation for paricipatory sensing data. In: PerCom, San Diego, CA, USA, March 2013
7. Boutsis, I., Kalogeraki, V.: On task assignment for real-time reliable crowdsourcing. In: ICDCS, Madrid, Spain, pp. 1–10, June 2014
8. Daly, E.M., Lecue, F., Bicer, V.: Westland row why so slow? fusing social media and linked data sources for understanding real-time traffic conditions. In: ACM IUI (2013)
9. Doraiswamy, H., Ferreira, N., Damoulas, T., Freire, J., Silva, C.T.: Using topological analysis to support event-guided exploration in urban data. IEEE Trans. Vis. Comput. Graph. **20**(12), 2634–2643 (2014)
10. Dou, A.J., Kalogeraki, V., Gunopulos, D., Mielikainen, T., Tuulos, V.: Scheduling for real-time mobile mapreduce systems. In: DEBS (2011)
11. DublinCity-FM: http://www.dublincityfm.ie/live-drive
12. Esper-Tech: http://www.espertech.com
13. Filipponi, L., Vitaletti, A., Landi, G., Memeo, V., Laura, G., Pucci, P.: Smart city: an event driven architecture for monitoring public spaces with heterogeneous sensors. In: 2010 Fourth International Conference on Sensor Technologies and Applications (SENSORCOMM), pp. 281–286. IEEE (2010)
14. Find&Fix-App: http://www.jra.org.za/index.php/find-and-fix-mobile-app
15. Gedik, B., Schneider, S., Hirzel, M., Wu, K.L.: Elastic scaling for data stream processing. IEEE Trans. Parallel Distrib. Syst. **25**(6), 1447–1463 (2014)
16. Hernández-Muñoz, J.M., Vercher, J.B., Muñoz, L., Galache, J.A., Presser, M., Hernández Gómez, L.A., Pettersson, J.: Smart cities at the forefront of the future internet. In: Domingue, J., Galis, A., Gavras, A., Zahariadis, T., Lambert, D., Cleary, F., Daras, P., Krco, S., Müller, H., Li, M.-S., Schaffers, H., Lotz, V., Alvarez, F., Stiller, B., Karnouskos, S., Avessta, S., Nilsson, M. (eds.) FIA 2011. LNCS, vol. 6656, pp. 447–462. Springer, Heidelberg (2011). doi:10.1007/978-3-642-20898-0_32
17. IBM-Streams: www-03.ibm.com/software/products/en/ibm-streams
18. INSIGHT-Project: http://insight-ict.eu/
19. Khazaei, H., Zareian, S., Veleda, R., Litoiu, M.: Sipresk: a big data analytic platform for smart transportation. In: EAI International Conference on Big Data and Analytics for Smart Cities (2015)
20. Lambda-Architecture: http://lambda-architecture.net/
21. Li, R., Lei, K.H., Khadiwala, R., Chang, K.C.C.: Tedas: a twitter-based event detection and analysis system. In: 2012 IEEE 28th International Conference on Data Engineering (ICDE), pp. 1273–1276. IEEE (2012)

22. Liao, L., Patterson, D.J., Fox, D., Kautz, H.: Learning and inferring transportation routines. Artif. Intell. **171**(5), 311–331 (2007)
23. Lucene: http://lucene.apache.org/
24. Ma, F., Li, Y., Li, Q., Qiu, M., Gao, J., Zhi, S., Su, L., Zhao, B., Ji, H., Han, J.: Faitcrowd: fine grained truth discovery for crowdsourced data aggregation. In: KDD, Sydney, Australia, pp. 745–754, August 2015
25. Mayer, R., Koldehofe, B., Rothermel, K.: Meeting predictable buffer limits in the parallel execution of event processing operators. In: Big Data, pp. 402–411 (2014)
26. Münz, G., Li, S., Carle, G.: Traffic anomaly detection using k-means clustering. In: GI/ITG Workshop MMBnet (2007)
27. Nguyen, H., Liu, W., Chen, F.: Discovering congestion propagation patterns in spatio-temporal traffic data
28. OpenStreetMap: https://www.openstreetmap.org
29. Pang, L.X., Chawla, S., Liu, W., Zheng, Y.: On detection of emerging anomalous traffic patterns using gps data. Data Knowl. Eng. **87**, 357–373 (2013)
30. Roy, S.B., Lykourentzou, I., Thirumuruganathan, S., Amer-Yahia, S., Das, G.: Task assignment optimization in knowledge-intensive crowdsourcing. VLDB J. **24**, 467–491 (2015)
31. Sakaki, T., Okazaki, M., Matsuo, Y.: Earthquake shakes twitter users: real-time event detection by social sensors. In: Proceedings of the 19th International Conference on World Wide Web, pp. 851–860. ACM (2010)
32. Saravanou, A., Valkanas, G., Gunopulos, D., Andrienko, G.: Twitter floods when it rains: a case study of the uk floods in early 2014. In: Proceedings of the 24th International Conference on World Wide Web Companion, pp. 1233–1238. International World Wide Web Conferences Steering Committee (2015)
33. da Silva, W.M., Alvaro, A., Tomas, G.H., Afonso, R.A., Dias, K.L., Garcia, V.C.: Smart cities software architectures: a survey. In: Proceedings of the 28th Annual ACM Symposium on Applied Computing, pp. 1722–1727. ACM (2013)
34. Stenneth, L., Wolfson, O., Yu, P.S., Xu, B.: Transportation mode detection using mobile phones and gis information. In: Proceedings of the 19th ACM SIGSPATIAL International Conference on Advances in Geographic Information Systems, pp. 54–63. ACM (2011)
35. Twitter-API: https://dev.twitter.com/rest/public/rate-limiting
36. Waze: https://www.waze.com/
37. Zacheilas, N., Kalogeraki, V., Zygouras, N., Panagiotou, N., Gunopulos, D.: Elastic complex event processing exploiting prediction. In: Big Data, Santa Clara, CA, USA. IEEE, October 2015
38. Zaharia, M., Das, T., Li, H., Hunter, T., Shenker, S., Stoica, I.: Discretized streams: fault-tolerant streaming computation at scale. In: Proceedings of the Twenty-Fourth ACM Symposium on Operating Systems Principles, pp. 423–438. ACM (2013)
39. Zygiaris, S.: Smart city reference model: assisting planners to conceptualize the building of smart city innovation ecosystems. J. Knowl. Econ. **4**(2), 217–231 (2013)
40. Zygouras, N., Panagiotou, N., Zacheilas, N., Boutsis, I., Kalogeraki, V., Katakis, I., Gunopulos, D.: Towards detection of faulty traffic sensors in real-time. In: MUD2, pp. 53–62 (2015)
41. Zygouras, N., Zacheilas, N., Kalogeraki, V., Kinane, D., Gunopulos, D.: Insights on a scalable and dynamic traffic management system. In: EDBT, Brussels, Belgium, pp. 653–664, March 2015

Automatic Detection of Non-Biological Artifacts in ECGs Acquired During Cardiac Computed Tomography

Rustem Bekmukhametov[1], Sebastian Pölsterl[1]([✉]), Thomas Allmendinger[2], Minh-Duc Doan[2], and Nassir Navab[1,3]

[1] Computer Aided Medical Procedures, Technische Universität München, Munich, Germany
{r.bekmukhametov,sebastian.poelsterl,nassir.navab}@tum.de
[2] Diagnostic Imaging and Computed Tomography, Siemens Healthcare GmbH, Forchheim, Germany
{thomas.allmendinger,minh-duc.doan}@siemens.com
[3] Johns Hopkins University, Baltimore, MD, USA

Abstract. Cardiac computed tomography is a non-invasive technique to image the beating heart. One of the main concerns during the procedure is the total radiation dose imposed on the patient. Prospective electrocardiographic (ECG) gating methods may notably reduce the radiation exposure. However, very few investigations address accompanying problems encountered in practice. Several types of unique non-biological factors, such as the dynamic electrical field induced by rotating components in the scanner, influence the ECG and can result in artifacts that can ultimately cause prospective ECG gating algorithms to fail. In this paper, we present an approach to automatically detect non-biological artifacts within ECG signals, acquired in this context. Our solution adapts discord discovery, robust PCA, and signal processing methods for detecting such disturbances. It achieved an average area under the precision-recall curve (AUPRC) and receiver operating characteristics curve (AUROC) of 0.996 and 0.997 in our cross-validation experiments based on 2,581 ECGs. External validation on a separate hold-out dataset of 150 ECGs, annotated by two domain experts (88 % inter-expert agreement), yielded average AUPRC and AUROC scores of 0.890 and 0.920. Our solution is deployed to automatically detect non-biological anomalies within a continuously updated database, currently holding over 120,000 ECGs.

Keywords: Anomaly detection · Cardiac computed tomography · Electrocardiography · Prospective ECG gating

1 Introduction

Computed tomography (CT) is a non-invasive imaging technique, where a number of X-ray projections, taken from different angles, form a volumetric image of an area inside the body. Here, we focus on images of the heart, i.e., cardiac CT, which is often used to detect coronary artery disease or to evaluate the heart's

© Springer International Publishing AG 2016
B. Berendt et al. (Eds.): ECML PKDD 2016, Part III, LNAI 9853, pp. 193–208, 2016.
DOI: 10.1007/978-3-319-46131-1_24

function and morphology [9]. Due to constant beating of the heart, cardiac CT is particularly challenging: to ensure sharp motion-free images, multiple X-ray projections need to be taken at the same cardiac phase. In addition, the imaging protocol needs to be optimized to reduce the total radiation dose a patient is exposed to, thereby lowering the risk of radiation-induced cancer [9]. Hence, keeping a proper balance between low radiation exposure and image quality is one of the major trade-offs in a cardiac CT [9].

One of the most effective imaging techniques in this field is based on prospective ECG gating [10], the central idea of which is to activate the X-ray source only at the "right" time windows, namely during the cardiac phases of interest. Such gating algorithms reduce radiation by over 70 %, while maintaining high image quality [9]. On the other hand, relying on ECG makes the whole cardiac CT workflow highly dependent on the quality of the ECG signal, which is influenced by various factors specific to a patient, hospital, and physician. If the ECG signal is corrupted by noise or artifacts, prospective ECG gating is prone to fail and the resulting image of the heart will be of poor quality. In some cases, the scan has to be repeated, which offsets the advantage of prospective ECG gating in reducing radiation dose. We describe typical non-biological artifacts that may disrupt the imaging workflow in Sect. 2.

While the field of ECG analysis is well-established, it addresses problems distinct from ours. Common use case are clinical decision support and patient monitoring, both of which use ECG to assess a patient's health status. Consequently, anomalies of biological origin are the primary focus. In contrast, this work aims to accurately identify ECG signals that are corrupted by various non-biological artifacts, disregarding any medical conditions a patient might have. In addition, the characteristics of ECG signals and artifacts encountered in cardiac CT differ from those encountered in clinical diagnosis (see Sects. 2 and 3). To the best of our knowledge, this is the first scientific work that thoroughly investigates methods to automatically identify anomalies occurring in the context of cardiac CT.

We developed a system that can process large pools of data from multiple medical centers across the world and automatically identify CT scanners experiencing anomalous behavior. Our approach has several advantages. First, it dramatically reduces the time and effort of identifying problems compared to a human analyst, which leaves more time to fix a particular problem. Second, our customers benefit by reduced response times to an incident. Third, we expect that our system helps to increase the rate of high quality cardiac CT images, while maintaining a low radiation exposure. Our solution utilizes existing techniques used in ECG analysis and incorporates two feature extraction methods, which are based on robust PCA [3] and a discord discovery algorithm [13]. We retrospectively analyzed 2,581 cardiac CT scans from 60 medical centers from 18 countries. We evaluated our solution by cross-validation and by comparing its predictions to annotations of two domain experts on a hold-out set of 150 scans. The results demonstrate that our system is highly discriminatory and allows processing thousands of ECGs with minimal human interaction.

The paper is structured as follows. In Sect. 2 and 3 we will describe the most prevalent noise patterns encountered in the context of cardiac CT and our dataset. Section 4 describes our system. Next, we present our evaluation results in Sect. 5 and end with concluding remarks in Sect. 6.

2 Noise Patterns in Cardiac CT

In this section, we will describe the most prevalent noise patterns encountered in cardiac CT. But first, let us provide a brief insight into the ECG signal's morphology. The letters P, Q, R, S, and T name the key features of an ECG waveform. A typical heartbeat starts with a so-called P wave, continues with a QRS complex – characterized by a narrow spike called R peak – and ends with a T wave (see Fig. 1). Each feature corresponds to a particular phase in the cardiac cycle.

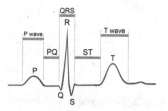

Fig. 1. Typical ECG waveform describing the heartbeart of a healthy patient.

Prospective ECG gating relies on detection of R peaks to predict the time of future R peaks. Since the R peak occurs at a distinct phase during the cardiac cycle, its detection enables imaging the heart in a predefined cardiac phase [9]. However, the presence of noise or non-biological artifacts in the ECG signal may result in false positive R peaks, which, in turn, may cause desynchronization of the whole workflow, resulting in a low quality image and the need for a repeat scan.

Typical non-biological artifacts observed in an ECG during cardiac CT can be classified into the following 6 categories:

- **Powerline noise** is caused by the interference of an ECG signal with an external power supply (Fig. 2a), for instance, if a power cord is placed across the patient or close to an ECG electrode.
- **Baseline wandering** is typically caused by breathing and movement of the patient, and becomes particularly strong when cardiograph's electrodes are unreliably connected to the body (Fig. 2b).
- **Rotational noise** is caused by an electrostatic charge near to or within the scanning area, which results in a rapid change of the electric field formed by the local static charge and rotating high-voltage generators of the CT scanner (Fig. 2c). Note that the noise is eliminated once the scanning process begins, because the X-ray leads to a discharge.
- **X-ray artifacts** are usually due to an X-ray beam hitting a piece of metal. This may happen when an electrode moves in the scanning area or the patient has one or more implants (Fig. 2e,f).
- **Table motion artifacts** are characterized by a noticeable fall of the ECG signal quality while the examination table is moving. Localized baseline and high frequency disturbances are sometimes observed after the table starts moving due to movements of the patient, improper wiring of the electrodes, or other reasons (Fig. 2g).

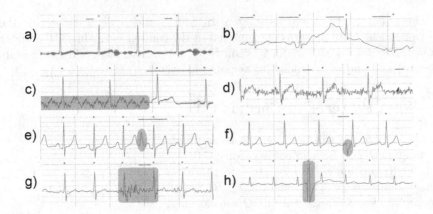

Fig. 2. Noise patterns observed during a cardiac CT scan. (a) powerline interference, (b) baseline wandering, (c) rotational noise, (d) other ubiquitous noise, (e-f) X-ray artifacts, (g) table motion artifact, and (h) localized disturbance. The purple lines indicate time intervals, during which the X-ray scanner was active.

- A wide range of **other noise** types due to a variety of reasons that are either unknown or do not fall into the abovelisted categories (Fig. 2d, h).

Although noise can be minimized by calibrating the CT equipment, not all medical centers may follow the best practices. By proactively identifying potentially unsuccessful scans, we mitigate the aforementioned health concerns and improve customer experience.

3   Dataset

Our dataset comprised 2,581 ECG signals from 60 medical centers from 18 countries annotated by a human expert as either "good" or contaminated with one or more of the noise patterns described above. Each ECG signal was sampled at a frequency of 100 Hz and on average ranged between 30 and 40 seconds in duration. In addition, each trace contained information about the time intervals, where the X-ray source was activated and the positions of QRS complexes, estimated by a proprietary R peak detection algorithm during image acquisition. Analyzing the dataset was challenging due to the following properties:

- ECG signals were highly heterogeneous due to different equipment used, different physicians performing the scan, and different technical and professional standards among countries.
- The ECG signal consisted only of the recording from a single lead, in contrast the conventional 12 lead ECG for clinical diagnosis.
- In cardiac CT, electrodes are placed outside of the patient's chest to not interfere with the X-ray scanner, which often results in atypical ECG waveforms, where only the R peak can be identified reliably.

– Relatively short ECG recordings, with the average length of 23 cardiac cycles and the minimal length of only 5 cycles.

4   Methods

In this section, we present our system to automatically quantify non-biological artifacts and noise in ECG signals. First, we present a high-level overview of our system. Subsequently, we explain our feature extraction technique for describing anomalous ECG signals. Finally, we illustrate how these features were incorporated into an ensemble of classification models.

4.1   High-Level Overview

Our general strategy is to first split the overall problem into multiple subproblems and address each individually, before combining our separate solutions into a unified system, which yields probabilistic scores representing the magnitude of noise in a given ECG trace. We can formulate two subproblems based on characteristics of the noise patterns depicted in Fig. 2:

1. *Global noise patterns* comprise disturbances that, once present, tend to contaminate the whole signal. This category includes baseline wandering, powerline interference, rotational noise caused by electromagnetic interference, and a subset of other noise types (Fig. 2a-d).
2. *Localized noise patterns* comprise non-biological artifacts that affect the ECG signal only within certain, relatively short, time intervals. It includes X-ray artifacts, disturbances related to movement of the examination table, and other miscellaneous localized disturbances (Fig. 2e-h).

For each category, we develop a feature extraction method tailored to that particular subproblem and train an ensemble of classification models on top of the extracted features to distinguish anomalous ECGs from normal ECGs and to quantify the extent of noise in a trace (see Fig. 3). Our approach can be summarized as follows.

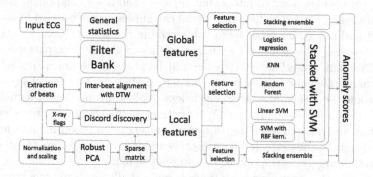

Fig. 3. High-level overview of our system.

1. A filter bank extracts features describing global noise patterns.
2. Each ECG trace is decomposed into a set of non-overlapping intervals constituting a full cardiac cycle – referred to as *beat* – based on the provided locations of QRS complexes.
3. Each beat is analyzed by a modified discord discovery algorithm [13], which identifies the most unusual beat based on the dynamic time warping distance [17] and a test for outliers [16].
4. At the same time, beats are combined into an inter-beat matrix, which is supplied to robust PCA [3] to detect anomalous patterns within this matrix.
5. Next, we compute features describing localized noise patterns based on the output of the previous two steps, i.e., discord discovery and robust PCA.
6. Finally, we use the features describing global and localized noise patterns to train three different ensembles of various classification models, each yielding an anomaly score in the interval $[0; 1]$:
   - The 1st model is trained to exclusively recognize global noise patterns. Its score represents the extent of global noise in a trace.
   - The 2nd model quantifies the extent of localized noise patterns in a trace.
   - The 3rd model, called *unified model*, is trained on the union of features describing global and localized noise patterns. It ought to quantify the overall amount of noise, disregarding the category of noise.

Let us now present individual steps in more detail.

4.2   Global Noise Patterns: Filter Bank Approach

Global noise patterns (Fig. 2a-d), by definition, should be detectable by looking at general properties of the signal. A straightforward approach would consider the signal-to-noise ratio of an ECG signal. Typically, it is estimated as the ratio of the signal's power (P_{signal}) to the power of the noise (P_{noise}):

$$\text{SNR} = 20 \cdot \log_{10}\left(P_{signal}/P_{noise}\right).$$

Obviously, we are unable to estimate the SNR in such a straightforward manner, as we do not know the noise component or the reference signal in advance. Instead, we develop a set of filters that separate the noise component from the observed signal. The extracted noise signal can subsequently be used to compute the SNR and to extract other features describing the signal. Next, we compose a set of features that describe the characteristics of global noise patterns.

To filter out powerline interference and baseline wandering (Fig. 2a,b), we utilize that both noise patterns are characterized by certain frequency bands, which would be either absent or much less explicit in unaffected signals. We employ a two-pass median filter [5] to extract noise stemming from baseline wandering, and a notch filter [18] to capture noise due to powerline interference. For clean signals, the extracted noise signal would be negligible and the denoised signal would largely correspond to the original signal.

Separating the remaining noise types is more challenging due to a large overlap between frequencies of the true (biological) signal and the noise. Standard band-pass filters affect both the noise and the actual signal, thereby distorting the ECG waveform, in particular the QRS complexes. We observed that many artifacts in the rotational noise and "other global noise" category (Fig. 2c,d) resembled white Gaussian noise, which can be filtered efficiently by utilizing the wavelet shrinkage technique [6], which works as follows. First, using the discrete wavelet transform [15], we represent the signal as a weighted sum of basis functions with different time and frequency resolutions. The weights or coefficients of basis functions corresponding to high frequency signals tend to capture the white Gaussian noise, which can be eliminated by applying the soft thresholding operator to the wavelet coefficients and reconstructing the signal via the inverse transform [6]. The result is a denoised version of the input signal with well preserved morphological features of the ECG waveform, in particular R peaks. In addition, we compute the median absolute deviation of wavelet coefficients at the highest resolution level, which quantifies noise based on the wavelet coefficients itself [6]. The features derived from wavelet coefficients and the denoised signal ought to differentiate clean signals from signals affected by rotational noise and various global noise patterns (Fig. 2c,d).

Up to this point, we addressed baseline wandering, powerline interference, rotational noise, and other types of ubiquitous noise independently. We combine these individual approaches into a filter bank: the separated noise and signal components are used to estimate the SNR and the original signal and its frequency domain representation to compute a number of statistics (normalized max. and min. amplitudes, mean, variance, skewness, kurtosis, and entropy). In total, we compute 65 features describing global noise patterns.

4.3 Localized Noise Patterns: Considerations

In contrast to global noise patterns, localized noise patterns (Fig. 2e-h) are characterized by pointwise, temporal changes in the signal, which requires methods operating at a high temporal resolution. Most existing work on ECG analysis is related to clinical diagnosis [11] and human identification [1]. For clinical diagnosis, feature extraction should focus on aspects that characterize a disease and at the same time account for the natural variability of ECG waveforms and heart rhythms across patients. For human identification, features need to differentiate individuals, while mitigating factors that vary across multiple measurements for the same individual, such as heart rate and signal quality. In both applications, the key morphological features of the ECG waveform, such as P wave, T wave, QRS complex, and so forth, often convey sufficient information about diseases and individuals. In our case, we require features that are robust to variations across CT scanners, imaging protocols, and individuals and their diseases. Most importantly, the source of noise patterns considered here is almost always independent from the individual and her heartbeat characteristics. Consequently, standard features of an ECG waveform may not be reliable in our context.

The notion of a localized noise pattern implies that there is a part of the signal, which notably deviates from the rest of the ECG. This suggests to first identify the most anomalous subsequences within the signal and then to assess

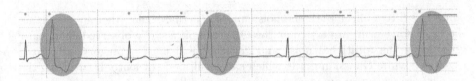

Fig. 4. ECG of a patient with a preliminary ventricular contraction (PVC).

type and degree of these anomalies. One of the key challenges is that such anomalies can be caused by technological as well as biological factors. A biological anomaly in the ECG signal is due to a physiological condition of the heart, such as preliminary ventricular contraction (PVC) depicted in Fig. 4. Therefore, artifacts of biological origin usually occur during specific cardiac phases and repeat themselves over time. In contrast, a technological anomaly, such as a sudden discharge caused by an X-ray, is not associated with a specific cardiac phase, instead, it can occur during any phase. Furthermore, the unique waveform of a technological artifact rarely occurs more than once in the same ECG trace, i.e., it is an outlier. Only in severe cases, when the noise results in a falsely detected R peak, we observe not one, but two beats with an unusual morphology.

Next, we present two feature extraction methods that detect anomalous structures within a signal, while mitigating the natural variability across diseases, patients, and medical centers. The first approach is based on a discord discovery algorithm, which finds the most unusual subsequence within a time series. The second approach utilizes robust PCA for identifying anomalous structures within a signal. These methods operate at the beat level, i.e., the time between two R peaks. We argue that this is the most reasonable level of detail for three reasons: (1) it provides necessary and sufficient information to identify repetitive and anomalous subsequences; (2) as mentioned in Sect. 3, only R peaks are well preserved across all signals; and (3) those traces, where an R peak was misdetected, are usually contaminated with non-biological artifacts and we have ways to recognize them, which we will describe next.

4.4 Localized Noise Patterns: Discord Beat Discovery

Discord beat discovery (DBD) performs a series of comparisons of ECG beats to identify the most anomalous beats. First, the ECG signal is decomposed into multiple beats based on the detected QRS complexes. Next, the beats are normalized to uniform length and compared with each other using a suitable distance measure. The result is an inter-beat dissimilarity matrix (see Fig. 5).

One of the key aspects of this approach is the choice of an appropriate distance measure. The two primary criteria for choosing a distance metric are the ability to handle ECGs with variable waveform morphology and its runtime performance. The latter criteria is crucial, because we require $O(B^2)$ comparisons for each ECG signal, where B is the number of beats in the signal, and we want to analyze thousands of ECG traces in a short amount of time. The Euclidean

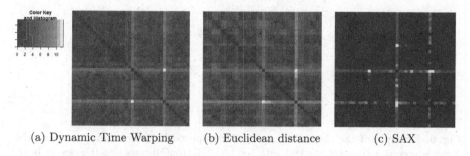

(a) Dynamic Time Warping   (b) Euclidean distance   (c) SAX

Fig. 5. Inter-beat dissimilarity matrices based on different distance measures. The yellow bands correspond to ECG beats contaminated with an X-ray artifact. SAX: symbolic aggregate approximation [14].

distance is fast to compute, but by definition is not tolerant to temporal inconsistencies within time series [17]. The dynamic time warping (DTW) algorithm [17] accounts for such differences and thereby mitigates natural morphological variabilities within the ECG waveform. The symbolic aggregate approximation (SAX) metric performs dimensionality reduction prior to distance comparison and is robust to changes in waveform morphology too [14]. Empirically, we have found that DTW provides the most optimal trade-off between accuracy and computational complexity for our purposes.

Our DBD approach can be considered as a modification of the brute force discord discovery (BFDD) algorithm [13]. The BFDD algorithm often performs well in finding the most unusual part of a time series, but has a runtime complexity of $O(L^2)$, where L denotes the length of the time series. Moreover, it requires us to specify the length of the anomaly, which is rarely known in advance. Instead, we adapt the BFDD algorithm with the following change: instead of comparing all possible subtraces of a fixed length with all others, we split the signal into non-overlapping beats first, and only compare beats with each other. This modification has the following consequences:

1. Focusing on the comparison between beats eliminates the need to specify a fixed window length and better suits the ECG analysis context.
2. Significantly faster runtime of $O(B^2)$ – the number of beats B is about two orders of magnitude smaller than the number of samples L in a signal.
3. It allows for an integration of domain knowledge in the form of predefined patterns (discussed below).

Identifying localized noise patterns can be challenging when the ECG signal contains both technological and biological anomalies, such as PVC beats (Fig. 4). Our DBD approach accounts for the presence of biological anomalies by utilizing that they tend to reappear over time. Therefore, for each beat – regular or biologically abnormal – we can find another beat within the trace whose similar (its distance is small). In contrast, a beat corrupted by a technological artifact possesses a unique waveform – it will have a large distance to all other beats in the trace. The DBD approach is suitable, given the following two conditions:

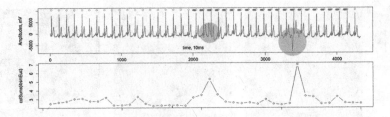

Fig. 6. Example of the discord beat discovery approach. Time intervals, where the X-ray source is active, are marked with red bars. Bottom: distance to the closest beat.

(1) there are at least two biologically abnormal beats, and (2) at least one of them occurs outside of time intervals where the X-ray source was active. These preconditions are met in most scenarios. In a few rare cases, it may happen that there is only one biological anomaly and it appears exactly under an X-ray region, in which case we might assume the presence of a non-biological X-ray artifact. Our solution for such cases is to maintain a set of patterns of typical biological anomalies. Each discord with a statistically significant deviation [16] from its closest beat should be additionally aligned with these patterns. In case a close match is found, the anomaly is likely of biological origin; otherwise, the observed discord is either a technological anomaly or a novelty, i.e., an unexpected form of a biological artifact. In either case, the trace would be of interest for the analyst. Considering the computational overhead of this approach and the rarity of these cases, we decided not to include predefined patterns in our deployed system, but this remains as a potential future improvement.

Figure 6 illustrates the DBD algorithm on an ECG contaminated with several X-ray artifacts. The main disadvantage of the algorithm is that it can only determine the presence of an anomalous beat, but not the exact location and structure of the anomaly. In the next section, we describe a technique that overcomes these limitations.

4.5   Localized Noise: Robust PCA

In this section, we present a technique based on robust PCA [3] that allows for a very precise localization of an anomaly at the sub-beat level, which is particularly useful for capturing X-ray artifacts (Fig. 2e,f). Moreover, this approach allows extracting anomalous structures and reconstructing the true, noise-free signal.

Robust PCA [3] is a modification of classical PCA designed to handle strong outliers. It seeks a decomposition of a matrix X into two components, $X = L + S$, such that L is a low-rank matrix that comprises regular patterns within the data, and S is a sparse matrix, which captures irregular structures. There are no strong assumptions about the irregularities – the only requirement is that they appear unusual with respect to the rest of the data, such that the sparsity condition of the S matrix holds. This enables us capturing a wide range of anomalies. The objective function of robust PCA consists of two terms: the nuclear norm

$\|X\|_* = \sum_i \sigma_i(X)$, with $\sigma_i(X)$ denoting the i-th singular value of X; and the L_1 norm $\|X\|_1 = \sum_{ij} |X_{ij}|$. The resulting optimization problem has the form:

$$\min_{L,S} \quad \|L\|_* + \lambda \|S\|_1, \qquad \text{subject to} \quad L + S = X, \lambda > 0.$$

In many applications, the data can be modeled as a sum of such low-rank and sparse components [3]. In the context of cardiac CT, we want L to capture the biological signal and S arbitrary localized non-biological anomalies.

Here, the rows of matrix X correspond to the beats in a single ECG trace: first, we segment the ECG into beats and scale individual beats to uniform length; next, the beats are stacked to form the inter-beat matrix X. Applying the robust PCA procedure to X yields the decomposition into the matrices L and S (see Fig. 7 a and b). An ECG waveform that is severely corrupted by a localized noise pattern, such as an X-ray artifact, can have its true waveform captured by the L matrix and the non-biological anomaly by the S matrix. As a result, the information we are interested in tends to accumulate in the S matrix.

We use S to answer two questions: (1) whether the ECG contains a significant anomaly and (2) whether this anomaly only occurs when the X-ray source is active. First, we apply a test for outliers [16] to the values of S and compute noise quantification measures such as the median absolute deviation. Next, we build a binary X-ray matrix M by splitting the vector of X-ray flags according to R peaks (Fig. 7), and compute the correlation between values in S and the positive flags in M. Finally, we divide S into groups corresponding to different values of the binary X-ray matrix M and perform t-tests to determine whether their mean significantly differs from each other.

We identified three requirements for this approach to yield good results:

1. There are enough beats within the trace to infer repetitive structures (15 beats are usually sufficient).
2. There is either a single beat corrupted by a non-biological artifact or multiple corrupted beats, each with its own unique waveform.
3. Artifacts of biological origin, if present, do repeat over time.

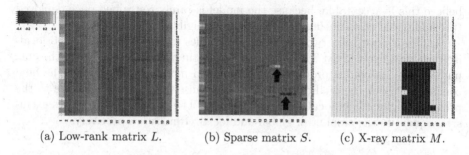

(a) Low-rank matrix L.   (b) Sparse matrix S.   (c) X-ray matrix M.

Fig. 7. The low rank (a) and the sparse (b) components produced by the robust PCA procedure applied to the ECG signal in Fig. 6, as well as the binary matrix of X-ray flags (c). Two anomalies are captured by the S matrix (marked by arrows). Overlaying (b) and (c) reveals that both anomalies are X-ray artifacts.

Our empirical findings and cross-validation results, which will be presented below, suggest that these three conditions hold for most of the noise patterns. Considering the third condition, in some cases an ECG signal contains few biologically abnormal beats (<15 %) such that the robust PCA mistakes them for outliers, i.e., they are captured by the S matrix. To address this ambiguity, we use our discord beat discovery approach described above, which can handle cases with two to three ectopic beats. By combining features extracted during discord beat discovery and robust PCA, we end up with over 100 features describing localized noise patterns.

4.6   ECG Trace Classification

After developing features describing global and localized noise patterns, respectively, we obtained two sets of features. These sets adequately describe the noise patterns in Fig. 2, but are applicable to their respective domain only and only some of them, such as SNR, directly provide information about the magnitude of noise in a given ECG trace. Therefore, we employ a classification model that utilizes all 181 extracted features to yield a probabilistic score representing the magnitude of noise, global or local. We refer to this model as the *unified model*.

Moreover, we noticed a considerable redundancy in the feature set and that some features contribute little to the overall model. Thus, prior to training, we rank features by importance – using the improvement in out-of-bag error estimated from a random forest (RF) model [2] – and retain all features in the top half. Subsequently, multiple classification models are trained on the selected features to distinguish good from anomalous ECGs. We use RF [2], linear SVM, SVM with RBF kernel [4], k nearest neighbors classification [8], and logistic regression. Although a single model trained on the selected feature set can provide satisfactory results (see Table 1), each model has its own biases determined by its learning principle and its hyper-parameter configuration. Thus, to further raise the reliability of our system, we construct an ensemble of the above mentioned models using the *model stacking* technique [19], where an SVM with RBF kernel is used as meta-model. This increases the complexity of training, but we believe this is acceptable, because the model is rarely re-trained once deployed (the additional costs during prediction are negligible).

Analogous, we train two additional ensembles to recognize global and localized patterns exclusively. Three anomaly scores in the range [0; 1] form the output. The main score is produced by the unified model and represents the final conclusion about the quality of the ECG, because it is equally sensitive to the presence of global and localized noise patterns. The remaining two scores exclusively quantify the amount of global and localized noise, respectively.

5   Evaluation

We evaluated our solution using cross-validation and a hold-out set consisting of annotations from two domain experts. The system's performance was measured by the area under the precision-recall curve (AUPRC) and the area under the receiver operating characteristics curve (AUROC). It is important to mention that our deployed system allows choosing a user-defined threshold on the predicted probabilities, because users have different requirements regarding precision and recall (see Sect. 5.2). Hence, we did not optimize the choice of a threshold, but provide accuracy, precision, and recall for a threshold of 0.6 for illustration.

5.1 Cross-Validation

Cross-validation was based on the dataset consisting of 2,581 cardiac CT scans presented in Sect. 3. It contained 1,733 "good" ECGs, 501 corrupted with global noise, and 391 with localized noise. Note that many traces were contaminated with multiple noise patterns of both categories. Experimental results with respect to the global, localized, and unified model are summarized in Table 1.

The results demonstrate that even individual models achieve high performance scores, with a negligible difference in AUPRC and AUROC between RF and SVM, but a slight advantage for RF with respect to precision. We achieved an additional improvement in precision and recall when combining several models into an ensemble. Although the improvement may seem minor, it becomes relevant when considering >10,000 traces. In production, the cost of not identifying a problem, i.e., a false negative, is generally higher than the cost of a false positive. Moreover, our primary objective is to assist technicians in identifying anomalous cardiac CT scanners and not individual ECGs, which results in Table 1 show. Therefore, multiple corrupted ECGs obtained from the same device need to be identified before action is taken, which justifies trading a higher recall for a lower precision – a corrupted ECG should not be missed. We allow the user to individually adjust the threshold, because the trade-off between precision and recall is often situational.

5.2 External Validation

We deployed our system at Siemens Healthcare, where it is used to automatically analyze previously unobserved ECG traces in a real world setting. Our ensemble

Table 1. Cross-validation results for global noise patterns, localized noise patterns, and both types of noise patterns (All) as defined in Sect. 4.1. Accuracy, precision, and recall were computed at a threshold of 0.6.

Metric	Global(RF)	Localized(RF)	All(SVM)	All(RF)	All(Ensemble)
mean AUROC	0.998	0.996	0.996	0.997	0.997
mean AUPRC	0.997	0.989	0.993	0.994	0.996
mean accuracy	0.990	0.981	0.973	0.978	0.983
mean precision	0.990	0.963	0.964	0.979	0.985
mean recall	0.970	0.934	0.952	0.954	0.964

of binary classification models was trained on all 2,581 ECG traces before deployment and processed 150 ECG traces during our evaluation period. Two domain experts independently analyzed these traces manually and assigned each trace to one of 5 categories: (1) perfect (no artifact), (2) good (only very minor artifacts), (3) corrupted (considerable amount of artifacts), (4) strongly corrupted, and (5) extremely corrupted. Overall, the inter-expert agreement was high, as indicated by Kendall's coefficient of concordance ($W = 0.938$, $P < 0.001$, corrected for ties) [12]. Most disagreements (24 of 53; 45.3 %) were due to traces of the 3rd category being assigned to the 2nd (15) or 4th category (9) instead, which indicates that it is difficult, even for experts, to draw a sharp line between clean and corrupted ECGs (see Table 2). We evaluated our system based on AUPRC, AUROC, and Kendall's coefficient of concordance [12], which measures the degree of agreement between our system and the experts on the five-level Likert-scale.

First, we treated categories 3, 4 and 5 as positive class and the remainder as negative class to allow comparison to our cross-validation results. We obtained an AUPRC and AUROC score of 0.875 and 0.898 with respect to expert 1 and 0.905 and 0.942 with respect to expert 2. Although performance scores dropped compared to our cross-validation experiment, it is noteworthy that the expert, who annotated the training set, did not participate in annotating the hold-out set. Thus, corner cases between "good" and corrupted signals are likely biased. The AUROC score still indicates a highly discriminatory model ($\gtrsim 0.9$) and the drop in AURPC can be attributed to a decrease in precision. Note that expert 1 assigned more ECGs to the positive class (cat. 3-5) than expert 2, thus only 80 % (56/70) of positive annotations of expert 1 match that of expert 2. In contrast, 93 % (56/60) of positive annotations of expert 2 match that of expert 1 (cf. Table 2). Consequently, we would expect that the AUPRC, or average precision, of our system would be around 0.9 at best. We obtained AUPRC scores of 0.875 and 0.905, indicating a highly discriminatory model.

When considering in which range our predicted probabilities fell, we noticed that predicted probabilities of ECGs belonging to categories 2 and 3 were inconsistent. Table 3 shows confusion matrices obtained after dividing predicted probabilities into 5 equally spaced bins ($[0; 0.2[, [0.2; 0.4[, \dots$). The table reveals that ECGs of categories 2 and 3 have been assigned probabilities in the whole interval $[0; 1]$, which results in a low precision. The recall remains high when disregarding

Table 2. Confusion matrix illustrating inter-expert agreement.

		Expert 1			
	1	2	3	4	5
Expert 2 1	32	21			
2		23	11	3	
3		4	23	6	
4			3	12	
5				5	7

Table 3. Confusion matrices demonstrating results of external validation.

		Expert 1					Expert 2			
	1	2	3	4	5	1	2	3	4	5
Predicted 1	27	32	6		1	45	19	2		
2		3	5	7			5	5	5	
3		5	4					6	3	
4			2	3	5		5	3	1	1
5	2	4	17	20	7	3	2	20	14	11

category 3: the system recognizes 59 out of 60 ECGs of categories 4 and 5 by predicting a probability above 0.6, whereas the recall drops to 0.838 (109/130) when including category 3. At the same time the precision increases from merely 0.492 for categories 4 and 5, to 0.790 for categories 3-5 due to less false positives. We concluded that predicted probabilities are not well calibrated, because very high and very low probabilities are over-represented. In fact, this is a problem for many machine learning methods, which can perform well by means of standard metrics for classification, but yield poorly calibrated probabilistic scores, or vice versa [7]. Alternate learning regimes, such as ordinal regression and learning-to-rank, could remedy this problem. However, in contrast to classification, richer annotations are required, which places more burden on human annotators and makes obtaining labels prohibitively costly in our case.

Next, we compared the model's predictions to the five-level Likert scale, which resulted in Kendall's coefficient of concordance of 0.863 ($P < 0.001$) based on the two expert annotations and the predicted probabilities (corrected for ties). The results demonstrate that most predictions were concordant with the experts' annotations (87.5 % and 82.8 %, excluding ties), thus the agreement between predicted probabilities and expert annotations is substantial.

Although results of the external validation suggest a less discriminatory system, compared to our cross-validation results, the overall performance of the system is still high. Moreover, we allow the user to individually adjust the threshold to identify only severe cases (categories 4 and 5) with very high recall but moderate precision, or all cases (categories 3–5), which increases precision. Overall, the external validation confirmed the practical applicability of our system. Most importantly, the automated analysis operates at a speed that allows processing over thousand ECGs per hour (single-threaded), compared to a few hundred per day of a human analyst.

6   Conclusion

The main goal of this work was to develop a system to automatically detect various non-biological artifacts and noise patterns in ECG signals acquired during cardiac CT. We adapted a discord discovery technique for detecting the most abnormal heartbeats and applied robust PCA for a more precise localization of non-biological anomalies. As a result, we produced a feature set that captures differentiating properties of various global and local noise patterns and used it to train an ensemble of classification models. We validated our system internally via cross-validation and externally in a real world setting. The results demonstrate that our system is highly discriminatory and allows processing thousands of ECGs with minimal human interaction. In the future, we would like to improve our model with regard to calibration, such that predicted scores accurately reflect the true severity of an artifact. Our system is currently deployed at Siemens Healthcare, where it continuously analyzes cardiac CT scans collected from various medical centers. The ultimate benefit of our work can be determined retrospectively as time passes, based on the overall reduction of reported problems and the time needed to resolve them.

Acknowledgments. This work was supported by Siemens Healthcare GmbH.

References

1. Biel, L., Pettersson, O., Philipson, L., Wide, P.: ECG analysis: a new approach in human identification. IEEE Trans. Instrum. Meas. **50**(3), 808–812 (2001)
2. Breiman, L.: Random forests. Mach. Learn. **45**(1), 5–32 (2001)
3. Candes, E.J., Li, X., Ma, Y., Wright, J.: Robust principal component analysis? J. ACM **58**(3) (2011). Article Number 11
4. Cortes, C., Vapnik, V.: Support-vector networks. Mach. Learn. **20**(3), 273–297 (1995)
5. De Chazal, P., Dwyer, M.O., Reilly, R.B.: Automatic classification of heartbeats using ECG morphology and heartbeat interval features. IEEE Trans. Biomed. Eng. **51**(7), 1196–1206 (2004)
6. Donoho, D.L.: De-noising by soft-thresholding. IEEE Trans. Inf. Theor. **41**(3), 613–627 (1995)
7. Esarey, J., Pierce, A.: Assessing fit quality and testing for misspecification in binary-dependent variable models. Polit. Anal. **20**(4), 480–500 (2012)
8. Fix, E., Hodges, J.L.: Discriminatory analysis - nonparametric discrimination: consistency properties. Technical report, USAF School of Aviation Medicine, Randolph Field, TX (1951)
9. Hausleiter, J., Meyer, T.S., Martuscelli, E., Spagnolo, P., Yamamoto, H., Carrascosa, P., Anger, T., Lehmkuhl, L., Alkadhi, H., Martinoff, S., et al.: Image quality and radiation exposure with prospectively ECG-triggered axial scanning for coronary CT angiography: the multicenter, multivendor, randomized PROTECTION-III study. JACC Cardiovasc. Imaging **5**(5), 484–493 (2012)
10. Hsieh, J., Londt, J., Vass, M., Li, J., Tang, X., Okerlund, D.: Step-and-shoot data acquisition and reconstruction for cardiac x-ray computed tomography. Med. Phys. **33**(11), 4236–4248 (2006)
11. Karpagachelvi, S., Arthanari, M., Sivakumar, M.: ECG feature extraction techniques - a survey approach. Int. J. Comput. Sci. Inf. Secur. **8**(1), 76–80 (2010)
12. Kendall, M.G., Smith, B.B.: The problem of m rankings. Ann. Math. Stat. **10**(3), 275–287 (1939)
13. Keogh, E., Lin, J., Fum, A.: Hot SAX: Efficiently finding the most unusual time series subsequence. In: 5th IEEE International Conference Data Mining, pp. 226–233. IEEE Computer Society, Washington, DC (2005)
14. Lin, J., Keogh, E., Wei, L., Lonardi, S.: Experiencing SAX: a novel symbolic representation of time series. Data Min. Knowl. Discov. **15**(2), 107–144 (2007)
15. Mallat, S.G.: A theory for multiresolution signal decomposition: the wavelet representation. IEEE Trans. Pattern Anal. Mach. Intell. **11**(7), 674–693 (1989)
16. Rosner, B.: Percentage points for a generalized ESD many-outlier procedure. Technometrics **25**(2), 165–172 (1983)
17. Vintsyuk, T.K.: Speech discrimination by dynamic programming. Cybern Syst. Anal. **4**(1), 52–57 (1968)
18. Widrow, B., Glover, J.R., McCool, J.M., Kaunitz, J., Williams, C.S., Hearn, R.H., Zeidler, J.R., Dong, E., Goodlin, R.C.: Adaptive noise cancelling: Principles and applications. Proc. IEEE **63**(12), 1692–1716 (1975)
19. Wolpert, D.: Stacked generalization. Neural Netw. **5**(2), 241–260 (1992)

Active Learning with Rationales for Identifying Operationally Significant Anomalies in Aviation

Manali Sharma[1]([⊠]), Kamalika Das[2], Mustafa Bilgic[1], Bryan Matthews[3],
David Nielsen[4], and Nikunj Oza[5]

[1] Illinois Institute of Technology, Chicago, USA
msharm11@hawk.iit.edu, mbilgic@iit.edu
[2] UARC, NASA Ames, Moffett Field, CA, USA
kamalika.das@nasa.gov
[3] SGT, Inc., NASA Ames, Moffett Field, CA, USA
bryan.l.matthews@nasa.gov
[4] MORi Associates, NASA Ames, Moffett Field, CA, USA
david.l.nielsen@nasa.gov
[5] NASA Ames, Moffett Field, CA, USA
nikunj.c.oza@nasa.gov

Abstract. A major focus of the commercial aviation community is discovery of unknown safety events in flight operations data. Data-driven unsupervised anomaly detection methods are better at capturing unknown safety events compared to rule-based methods which only look for known violations. However, not all statistical anomalies that are discovered by these unsupervised anomaly detection methods are operationally significant (e.g., represent a safety concern). Subject Matter Experts (SMEs) have to spend significant time reviewing these statistical anomalies individually to identify a few operationally significant ones. In this paper we propose an active learning algorithm that incorporates SME feedback in the form of rationales to build a classifier that can distinguish between uninteresting and operationally significant anomalies. Experimental evaluation on real aviation data shows that our approach improves detection of operationally significant events by as much as 75 % compared to the state-of-the-art. The learnt classifier also generalizes well to additional validation data sets.

1 Introduction

As new technologies are developed to handle complexities of the Next Generation Air Transportation System (NextGen), it is increasingly important to address both current and future safety concerns along with the operational, environmental, and efficiency issues within the National Airspace System (NAS). NASA, in partnership with the Federal Aviation Administration (FAA) and industry is continuing to develop new technologies to identify previously undiscovered safety events through data mining of large heterogeneous aviation data sets that are collected on a regular basis. These techniques have the potential to discover new safety risks in the existing system or risks that did not exist previously but

© Springer International Publishing AG 2016
B. Berendt et al. (Eds.): ECML PKDD 2016, Part III, LNAI 9853, pp. 209–225, 2016.
DOI: 10.1007/978-3-319-46131-1_25

are a result of the implementation of the NextGen concepts. Combined with more traditional monitoring of safety, the Aviation Safety program at NASA has invested significant resources for development and use of data mining methods for identification of unknown safety and other events in Flight Operations Quality Assurance (FOQA) data [6].

Several unsupervised anomaly detection methods have been developed to identify anomalies in commercial flight-recorded data. In the absence of knowledge regarding the types of safety events that are present in the data, and absence of labels, unsupervised techniques are the only ones that have the unique ability to find previously unknown anomalies; however, they do so only in the statistical sense—the anomalies found are not always operationally significant (e.g., represent a safety concern). After an algorithm produces a list of statistical anomalies, a Subject Matter Expert (SME) must go through that list to identify those that are operationally relevant for further investigation. A very small fraction of statistical anomalies (less than 1 %) turns out to be operationally relevant, so substantial time and effort is spent by SMEs in examining anomalies that are not of interest.

The goal of this work is to semi-automate the process of distinguishing between operationally significant anomalies and uninteresting statistical anomalies through use of supervised learning approaches, which require labeled instances. We propose to use active learning for training a classifier, so that SME time and effort is spent on only the most informative and critical anomaly instances. In this process, first an unsupervised anomaly detection algorithm is run on all the flight data to generate a ranked list of statistically significant anomalies. A very small percentage of these are presented to SMEs to bootstrap the active learning process. The SME provides labels for each of these instances along with an explanation about the label. A positive label indicates an operationally significant safety event whereas a negative label indicates otherwise. Based on these few labels we build an active learning system that (i) utilizes the SME's time in the most effective manner by iteratively asking for labels for few informative instances, (ii) elicits rationales/explanations from the SME for why s/he assigns a certain label to an instance, and (iii) constructs new features, based on rationales, that are incorporated in future iterations of active learning and classifier training.

Active learning for anomaly detection has been studied in the past with the goal of finding *useful* anomalies as opposed to statistical anomalies [7] where a priori knowledge of the number of rare event classes is assumed. In our application the number of types of anomalies encountered is unknown and therefore, the assumption does not hold true. Recent work in active learning has focused on eliciting richer feedback from the experts in addition to labels, to speed up the annotation process. For example, experts are asked to annotate features as relevant/irrelevant for a specific task [1, 15]. Similarly, several researchers have investigated eliciting rationales, which often correspond to highlighting a piece of text in text classification or highlighting feature values in feature-valued representations, and incorporated them into the training of classifier [14, 18]. In this work, we build on the rationale framework by allowing the domain experts to provide rationales for their classification. The main difference between our work and existing work is that in this paper we enrich the representation by creating additional features that are combinations of existing features rather than focusing on feature value distribution.

The advantages of this method are twofold: (i) it dramatically minimizes the time an SME needs to spend to find operationally significant anomalies from the long list of statistical anomalies output by any unsupervised anomaly detection method, and (ii) at the end of training, we have a classifier that can be run on the original flight operations data set to uncover many more operationally significant safety events that might have been missed in the original anomaly detection process due to the presence of overwhelming number of statistically significant, but uninteresting, anomalies. Our experiments with real aviation data show that using active learning with rationales improves *precision*@5 (defined as number of positive instances in top 5 instances ranked according to their distance from the decision boundary) results by as much as 75 % compared to the state-of-the-art.

The rest of the paper is organized as follows. Section 2 discusses the data setup and the existing unsupervised anomaly detection framework. Section 3 discusses our proposed active learning algorithm and its performance is analyzed in Sect. 4. Section 5 discusses deployment plans. Section 6 concludes the paper.

2 Background

In this section we describe the state-of-the-art unsupervised anomaly detection method used for identifying statistical anomalies in flight operations data, followed by description of the data used in this study.

2.1 Multiple Kernel Anomaly Detection

The unsupervised anomaly detection algorithm that is currently used in the aviation safety community most frequently is Multiple Kernel Anomaly Detection (MKAD)[1] [5]. The MKAD algorithm is designed to run on heterogeneous data sets consisting of multiple attribute types including discrete and continuous. MKAD is a "multiple kernel" [2] based approach where the major advantage is the method's ability to combine information from multiple heterogeneous data sources. The heart of MKAD is a one-class SVM model that constructs an optimal hyperplane in the high dimensional feature space to separate the abnormal (or unseen) patterns from the normal (or frequently seen) ones. This is done by solving the following optimization problem [10]:

$$\min \quad Q = \frac{1}{2} \sum_{i,j} \alpha_i \alpha_j K\left(\mathbf{x}_i, \mathbf{x}_j\right) \tag{1}$$

$$\text{subject to} \quad 0 \leq \alpha_i \leq \frac{1}{\ell \nu}, \sum_i \alpha_i = 1, \rho \geq 0, \quad \nu \in [0, 1]$$

---

[1] http://ti.arc.nasa.gov/opensource/projects/mkad/.

where α_i's are Lagrange multipliers, ℓ is the number of data tuples in the training set, ν is a user-specified parameter that defines the upper bound on the training error, and also the lower bound on the fraction of training points that are support vectors, ρ is a bias term, and K is the kernel matrix. Once this optimization problem is solved, at least $\nu\ell$ training points with non-zero Lagrangian multipliers (α) are obtained and the points for which $\{\mathbf{x}_i : i \in [\ell], \alpha_i > 0\}$ are called the support vectors. The decision function is:

$$f(\mathbf{z}) = sign \left(\sum_i \alpha_i \sum_p \eta_p K_p(\mathbf{x}_i, \mathbf{z}) - \rho \right)$$

which predicts positive or negative label for a given test vector $\mathbf{z}$. Instances with negative labels are categorized as outliers.

The classifier that we learn using active learning for differentiating between operationally significant and uninteresting anomalies is a two-class support vector machine using multiple kernels. Therefore, it differs from MKAD in the fact that it is not based on a one-class SVM like MKAD, but has the same kernel structure as MKAD. The dual objective function for the two-class problem is:

$$\max_\alpha \sum_{i=1}^{\ell} \alpha_i - \frac{1}{2} \sum_{i,j} \alpha_i \alpha_j y_i y_j K(\mathbf{x}_i, \mathbf{x}_j)$$

where $(\mathbf{x}_i, y_i)$'s are the data tuples for $i = 1, \ldots, \ell$. Here $\mathbf{x}_i$ and y_i are the input data points and class labels respectively. In the supervised classification case, the $\mathbf{x}_i$'s correspond to the anomalies found by the MKAD algorithm as discussed above and y_i's correspond to the labels provided by the SMEs. For identifying operationally significant anomalies, this classifier is used to rank the test instances based on their distance from the hyperplane.

2.2 Data Preparation

The surveillance data used in this study comes from combining two Air Traffic Control (ATC) facilities — Denver Terminal Radar Approach Control (D01) and the Denver Air Route Traffic Control Center (ZDV). The objective of this work is to develop a process that automatically discovers previously unmonitored, operationally significant, flight trajectories representing a safety risk to the airspace. The end goal is to produce a tool that can rank these anomalous flights for controllers to review and help make mitigating decisions about the safety of the airspace. The types of anomalies that are being targeted in this study are unusual trajectories from 30 nautical miles (NM) on approach to landing. These can include strange vectoring that do not conform to standard operating procedures, significant overshooting of the final approach fix, or high altitude and speed profiles that can lead to unstable approaches. Figure 1 illustrates the data processing flow from data collection through merging, filtering, unsupervised anomaly detection, and SME feedback incorporation for classification of anomalies into operationally significant and uninteresting categories. Data collection refers to the process of recording the relevant data that is used in this study (done by the PDARS program responsible for collection, processing, and reporting of aviation data from multiple sources). NASA was given access to PDARS data for the 2014 and 2015 calendar years. Approximately 25,000 flights are available to us from 2014, of which approximately 2400 flights for a particular month are being analyzed as part of our safety study for Denver for 2014. The 2015 flights are only used for validation of results. For each trajectory, from 30 NM out from the destination airport, the minimum separation is found and used to create four-dimensional trajectories: latitude, longitude, altitude and distance to nearest flight. These four features are then averaged over half NM intervals from 30 NM to the runway threshold based on distance traveled and are partitioned by runway and destination airport sets on each day. This results in trajectories with fixed vector lengths because of the half-mile binning and the fixed 30 NM distance traveled, which are then used to create similarity kernels. We also use the PDARS turn-to-final (TTF) reports that provide specific characteristics of how the aircraft performed the turn on to the final approach within 20 NM of a runway. All deviations are calculated with respect to the intercept, which is the point at which the flight trajectory crosses the extended runway centerline before making its final approach. These deviations include intercept distance, angle of intercept, altitude deviation, distance deviation, and speed. Maximum overshoot and aircraft size (categorical feature indicating one of four weight categories) are two additional features from this source. In addition, three binary parameters are derived based on the characteristics of the flight identified as the nearest neighbor for each time step. These features are designed to provide domain context since flights on parallel runways or flights in the same flow are allowed to encroach within the standard separation threshold, whereas flights on the same runway should not fall below the separation threshold. These parameters indicate whether two nearest neighboring flights are on the same runway, parallel runway, or are part of the same flow. An additional derived feature called separation is constructed as the 3-d separation between two flights based on the l_2 norm of the horizontal and vertical separation. It should be noted here that all of these (raw and derived) features together constitute the original feature set for our study. The data is heterogeneous in the sense that some of these features are time-series data while others are a single-point feature and some are continuous whereas others are discrete, nominal, or binary.

The data mining block in Fig. 1 consists of the next steps of unsupervised anomaly detection followed by SME review and labeling, and finally, classifier learning for distinguishing between operationally significant anomalies and uninteresting anomalies. Depending on the size of the input data set, MKAD algorithm may discover hundreds to thousands of ranked anomalies, making it difficult for domain experts to validate all of them. Therefore, we use active learning to learn a classifier using very few labeled instances for this purpose. Each time an SME is provided an instance to be classified, the SME provides the label, along with an explanation/rationale for his/her decision. This rationale, whenever possible, is converted into a new additional feature, which is then incorporated into

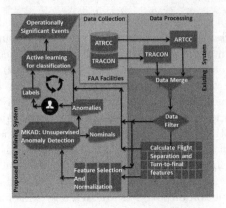

Fig. 1. System setup: Data collection, processing, and mining.

the classifier training through the creation of a new kernel. The details of this process and approach are described in the next section.

3   Active Learning with Rationales

Active learning algorithms iteratively select informative instances for labeling to save annotation time, cost, and effort [11]. For skewed data sets with minority class distribution much less than the majority class, a common and simple approach for selecting informative instances is to maximize the chances of retrieving positive instances [4]. Most-likely positive (MLP) strategy aims to add more positive instances into the labeled training set. The objective is:

$$\mathbf{x}^* = \arg\max_{\mathbf{x} \in \mathcal{U}} P_\theta(\hat{\mathbf{y}}^+ | \mathbf{x})$$

where $\hat{\mathbf{y}}^+$ represents the predicted positive label. Intuitively, MLP is a way of over-sampling the minority class to address the imbalanced class distributions. Examples of other strategies include query-by-committee [12], uncertainty sampling [16], expected error reduction [9], evidence-based uncertainty sampling [13], and more. Ramirez-Loaiza et al. [8] provide an empirical evaluation of common active learning strategies. Recent active learning work has looked at eliciting domain knowledge in form of rationales [14] and feature annotations [1] from the SMEs instead of just the labels of instances. In learning with rationales approach, SMEs provide rationales in the form of features that they think are responsible for classifying an instance into a particular class. In this paper, we elicit the rationales from SMEs and incorporate them into the learning process. The main difference between previous work on incorporating rationales and our work is that we create new features based on the rationales provided by the SMEs.

For training our classifier using active learning, we work with the list of anomalies produced by running the unsupervised anomaly detection algorithm,

MKAD, on the data described in Sect. 2.2. For each flight, MKAD returns an anomaly score, which is the flight's distance from the hyperplane of a one-class SVM model. Flights with a negative score are considered as anomalous and flights with a positive score are considered as not anomalous. The SMEs are asked to provide labels for top 5 % anomalous flights based on whether they think the anomaly is operationally significant (OS/positive labels) or not (NOS/negative labels). They are also asked to provide a rationale for the chosen label. Since labels and rationales are subjective opinions of each SME, we consolidate the labels and rationales from two SMEs by resolving conflicts (by reviewing each others' labels and rationales) whenever there is one, to get gold standard labels and rationales for our study.

3.1   Creating Rationales

When the SMEs identify a flight as an OS flight, they provide rationales in the form of either domain knowledge or using existing features and thresholds. However, when the SMEs identify a flight as NOS, they only provide acknowledgment of certain characteristics of the flight (e.g., a little overshoot, speed not a factor, small deviations on final). In anomaly detection tasks, it is easy to provide a rationale for why a particular instance is anomalous, but it is often difficult, if not impossible, to provide a rationale for why an instance is not anomalous. Therefore, we use the rationales for only the OS flights to create new features and use them to extend the feature representation. Note that the rationales provided by SMEs are often in terms of the original features that are already captured by PDARS. Some rationales talk about two or more features whereas some highlight only one feature.

In our training set, most OS anomalies could be explained by one or more of three different rationales. The first rationale provided for operational significance is loss-of-separation, which the domain experts define as 'horizontal separation is less than 3 miles and vertical separation is less than a 1000 feet, and the nearest neighboring flight is not on parallel runways and not part of the same flow'. When a loss-of-separation rationale is provided, we create a new feature that checks whether the criteria 'horizontal separation less than 3 miles and vertical separation less than 1000 feet' and 'the nearest neighboring flight is not on parallel runway and not in the same flow' hold and incorporate it as a new binary feature in our training set.

The second rationale provided by the SMEs is for large overshoots where an overshoot is defined as going past a certain point in the landing trajectory against standard operational procedures. For rationales such as 'maximum overshoot is too large', we create a new feature that checks whether the overshoot is greater than a threshold. The threshold can be either chosen manually based on domain knowledge or based on the values of the overshoot feature for the labeled OS flights with overshoot rationale observed until that point, and updated iteratively.

The third rationale provided by the SMEs is for unusual flight path. Since this rationale is more qualitative than quantitative, and none of the original features

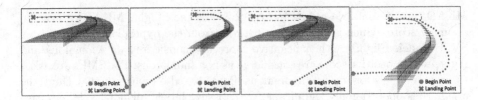

Fig. 2. Expected flight path and deviation from it for 4 flights. The first three flights are NOS. The last flight is an OS flight. (Color figure online)

represent an 'unusual flight path', we compute a new feature as follows. For each runway, using latitude and longitude features, we compute expected flight trajectory as the average trajectory of all flights that land on a runway. Then we create a new feature that captures the overall deviation of each flight from its expected flight trajectory over the last 10 points in the trajectory. Figure 2 shows the plots for a few trajectories. It can be seen that for the first three flights in Fig. 2, the red dots align well with the expected trajectory (highlighted using the red box), whereas for the last flight there is significant deviation from the expected trajectory. This can have severe safety implications and is therefore considered an operationally significant safety event.

3.2   Active Learning with Rationales Algorithm

Algorithm 1 describes our approach for incorporating rationales into active learning. Active learning algorithm starts with a small set of labeled flights, $\mathcal{L}$, and finds the most informative flight, $\mathbf{x}^\star$, from the unlabeled set, $\mathcal{U}$. The most informative flight is the one that provides the classifier maximum information in terms of the decision boundary, or, in other words, one that has the maximum *utility*. The flight $\mathbf{x}^\star$ is then presented to the SME, who provides its label $\mathbf{y}^\star$. For every flight we present to the SME, in addition to a label, we also request for a rationale $R(\mathbf{x}^\star)$ describing why s/he labeled the flight as OS or NOS. If the label is OS, we create a new feature, $f_r^\star$, if possible, for the rationale $R(\mathbf{x}^\star)$ and add it into our existing feature representation: $\mathbf{f} = \langle f_1, f_2, \cdots f_n \rangle \bigcup \langle f_r \rangle$. We assign weight w_o for the original feature kernels and weight w_r for the rationale feature kernels, where $w_r \geq w_o$, since intuitively the rationale features are the ones that have the highest power to separate the OS flights from the NOS ones. However, to satisfy Mercer's condition, we need to ensure that it is a convex combination of the kernels. Therefore, we normalize each weight by the sum of the weights $w = w_o \times n + w_r \times p$, where n and p denote the number of original and rationale features respectively. Let η denote the normalized kernel weights for the enhanced feature set. Note that the kernel weights for original features $\langle \eta_1, \eta_2, \cdots \eta_n \rangle$ are uniform and hence the kernel weight for each original feature will be η_o, which is computed in Step 10 of Algorithm 1. Similarly, the kernel weight for the rationale feature set $\langle \eta_{n+1}, \eta_{n+2}, \cdots \eta_{n+p} \rangle$ is η_r and is computed in Step 11 of Algorithm 1. The final kernel is computed using the updated set of

Algorithm 1. Active Learning with Rationales for Identifying Operationally Significant Anomalies in Aviation

1: **Input:** $\mathcal{U}$ - unlabeled flights, $\mathcal{L}$ - labeled flights, $\mathcal{T}$ - test flights, $\mathbf{f} = \langle f_1, f_2, \cdots f_n \rangle$ - current set of features, $\eta = \langle \eta_1, \eta_2, \cdots \eta_n, \eta_{n+1}, \eta_{n+2}, \cdots \eta_{n+p} \rangle$ - normalized kernel weights for enhanced feature set, θ - underlying classification model, B - budget

2: **repeat**

3:    $\mathbf{x}^* = \arg \max\limits_{\mathbf{x}^i \in \mathcal{U}} utility(\mathbf{x}^i | \theta)$

4:    request label $\mathbf{y}^*$ for the flight $\mathbf{x}^*$

5:    **if** $\mathbf{y}^* ==$ OS **then**

6:       request SME to provide a rationale $R(\mathbf{x}^*)$ for why the flight is operationally significant

7:       **if** rationale $\neq \phi$ **then**

8:          create feature f_r^* for $R(\mathbf{x}^*)$

9:          add f_r^* to $\mathcal{U}, \mathcal{L}$, and $\mathcal{T}$

10:          $\eta_o = \frac{w_o}{\sum_{i=1}^{n} \eta_o + \sum_{j=1}^{p} \eta_r}$

11:          $\eta_r = \frac{w_r}{\sum_{i=1}^{n} \eta_o + \sum_{j=1}^{p} \eta_r}$

12:          $\eta = \langle \eta_1, \eta_2, \cdots \eta_n \rangle \bigcup \langle \eta_r \rangle$

13:          $\mathbf{f} = \langle f_1, f_2, \cdots f_n \rangle \bigcup \langle f_r \rangle$

14:       **end if**

15:    **end if**

16:    $\mathcal{L} \leftarrow \mathcal{L} \cup \{\langle \mathbf{x}^*, \mathbf{y}^*, R(\mathbf{x}^*) \rangle\}$

17:    $\mathcal{U} \leftarrow \mathcal{U} \setminus \{\langle \mathbf{x}^* \rangle\}$

18:    Train θ on $\mathcal{L}$

19: **until** Budget B is exhausted; e.g., $|\mathcal{L}| = B$

kernel weights η containing normalized weights η_o for the original feature kernels and the normalized weights η_r for the rationale feature kernels for the enhanced feature set $\mathbf{f}$.

Possible enhancements: Based on the training data and the rationales provided by the SMEs, in this paper, we created three features that encompass a significant number of OS safety scenarios. However, this set is far from complete as there can be a huge variety of other explanations that can come from SMEs. So the set of rationale features is always expanding. As the set of features grows based on rationales, there might be a need to consolidate features into conjunctions and disjunctions depending on redundancy. For example, two common rationales in our study are loss-of-separation and large overshoot. However, not all OS flights have both reasons for being labeled OS. Some flights are OS because of loss-of-separation, but they might have perfectly acceptable overshoot values, whereas other OS flights might not have a loss-of-separation but might have large overshoot values. Current framework creates one feature per rationale. An alternative approach is to create one indicator feature and keep revising it by adding the new rationales as disjunctions. Also, once a classifier is trained using this framework, our goal is to find operationally significant events in the original flight data. However, since the classifier is trained on only the anomalies, the feature distribution does not necessarily match that of the overall data set. This unaccounted bias can be handled by sub-sampling some of the flights that are not signaled by MKAD and adding them to the training with NOS (negative) labels. Selecting flights that are ranked lowest by MKAD, for this purpose, can ensure with a high probability that the flights which are most certainly nominal are being used as NOS samples.

4   Empirical Evaluation

Experimental Setup: The data set used for training the classifier using active learning corresponds to PDARS data from the Denver Airport for August 2014, containing approximately 2400 flights out of which 153 flights are marked anomalous by MKAD. These 153 flights are reviewed by two SMEs independently (with conflict resolutions as needed) to provide labels and explanations. In these 153 flights, 26 are marked OS (positive) and the remaining 127 are marked NOS. The original data set contains 16 features as described in Sect. 2.2. Additionally, we construct 3 rationale features supporting the explanations for the OS flights during the active learning iterations, when OS flights with one or more rationales provided in Sect. 3.1 are encountered.

Our proposed active learning strategy, MLP_w/Rationales, selects most-likely positive (MLP) instances for labeling at each iteration of training and creates (or updates) rationale features whenever an appropriate new instance is encountered. We compare our algorithm's performance with three baselines: (i) random strategy (RND) where random instances are picked from the unlabeled pool and given to the SME for labeling, (ii) most-likely positive strategy (MLP) that selects more of the positive instances for labeling at each iteration, but does not add new features (or rationales), and (iii) MKAD-Sampling strategy where flights are given to the SME for labeling in the order of their MKAD anomaly ranking (higher the anomaly rank, the more informative it is for labeling).

We evaluate all strategies using *precision@k* measure which can be defined as the number of positive instances in top k instances ranked by the classifier. This measure is most suitable for our application because the SMEs go through a list of anomalies to identify those that are operationally significant for further investigation, and improving *precision@k* means that the SMEs would analyze more of the OS flights compared to the NOS flights. We chose *precision@5* and *precision@10* for evaluation since they are the most frequently used in the literature measures to use (e.g., [3,17]). We bootstrap the classifier using an initially labeled set containing one OS flight and one NOS flight, and at each round of active learning the learner picks a new flight for labeling. We evaluate all strategies using 2-fold cross validation and repeat each experiment 10 times per fold starting with a different bootstrap, and present average results over 20 different runs. We set the budget (B) in our experiments to 45 flights, as most learning curves flatten out after about 35 flights. Since each learning curve is an average over 20 runs, for each learning curve, we report error bars for standard error of the mean (SEM), which is computed as standard deviation divided by the square root of sample size ($SEM = \frac{s}{\sqrt{n}}$).

4.1 Results

Figure 3 presents the learning curves comparing RND, MKAD-Sampling, and MLP strategies for *precision*@5 and *precision*@10. MKAD-Sampling performs worse than RND for *precision*@5 and it outperforms RND for *precision*@10. However, MLP outperforms both RND and MKAD-Sampling for *precision*@5 and *precision*@10. We performed pairwise one-tailed t-tests under significance level of 0.05, where pairs are area under the learning curves for 20 runs of each method. If a method has higher average performance than a baseline with a significance level of 0.05 or better, it is a win, if it has significantly lower performance, it is a loss, and if the difference is not statistically significant, the result is a tie. The t-test results show that MKAD-Sampling statistically significantly loses to RND for *precision*@5 and significantly wins over RND for *precision*@10. MKAD-Sampling performs better than MLP at the very beginning of the learning curves, but t-test results show that overall, MLP statistically significantly wins over MKAD-Sampling for both *precision*@5 and *precision*@10. This justifies our choice of using MLP as the active learning strategy for training our classifier for a highly skewed distribution of class labels.

Table 1 presents a comparison of the number of labeled flights required by these methods to achieve a target value of *precision*@5 and *precision*@10.

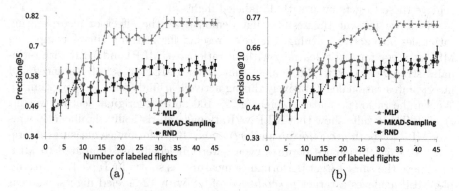

Fig. 3. MLP vs. RND and MKAD-Sampling. MLP significantly outperforms RND and MKAD-Sampling for both (a) *precision*@5 and (b) *precision*@10.

Table 1. Comparison of number of labeled flights required by various strategies to achieve a target performance measure. 'n/a' represents that the target performance cannot be achieved by a method even with 45 labeled flights.

Method	Target *precision*@5						Target *precision*@10					
	0.5	0.6	0.7	0.8	0.9	1.0	0.50	0.55	0.60	0.65	0.70	0.75
RND	6	25	n/a	n/a	n/a	n/a	12	18	33	n/a	n/a	n/a
MKAD-Sampling	4	6	n/a	n/a	n/a	n/a	4	6	13	n/a	n/a	n/a
MLP	5	10	16	32	n/a	n/a	8	12	15	16	23	34
MLP_w/Rationales	2	2	2	8	10	29	2	5	7	11	19	29

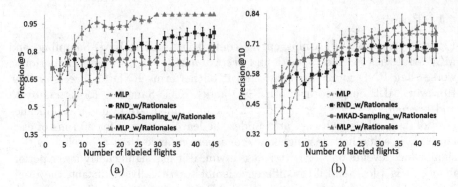

Fig. 4. MLP_w/Rationales vs. MLP. Incorporating rationales further improves performance over MLP for both (a) *precision*@5 and (b) *precision*@10.

The maximum target for each metric is chosen based on the best performance observed in the learning curves for each of the strategies. The results show that MLP often requires fewer labeled flights compared to RND and MKAD-Sampling. Moreover, MLP achieves a *precision*@5 of 0.7 and *precision*@10 of 0.65 with just 16 labeled flights, whereas RND and MKAD-Sampling could not achieve these targets even with 45 labeled flights.

Next, we present the results that demonstrate the effect of incorporating rationales into active learning. Figure 4 presents the learning curves comparing MLP strategy for active learning without rationales (MLP) and MLP with rationales strategy (MLP_w/Rationales) that utilizes MLP to select instances and incorporates rationales iteratively during active learning (refer to Algorithm 1). We set the rationale feature weight $w_r \cdot = 100$ and the original feature weight, $w_o = 1$. The results show that MLP_w/Rationales statistically significantly wins over MLP for both *precision*@5 and *precision*@10 performance measures. Moreover, MLP_w/Rationales requires even fewer labeled flights compared to MLP to achieve the same target performance measure, as shown in Table 1. For example, MLP achieves a target *precision*@5 of 0.8 with 32 labeled flights, whereas MLP_w/Rationales achieves this target with only 8 labeled flights, which is 75 % savings in the labeling effort over MLP.

Figure 4 also compares MLP_w/Rationales to RND_w/Rationales and MKAD-Sampling_w/Rationales. MKAD-Sampling_w/Rationales performs better than MLP_w/Rationales at the beginning for both *precision*@5 and *precision*@10, but after seeing approximately 10 labeled instances, MLP_w/Rationales outperforms MKAD-Sampling_w/Rationales. T-tests show that MLP_w/Rationales statistically significantly outperforms both MKAD-Sampling_w/Rationales and RND_w/Rationales for both *precision*@5 and *precision*@10.

Choice of rationale weights: We ran experiments to study the effect of weights w_r and w_o on the performance of our algorithm. We chose uniform weighting for the original feature kernels since all 16 of those were suggested by

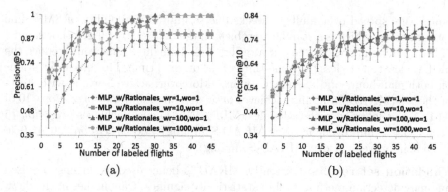

Fig. 5. Comparison of rationale features weights w_r for MLP_w/Rationales using (a) *precision*@5 and (b) *precision*@10

domain experts and were supposed to be important for this safety study. We fixed $w_o=1$ and experimented with four weight settings for w_r (1, 10, 100, or 1000). Figure 5 presents the learning curves for these four weight settings for MLP_w/Rationales. The results confirm our intuition that weighting rationale features higher than original features provides benefit to the active learner. The *precision*@5 results are significantly better with $w_r=100$ than other weights for w_r. For *precision*@10, setting higher weights for rationale features improves performance at the beginning of active learning, however, t-test results show that weights $w_r=1$, 10, and 100 statistically significantly tie with each other. In general, weighting rationale features higher than original features improves learning. The kernel weights for optimal performance can be obtained through multiple kernel learning.

Ideally, one would want to search for the best weights setting using cross validation, but given the limited number of anomalous instances that domain experts could review, it was not possible for us to perform cross validation over the training set. Based on the performance observed for these four weight settings, we chose $w_o=1$ and $w_r=100$ for all our experiments.

Scalability: Active learning methods are typically computationally expensive, since they need to build a classifier at each iteration of learning and evaluate the utility score for every instance in the unlabeled pool. However, in our setting, when active learning is used on the output of an unsupervised anomaly detection algorithm, the unlabeled pool is much smaller in size compared to the entire set of raw instances. Therefore, utilizing this framework in a practical setting is easily viable, without the iterative nature of active learning being a performance bottleneck.

4.2 Performance Benefits

In the absence of active learning framework, our SMEs took approximately 33 hours to review the entire set of 153 anomalies produced by MKAD. These 33 hours were spread over multiple weeks due to limited availability of SME time for such tasks, which is a standard problem in the industry. As Fig. 4 shows, most of the learning curves flatten out after labeling 35 flights. This would reduce the SME review time to less than one-third of the original time. This has implications on both man-hours and monetary savings. Moreover, active learning with state-of-the-art (MKAD-Sampling) achieves *precision*@5 of 0.57 and *precision*@10 of 0.61. Active learning with rationales (MLP_w/Rationales) achieves *precision*@5 of 1 (75.4 % improvement over MKAD-Sampling) and *precision*@10 of 0.76 (24.6 % improvement over MKAD-Sampling).

Validation set results: Currently, MKAD is being used as an unsupervised anomaly detection method to find statistically significant anomalies in the data. We compare performance benefits that active learning with rationales framework (MLP_w/Rationales) provides over the MKAD based classifier for finding OS anomalies in two external validation data sets, July 2014 and July 2015 data sets for the Denver airport. The July 2014 data set has 149 labeled flights with 24 OS anomalies and July 2015 data set has 257 labeled flights with 84 OS anomalies, as determined by the SMEs. Both *precision*@5 and *precision*@10 values for MKAD are 0.4 for the July 2014 data set, and 0.2 for the July 2015 data set. Using our (MLP_w/Rationales) framework, *precision*@5 improves by 15 % for July 2014 data set and by 50 % for July 2015 data set. On the other hand, *precision*@10 improves by 25 % and 110 % for the July 2014 and July 2015 data sets, respectively.

It should be noted that MKAD performs very poorly for the July 2015 data set. This is because the data set is expected to evolve significantly over the years (due to change in landing procedures and other regulation changes) and the MKAD classifier does not capture the signatures of the OS flights, but rather focuses on finding statistically different data points which can vary over time due to a change in the underlying distribution. However, the nature of the operationally significant anomalies still remains consistent and therefore MLP_w/Rationales can identify those types of anomalies much better than MKAD. These results show how active learning with rationales framework can help in building a classifier that is robust to changing distribution of statistically significant anomalies and can, therefore, be used on new data sets without further labeling needs.

5   Towards Deployment

The active learning framework improves over traditional learning, and incorporating rationales further improves learning, utilizing the SME's time much more efficiently. The classifier that is trained through this framework is focused on finding operationally significant anomalies, rather than simply statistically significant anomalies, and hence the flights that are signaled by the two-class classifier approach are of higher relevance to FAA.

This active learning framework has been developed as an extension to the anomaly detection framework that is currently used for detecting safety events.

We expect this framework to easily fit into the existing anomaly detection framework because the classifier training is part of the same data flow pipeline that can take the output of MKAD as input and can seamlessly plug-in new data sources as needed. Given that the new classifier reduces SME review time significantly while improving coverage and reducing false alarm rate, it seems to be the perfect addition to bolster the existing anomaly detection framework, especially since these safety studies are conducted on a regular basis on data that gets collected every month. We expect that this enhanced data processing pipeline with the active learning framework incorporated into it will make the review and detection system significantly more efficient. In our current setup, we provide our SMEs an excel sheet containing the list of anomalies returned by MKAD and the SMEs note down the annotations and rationales textually. This process is repeated iteratively for each round of labeling. The textual information is then converted into features in batches. The next step towards the deployment of our active learning with rationales framework is to fully automate this process where the SMEs can select appropriate rationales using a drop-down list of features by choosing the criteria that were satisfied or violated by the flight in question. The SMEs can choose multiple features for each flight and, therefore, create complex rationale conditions that can be used to create new complex discriminative features on the fly and those features can be immediately utilized for the next iteration of active learning. Figure 6 shows a diagrammatic representation of the software that we are currently developing for deploying as part of the existing framework. It shows the SME initial bootstrap instances for labeling by randomly selecting from the list of anomalies found by MKAD, along with the feature contributions and asks for labels and rationales using drop-down menus. As soon as the classifier has enough number of bootstrap samples, training begins for the classifier. After every iteration the most informative instance is populated in the table for the SME to label and rationalize and classifier training begins again. This iterative process is repeated until the budget B is exhausted or there is no further improvement in the classifier performance on a held-out set.

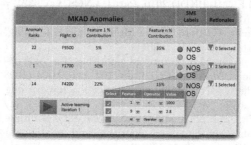

Fig. 6. Diagrammatic representation of the GUI for deployment of active learning as part of the anomaly detection framework

6   Conclusion

We present an active learning framework to build a classifier that can distinguish between operationally significant anomalies and uninteresting ones. Our proposed framework is novel in the sense that it incorporates SME feedback into the learning process in the form of new features constructed to support the labels. Experimental evaluation on real aviation data shows that our approach improves detection of operationally significant events by as much as 75 % compared to the state-of-the-art. The learnt classifier also generalizes well when tested on additional validation data sets. We also observe that our approach provides significant reduction in SME review time and labeling effort in order to achieve the same target performance using other baselines.

We are working toward deploying our framework as a daily reporting system that can reveal operationally significant anomalies to safety analysts with the goal of developing mitigation opportunities by changing standard operating procedures. The reduced false alarm rate of our framework compared to the unsupervised anomaly detection method is critical for domain experts to accept our reporting system and not just ignore the alarms, as has happened with other warning systems. Future work also includes developing richer rationales and ability to integrate multiple data sources for supporting those rationales for increased coverage of a wider range of operationally significant anomalies.

Acknowledgments. This research is supported by the NASA Airspace Operation and Safety Program. Manali Sharma and Mustafa Bilgic are supported by the National Science Foundation CAREER award no. IIS-1350337. The authors would also like to thank the SMEs: Steve Wyloge and Glenn Hilgedick for their insightful comments and perspective on the identified events.

References

1. Attenberg, J., Melville, P., Provost, F.: A unified approach to active dual supervision for labeling features and examples. In: Balcázar, J.L., Bonchi, F., Gionis, A., Sebag, M. (eds.) ECML PKDD 2010, Part I. LNCS, vol. 6321, pp. 40–55. Springer, Heidelberg (2010)
2. Bach, F., Lanckriet, G., Jordan, M.: Multiple kernel learning, conic duality, and the SMO algorithm. In: ICML (2004)
3. Bharat, K., Henzinger, M.R.: Improved algorithms for topic distillation in a hyperlinked environment. In: ACM SIGIR, pp. 104–111. ACM (1998)
4. Bilgic, M., Bennett, P.N.: Active query selection for learning rankers. In: ACM SIGIR, August 2012
5. Das, S., Matthews, B.L., Srivastava, A.N., Oza, N.C.: Multiple kernel learning for heterogeneous anomaly detection: algorithm and aviation safety case study. In: Proceedings of KDD, pp. 47–56 (2010)
6. National Research Council: Advancing Aeronautical Safety: A Review of NASA's Aviation Safety-Related Research Programs. The National Academies Press, Washington, DC (2010)
7. Pelleg, D., Moore, A.: Active learning for anomaly and rare-category detection. In: NIPS, December 2004

8. Ramirez-Loaiza, M.E., Sharma, M., Kumar, G., Bilgic, M.: Active learning: an empirical study of common baselines. Data Min. Knowl. Discov., 1–27 (2016)
9. Roy, N., McCallum, A.: Toward optimal active learning through sampling estimation of error reduction. In: ICML, pp. 441–448 (2001)
10. Schölkopf, B., Platt, J.C., Shawe-Taylor, J.C., Smola, A.J., Williamson, R.C.: Estimating the support of a high-dimensional distribution. Neural Comput. **13**(7), 1443–1471 (2001)
11. Settles, B.: Active Learning. Synthesis Lectures on Artificial Intelligence and Machine Learning. Morgan & Claypool Publishers (2012)
12. Seung, H.S., Opper, M., Sompolinsky, H.: Query by committee. In: ACM Annual Workshop on Computational Learning Theory, pp. 287–294 (1992)
13. Sharma, M., Bilgic, M.: Evidence-based uncertainty sampling for active learning. Data Min. Knowl. Discov., 1–39 (2016)
14. Sharma, M., Zhuang, D., Bilgic, M.: Active learning with rationales for text classification. In: NAACL-HLT, pp. 441–451 (2015)
15. Sindhwani, V., Melville, P., Lawrence, R.D.: Uncertainty sampling and transductive experimental design for active dual supervision. In: ICML, pp. 953–960 (2009)
16. Tong, S., Koller, D.: Support vector machine active learning with applications to text classification. JMLR **2**, 45–66 (2001)
17. Yu, C.N.J., Joachims, T.: Learning structural SVMS with latent variables. In: Proceedings of the 26th Annual International Conference on Machine Learning, pp. 1169–1176. ACM (2009)
18. Zaidan, O.F., Eisner, J., Piatko, C.: Machine learning with annotator rationales to reduce annotation cost. In: Proceedings of the NIPS* 2008 Workshop on Cost Sensitive Learning (2008)

Engine Misfire Detection with Pervasive Mobile Audio

Joshua Siegel[✉], Sumeet Kumar, Isaac Ehrenberg, and Sanjay Sarma

Massachusetts Institute of Technology, Cambridge, MA 02139, USA
{j_siegel,sumeetkr,yitzi,sesarma}@mit.edu

Abstract. We address the problem of detecting whether an engine is misfiring by using machine learning techniques on transformed audio data collected from a smartphone. We recorded audio samples in an uncontrolled environment and extracted Fourier, Wavelet and Mel-frequency Cepstrum features from normal and abnormal engines. We then implemented Fisher Score and Relief Score based variable ranking to obtain an informative reduced feature set for training and testing classification algorithms. Using this feature set, we were able to obtain a model accuracy of over 99 % using a linear SVM applied to outsample data. This application of machine learning to vehicle subsystem monitoring simplifies traditional engine diagnostics, aiding vehicle owners in the maintenance process and opening up new avenues for pervasive mobile sensing and automotive diagnostics.

Keywords: Pervasive sensing · Mobile phones · Sound classification · Audio processing · Fault detection · Machine learning

1 Introduction

People spend more time in their cars than ever before, and with growing miles traveled [25], hours spent in traffic [18], and an aging vehicle fleet in the United States and around the world [7], vehicle maintenance has become an increasingly critical part of vehicle ownership. Proactive or rapid-response maintenance saves significant cost over the life of a vehicle and reduces the likelihood of an unplanned breakdown. Anticipatory maintenance can further alleviate reliability concerns and increase the overall satisfaction of vehicle owners and operators through reduced fuel consumption, emissions, and improved comfort. For these reasons, the consumer-facing diagnostic market for vehicles has grown to include products intended to help vehicle owners maintain and supervise the operation of their vehicles without the assistance of a mechanic.

At the core of any vehicle's maintenance requirement is the engine, responsible for efficient and reliable propulsion. Automotive internal combustion engines require only three "ingredients" to run: a supply of fuel, intake air, and ignition sparks. Delivery of one or more of these elements can fail, as is the case when an air filter or fuel injector clogs or when an ignition coil is damaged. One common engine fault results from a weak or non-existent spark, causing the fuel

© Springer International Publishing AG 2016
B. Berendt et al. (Eds.): ECML PKDD 2016, Part III, LNAI 9853, pp. 226–241, 2016.
DOI: 10.1007/978-3-319-46131-1_26

in a cylinder to fail to combust. With one or more cylinders failing to explode and generate motive force, fuel efficiency and power output drop, with the engine operation increasing in noise, vibration, and harshness. This fault, called a "misfire," results in engine wear and leads to hesitation upon acceleration. A weak spark may be the result of neglected maintenance, such as fouled spark plugs, or component failure, such as an intermittently connected plug wire or an ignition coil pack stressed from powering improperly gapped spark plugs.

Per a 2011 CarMD "Vehicle Health Index" [2], misfires are severe faults and the most commonly occurring vehicle failure, representing 13.8 % of reported problems. Beyond the cost of damage resulting from inaction, misfires have the potential to incur significant additional fuel costs resulting from inefficient or incomplete combustion.

In modern vehicles, computer systems monitor combustion, misfires, and other emission-related functions through a system called "On Board Diagnostics" (OBD) [17]. While OBD systems are capable of detecting a misfire, they are slow to react, rely on proprietary and non-standard algorithms, and necessitate the use of a specialized interface device to provide human-readable information. In a survey we conducted of 15 drivers who had recent, active check-engine lights, we determined that owners left problems unaddressed an average of 3,500 miles [20]. Though OBD tools are available, we determined that they are underutilized by the average vehicle owner.

To better enable preventative maintenance, it is desirable to instead detect these faults passively, more reliably, and without specialized equipment, applying sensing from devices such as mobile phones and allowing location- and orientation-independent analysis. This would remove the barrier to entry posed by requiring a dedicated code-reading device and enable pervasive sensing to allow drivers to monitor the health of their vehicles with increasing frequency at no additional cost. Through improved early detection, the source of the misfire can be addressed easily and inexpensively with the replacement of a spark plug, wire, or ignition coil, before the failure takes a more costly toll on other components like the catalytic converter due to long-term rich fuel trim.

A concurrent proliferation in mobile devices, along with recent advances in sensing and computation, has made pervasive sensing a valuable field for exploration. The use of mobile phones as "automotive tricorders" capable of non-invasively detecting vehicle condition will encourage drivers to take an active role in vehicle maintenance through improved ease-of-use and widespread adoptability relative to current diagnostic offerings. Passive sensing will allow a shift from today's paradigm of reactive repair to one of proactive maintenance, with this technique having been used successfully for passive monitoring of wheels and tires [21, 22].

In this paper, we show that pervasive sensing may be used to differentiate normally operating engines from those operating with misfires. Because we lack a robust physical model describing misfire phenomena, we apply machine learning techniques to uncontrolled data collection and demonstrate an approach to misfire detection making use of extensive feature generation and set reduction to improve classification without a physically-derived hypothesis. We demonstrate that a mobile device may be used to generate data, create a set of features, reduce the size of that set, and apply machine learning to classify accurately and efficiently based on the reduced set.

This paper covers topics ranging from data collection to feature generation to classification. In Sect. 2, we consider prior art and how our method differs from the in-situ and externally sensed solutions before it, illustrating the opportunity space and motivating our work. Section 3 describes our approach to data generation, and how we minimize experimental setup in favor of more naturalistic and representative data collection capable of more easily translating to a consumer-friendly application. Section 4 explores the algorithms we use to generate a comprehensive feature vector. Our approach applies exhaustive feature generation because we have no prior art to distinguish what might be important to classify misfires from normally operating engines. We further discuss our approach to reducing feature set size using feature ranking techniques to facilitate lower computational and other resource overheads. In Subsect. 4.3, we briefly discuss the various classification algorithms we implemented and their relative merits, drawbacks, and efficacy. We conclude in Sect. 5 and show 99 % classification accuracy with 50 % outsample data, before Sect. 6 which discusses plans for future work in this area.

2 Prior Art

Engine misfires have been detected in a variety of ways. Under normal operating conditions, the crankshaft rotates through a fixed angular displacement between every cylinder firing attempt. A misfire detectably alters the precession of the crankshaft which is sensed by a crankshaft position sensor. Measuring a series of unexpected angular measurements within a time window prompts the illumination of a check engine light indicating that the engine is operating outside of specifications and malfunctioning. The use of an OBD scan tool may reveal which cylinder or cylinders are misfiring, but this information is of uncertain provenance and dubious value due to the use of proprietary classification schemes [14, 19]. Some direct-sensing alternatives to crankshaft position-based detection include sampling of the instantaneous exhaust gas pressure, measuring ionization current in the combustion chamber, or installing other sensors within [5, 27] or outside the combustion chamber [15, 26].

Other diagnostics have demonstrated the capacity to identify misfires through audio signal processing. Aside from less obviously discernible symptoms like increased fuel consumption or visual indications like an oily or white residue on the tip of the spark plug, misfires have a characteristic audible "pop" and cause the engine to vibrate as though it is unbalanced or otherwise "missing a beat". The sound emanating from an abnormally-firing engine can be captured at a distance by a microphone and analyzed in both the time and frequency domains for patterns indicative of cylinder misfires. Auto mechanics have long employed a form of auditory diagnosis, listening to engines and easily determining the presence of a misfire. The fact that physical models of the sound and vibration profiles produced by an engine misfire are complex, yet experienced mechanics can classify a firing abnormality by ear, lends credence to the idea that a machine learning approach to detection may be tenable.

Researchers have applied this sort of classification technique successfully. To acquire the audio signals, Dandare [4] and Sujono [23] made use of dedicated recording equipment to analyze the sound from automotive internal combustion engines in a laboratory environment. Engines were recorded during normal operation, as well as in the presence of different faults including cylinder misfire. In Dandare, an Artificial Neural Network classified faults with accuracies ranging from 85–95% overall. Kabiri and Ghaderi [8,9] introduced noise into their misfire measurements of over 300 single cylinder engines by moving outside the laboratory and into a garage. Principal Component Analysis and correlation-based feature selection in the time and frequency domains achieved an accuracy between 70 % and 85 % for these vehicles. Anami made similar recordings of motorcycles [1] to aid mechanics in the rapid classification of healthy versus faulty. Several hundred motorcycle engines were recorded from a distance of half a meter, with wavelet-based machine learning techniques distinguishing not only between healthy and faulty motorcycles, but also the category of fault present. This included, for example, whether the fault was in the engine or exhaust system. Experienced mechanics provided ground truth, with the classification system reporting > 85% accuracy relative to these uncertain reference values.

With the proliferation of smartphones among the car owning public, researchers have considered how these devices can be used to aid in vehicle diagnostics and more specifically engine misfire detection. Using the smartphone at the center of a remote maintenance system, Tse [24] installed sensors including accelerometers and laser encoders within a test vehicle. When a misfire or other engine events were detected, a message was sent via smartphone to inform the user. In Navea [16], the smart phone itself was used as the data collection and processing device, and was held 30 cm above the engine cover to record sounds of the engine and drive belt during startup, while idling, and at 1000 RPM. Thirty-five Honda Civics were used and recordings were taken at various locations and ambient conditions as input data for Fast-Fourier Transformed data based classification. Startup issues relating to the car battery, fuel supply and timing were recognized 100 % of the time, while a normal engine at idle or 1000 RPM was identified with a 33 % false positive rate. Pulley bearing defects or belt slips were properly diagnosed less than 50 % of the time, while valve clearance issues were more reliably detected.

Previous work has laid a strong foundation and shown great potential for using audio signals as a vehicle diagnostic technique, with the capacity for smartphones to serve as capable diagnostic tools within the reach of the general public. Indeed, in past studies we have utilized internal smartphone sensors for a variety of automotive applications, from wheel imbalance detection [22] to tire pressure monitoring [21]. Thanks to mobile computing and pervasive sensing, there is an opportunity to help vehicle owners passively supervise the operation and maintenance of their cars without environmental control or specialized equipment, yielding accuracy meeting or exceeding that of a trained and certified mechanic.

3   Data Collection

3.1   Experimental Design

The goal of the experiment was to collect audio in a manner that could reasonably be duplicated by typical vehicle owners with access to a smartphone. To that end, the procedure did not rely on fixed position or orientation of the vehicle or mobile phone, and the background environment was not controlled, which allowed ambient sources such as wind and other vehicles to add noise to both the training and testing data. In effect, we applied non-invasive and uncontrolled data collection.

To record the audio samples, each vehicle was warmed up for at least five minutes to ensure the engine was no longer running a "fast idle," which could provide unwanted audio artifacts. Then, the vehicle's hood was opened and propped up. Opening the hood allowed clearer audio signal capture during the proof-of-concept phase, and is something most drivers can easily complete without guidance or the use of tools.

For between two minutes and thirty seconds and six minutes, we used a mobile phone to record the engine idle sound as an uncompressed stereo. WAV file at 48000 Hz. During this time, mobile device was swept over the engine to provide a robust training set that incorporated noise from the engine intake, exhaust, belts, and other periodic signals present in the engine compartment. This relative motion is shown in Fig. 1.

With baseline testing completed, the procedure was repeated for anomalous engine operation and misfires. To simulate a misfire, the engine coil pack was

Fig. 1. The phone recorded as it was moved over the engine to provide background noise to test algorithm robustness. Engine covers were left on to minimize prep work and provide a better representative use case for in-situ monitoring.

Fig. 2. The supply to the ignition coil pack was disconnected in order to induce a complete misfire on individual cylinders.

disconnected with the engine turned off, removing the 12 V supply. This connector is shown for two vehicles in Fig. 2. Misfires induced in this manner manifest identically to misfires caused by coil failure, broken spark plug wires, fouled spark plugs, and improper grounding.

The engine was allowed to run for two minutes in this configuration prior to recording in order to allow the engine time to adapt to a cycle with periodic non-ignition. We selected two minutes as a lower limit because many engine control parameters, such as "long term fuel trim," reference only 30 seconds of driving history. In all cars, at least one cylinder was "deactivated" via induced misfire; in some cars, data were collected for multiple cylinders misfiring individually and in aggregate. After testing, the engine was shut off and the coil pack reconnected. If a check engine light had illuminated during testing, it was cleared using a standard ELM327-based automotive diagnostic tool.

Audio data were collected from multiple vehicles with different engine configurations over several days and in different parking locations (outdoor parking lots, a garage, and indoor parking structures). This allowed for the creation of a rich training set capable of providing in-data and out-data for testing. In the case of this experiment, the two engine configurations tested were a normally aspirated inline-four cylinder layout in a Kia Optima and a Ford Focus, as well as a normally aspirated V6 configuration in a Chevrolet Traverse and Nissan Frontier SUV. In cases where the engine cover had been removed to disconnect the coil pack, the cover was replaced prior to recording to better replicate a typical misfire condition wherein the engine's exterior features remain unperturbed.

4   Audio Analysis and Engine State Classification

Armed with audio samples from vehicle engines, we employed several data mining techniques in an attempt to detect and classify misfire occurrences. The detection task was formulated as a supervised learning problem, and to simplify initial algorithm development, the audio samples were classified over only two operational states (normal and anomalous) as opposed to three or more (normal and different cylinder misfire configurations).

4.1   Feature Construction

The 48 kHz audio samples were first assigned labels based on whether the engine was operating normally or abnormally (an engine operating with a single misfiring cylinder) during recording. These samples were merged from stereo into a single, mono channel via averaging and the averaged samples were then subdivided into 2.5s segments. The first 1s and the last 2s of each audio sample were discarded to reduce noisy edge effects and clips with poor signal strength caused as a result of manipulating the mobile device.

2.5s samples recorded at 48 kHz correspond to 120,000 discrete signal elements. The total number of samples in our data set was 992, out of which 373 corresponded to a normal engine. Figure 3 shows a segment of a normal engine audio signal along with that of a misfiring engine.

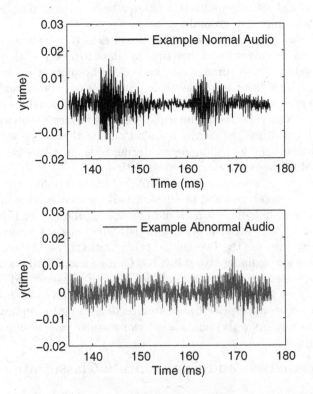

Fig. 3. Comparison of a segment of a normal audio sample with a misfiring audio sample.

To generate features for use in classification, each 120,000 discrete signal element was then converted into a feature vector. We sought to generate a range of features to allow classification without the need for targeted, hypothesis-driven feature creation. Three classes of feature construction were employed and concatenated to form a long feature vector. The three classes include binned Fourier Transform coefficients, Wavelet Transform coefficients, and Mel Frequency Cepstral coefficients.

Though dense feature generation is an intensive process, this approach was adopted to remove any preconceived bias on what features have good discriminative power and rather allow machine learning techniques to drive the solution towards a reduced size feature set. A reasonable feature set size will allow rapid computation using the programmable Digital Signal Processors (DSPs) on mobile devices. Use of such processors has been shown to minimize a classification algorithm's impact on battery life significantly, even allowing Cloud-free operation [11], though further studies are required to optimize DSP computation and data transmission to the cloud in order to minimize overall power consumption and enable pervasive sensing with minimal annoyance to drivers.

Binned Fourier Transform (FT) Coefficients. The discrete samples were first normalized based on power and detrended to remove bias and linear drift. The Fast Fourier Transform (FFT) was then applied to convert the detrended time-domain signals into the frequency domain. Frequencies $< 10\,\text{kHz}$ were divided into bins $10\,\text{Hz}$ wide. Higher frequencies were discarded as not providing additional differentiation because on average they comprised $< 25\,\%$ of the total energy and typically represented harmonics of lower frequencies. The average FT magnitude in each bin provided one feature. This process resulted in the creation of a feature vector of size 1000.

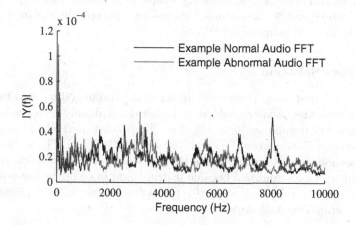

Fig. 4. Comparison of the spectral density of a normal audio sample with a misfiring audio sample.

Figure 4 shows an example comparison of the magnitude of the FT of a normal engine audio signal with a misfiring engine audio sample. We observe that several frequencies (in this particular example segment, around 2 kHz and 8 kHz) have a distinct pattern in the normal vs. abnormal cases. These frequencies that are statistically more powerful classifiers will be identified and used to classify a normal engine from a misfiring engine.

Discrete Wavelet Transform (DWT) Coefficients. In addition to the binned FT, we conducted a wavelet decomposition at level 10 on the power normalized, detrended discrete signal using Daubechies 4 wavelet. At each level of signal decomposition, mean, standard deviation and skewness was computed resulting in a 33-dimensional feature vector.

Mel Frequency Cepstral Coefficient (MFCC). The MFCC creates a spectral signature of short-term frames of the original signal that has been successfully applied to speech recognition [13]. We used a frame size of 1024 samples, with each frame incrementally shifted by 512 samples leading to a total number of 233 frames. For each frame 12 MFCC coefficients were extracted to form a feature vector of size 2796. We made use of the GNU-licensed Voicebox toolbox for MATLAB to conduct MFCC feature extraction.[1]

Concatenating the three sets of feature vectors from the FT, DWT, and MFCC resulted in a 3829-dimensional feature representation of the audio signal and a data matrix of size 992 × 3829. The data set was randomly divided into a 50 % training set and a 50 % test set. In most cases, samples of each state were drawn from different recording events. Rarely, segments of the same file may have been used in both training and testing. In such cases, the movement of the mobile device minimized the likelihood that samples were taken from similar locations and orientations, reducing sample dependence. After splitting the segments, subsequent work continued to develop appropriate feature reduction and classification techniques.

4.2   Feature Selection

To simplify computation, reduce redundancy and training time, and minimize overfitting, it was necessary to reduce the higher-dimensional feature vector using feature selection techniques [6,10,12]. Two filter-based methods were used for feature ranking: Fisher Score (FS) and Relief Score (RS) [10]. The use of feature ranking methods provides novelty over the state-of-the-art in audio classification for automotive faults, and will become instrumental in enabling low-power and resource-constrained devices to run this type of classification by eliminating the need to generate certain features.

---

[1] http://www.ee.ic.ac.uk/hp/staff/dmb/voicebox/voicebox.html.

The Fisher score [10] of a feature for binary classification is calculated using:

$$FS(f_i) = \frac{n_1(\mu_1^i - \mu^i)^2 + n_2(\mu_2^i - \mu^i)^2}{n_1\left(\sigma_1^i\right)^2 + n_2\left(\sigma_2^i\right)^2} \tag{1}$$

$$= \frac{1}{n}\frac{(\mu_1^i - \mu_2^i)^2}{\frac{\left(\sigma_1^i\right)^2}{n_1} + \frac{\left(\sigma_2^i\right)^2}{n_2}},$$

where n_j is the number of samples belonging to class j, $n = n_1 + n_2$, μ^i is the mean of the feature f^i, μ_j^i and σ_j^i are the mean and the standard deviation of f_i in class j. A larger value corresponds to a variable having higher discriminating power.

The Relief score [12] of a feature is computed by first randomly sampling m instances from the data and then using:

$$RS(f_i) = \frac{1}{2}\sum_{k=1}^{m} d\left(f_k^i - f_{NM(\mathbf{x}_k)}^i\right) - d\left(f_k^i - f_{NH(\mathbf{x}_k)}^i\right), \tag{2}$$

where f_k^i denotes the value of the feature f_i on the sample $\mathbf{x}_k$, $f_{NH(\mathbf{x}_k)}^i$ and $f_{NM(\mathbf{x}_k)}^i$ denote the values of the nearest points to $\mathbf{x}_k$ on the feature f_i with the same and different class label respectively, and $d(.)$ is a distance measure which was chosen to be the ℓ_2 norm. Here again, a larger score indicates a higher discriminating power of the variable.

Figure 5 shows the normalized score (scaled to $\in [0,1]$) computed by the two methods noted above for each of the generated features. Though there is a significant correlation between the weights of FS and RS (a linear correlation coefficient of 0.49), combining the information from the two methods may reduce the likelihood of overfitting. To achieve this, we take a simple average of the scores from the two methods, calculated by:

$$AS(f_i) = \frac{1}{2}\left(\frac{FS(f_i) - \min(FS(f_i))}{\max(FS(f_i)) - \min(FS(f_i))} + \frac{RS(f_i) - \min(RS(f_i))}{\max(RS(f_i)) - \min(RS(f_i))}\right). \tag{3}$$

With the features scored, we performed a systematic feature reduction study in order to identify a suitable subset of features. These feature subsets were parametrized by a variable p, with all features whose scores were in the top $(100 - p)^{th}$ percentile for discrimination were included in the subset. Figure 6 demonstrates how feature weighting varied with the FS, RS, and AS methods.

Figure 7 shows the variation in the 10−fold Misclassification Error Rate (MCR) on the training set using a linear Support Vector Machine (SVM), as well as the feature set size (#F) for different scoring schemes and the percentile cutoff p. We performed a grid search to find the optimal box constraint hyperparameter (C) for each of the feature subsets in the figure. From inspection, we identified a minimum MCR at $p = 90$ for each of the three feature scoring methods. Selection of a lower p results in a higher number of less informative features

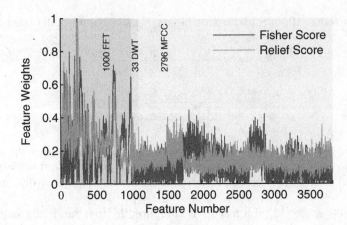

Fig. 5. Comparison of the feature score calculated by the Fisher and the Relief Score methodology.

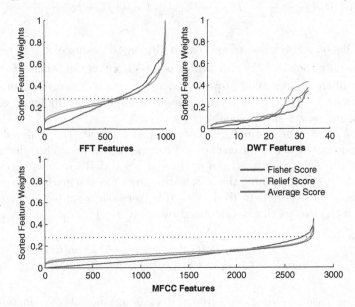

Fig. 6. Feature selection illustrated by method and type. For the AS method, $p = 90$ selection cutoff threshold is indicated as dotted black line at $w = 0.2784$.

in the subset, leading to overfitting and poorer cross-validation performance. Use of a higher p removes important features from the subset leading to a weaker model with decreased accuracy. We additionally observe that with the AS feature ranking the MCR increases less sharply after $p = 90$ when compared to FS or RS, likely due to variance reduction by model averaging. For these reasons, we selected AS with $p = 90$ as the optimal feature subset selection criterion.

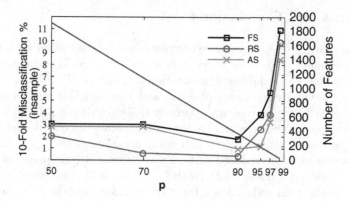

Fig. 7. Comparison of the 10−fold misclassification rate (MCR) and the feature set size with the variation in p.

The binned FT features alone result in a 10−fold misclassification rate of 0.8 %, while with the DWT the error is 36 % and the MFCC based features provide a 29 % error. Concatenating all the above features results in a misclassification rate of 2.6 % which is higher than FT alone. The minimum misclassification rate with FS, RS and AS scoring is 1.8 %, 0.4 % and 1.0 % respectively (Fig. 7).

The FT features have a higher discriminating power when compared to the other two classes of features. Simply combining the features from all three does not provide more discrimination than using the FT features. The ability to perform feature ranking and selecting the optimal subset improves the ratio of the discriminating power to the feature set size (i.e. (1-MCR)/#F) and therefore helps determine a small feature set with high discriminative power. It is also noted that the feature subset with AS and $p = 90$ has 358 FT features out of a total of 383 features, 5 DWT features and 20 MFCC features. Among the FT features selected from aggregate data, several were found to group around the 2.5 kHz and 7.5 kHz frequency bands.

4.3   Classification Algorithms

Using the chosen reduced feature set (AS feature weighting with $p = 90$ and $100 - p =$ the top 10th percentile of features selected), several classification algorithms were studied. The hyperparameters of the classification algorithms were optimized by conducting a grid search to minimize 10−fold cross-validation on the training data. The algorithms tested were k-Nearest Neighbor, Adaboost and SVM with linear, quadratic and RBF kernels. We found that for the SVM with the quadratic kernel all choices of the hyper parameter box-constraint cost (C) led to the same 10-fold misclassification error while for the RBF kernel the error sharply dropped from 38 % to 0 % around the optimal grid points (for finding C and γ). We therefore decided to remove the quadratic and RBF SVM from the final list of classifiers because we were unable to find a robust set of optimal hyperparameters.

5   Results and Conclusions

Table 1 summarizes the performance of the different classification algorithms on the 50 % outsample data. We observe that the linear SVM significantly outperforms the knn and Adabosst classification algorithms. With linear SVM, we obtained a misclassification rate of 1.0 % and the confusion matrix shown in Table 2. The 99 % accuracy of our approach well exceeds the prior art, indicating that our feature selection and reduction techniques are effective at not only improving algorithm efficiency, but increasing accuracy as well.

Considering that the reduced feature set is primarily comprised of the FT features, we trained a linear SVM (with C = 0.01) using only the FT features contained in the final reduced set from the previous section. The outsample misclassification rate with the top FT features was a slightly higher 1.2 % when compared to the results with using the top features of all types (see Table 1). This indicates that most of the discriminative information is contained in the FT features, with the DWT and MFCC features helping primarily differentiate edge cases. This presents an interesting trade off between computing cost and accuracy which will be relevant for designing a mobile application employing this technique. Current efficient implementations of FFT on smartphones [3] can be directly implemented for constructing the FT features in our reduced feature set, while there exist fewer algorithms to efficiently generate DWT and MFCC features.

Table 1. This table compares the classification accuracy (misclassification rate, reported-normal-when-abnormal false positive rate) for different tested algorithms.

Classifier Type	Optimal Hyperparameters	Misclassification Rate	False Positive Rate (Abnormal as Normal)
kNN (l_2 distance based)	Number of neighbors (n) = 11	25 %	27 %
Adaboost (learning rate = 0.3)	Tree depth = 7, Number of trees = 70	15 %	11 %
Linear SVM	Box-constraint cost (C) = 0.01	1.0 %	1.6 %

Table 2. The confusion matrix shows promising results for misfire detection, with 1.6 % false positives (reported normal when actually abnormal). We achieve similarly strong performance for false negatives (reporting abnormal when actually normal), potentially saving drivers money on unnecessary repairs.

		Predicted	
		Normal	**Abnormal**
Actual	**Normal**	100.0%	0.0%
	Abnormal	1.6%	98.4%

Finally, we note that in only one of the four vehicles did a "check engine" light come on at any point during testing, indicating that audio detection such as the one presented here with high accuracy and sensitivity may lend itself to the identification of a misfire prior to detection by an On-Board Diagnostic system. Early detection facilitates proactive response, and can help to lower vehicle maintenance and operating costs relative to drivers relying on the reactive diagnostic systems found in cars today.

6   Future Work

As a component of future work, we intend to explore the resource savings (computational and power) afforded by working with a reduced feature set. We have shown that feature ranking techniques facilitate the discarding of features with minimal loss in accuracy. These unused features need not be computed, enabling more efficient implementations of our feature generation algorithms suited to the limited resources found on mobile devices. Additionally, improving the off-line efficiency of these algorithms will allow us to develop an improved on-line approach, by minimizing bandwidth used for unnecessary data transmission and decreasing reference database size.

While this paper demonstrates promising results for the use of a mobile phone as a pervasive automotive diagnostic tool, the classification can be enriched and robustness improved to yield a more beneficial application, namely identification of the misfiring cylinder itself. That was difficult to discern in this study, as we suspect that information to be embedded within phase-based audio features, which are difficult to discern without a reliable indexing feature in the audio relative to engine component rotations. Other, non-combustion sounds are as of yet ill-defined (considering amplitude/frequency spread) and not available as a phase reference. Similarly, with the collected data it was not immediately feasible to distinguish among various anomalous misfire configurations, but we aim to study other techniques which may be used to improve differentiation among failed states. Such approaches may also improve classification of faults with lesser-defined signals, such as partial misfires due to lean conditions, and non-misfire faults such as clogged air filters or exhaust leaks.

To account for background noise, we intend to build a model to determine dependency of the audio waveform on the engine configuration (idle speed, cylinder count, aspiration, displacement, and firing order). Additionally, audio samples will be recorded from within the car to test whether the application can function from inside the vehicle.

Providing further data to enrich classification, the authors intend to develop algorithms for differential diagnosis: for example, measuring the sound near the air intake and exhaust to monitor airflow issues, identifying where in the airflow process an issue might be occurring. Finally, integrating audio data with information from the On-Board Diagnostic system may be possible, yielding richer fault information than is possible with either system alone.

References

1. Anami, B.S., Pagi, V.B.: Multi-stage acoustic fault diagnosis of motorcycles using wavelet packet energy distribution and ann. In: SERSC International Journal of Advanced Science and Technology (December 2012), International Journal of Advanced. Science and Technology **49**, 47–62 (2012)
2. CarMD: 2011 CarMD Vehicle Health Index. https://www.carmd.com/wp/vehicle-health-index-introduction/2011-carmd-vehicle-health-index/
3. de Carvalho Jr., A.D., Rosan, M., Bianchi, A., Queiroz, M.: Fft benchmark on android devices: Java versus jni. Nexus **7**, 1 (2013). http://compmus.ime.usp.br/sbcm/2013/pt/docs/pos_tec_4.pdf
4. Dandare, S.N.: Multiple fault detection in typical automobile engines: a soft computing approach. WSEAS Trans. Signal Process. **9**(3), 158–166 (2013)
5. Galloni, E.: Dynamic knock detection and quantification in a spark ignition engine by means of a pressure based method. Eng. Convers. Manag. **64**, 256–262 (2012)
6. Gheyas, I.A., Smith, L.S.: Feature subset selection in large dimensionality domains. Pattern Recogn. **43**(1), 5–13 (2010)
7. IHS Inc: Aging vehicle fleet continues to create new opportunity for automotive aftermarket, ihs says. http://press.ihs.com/press-release/automotive/aging-vehicle-fleet-continues-create-new-opportunity-automotive-aftermarket
8. Kabiri, P., Ghaderi, H.: Automobile independent fault detection based on acoustic emission using wavelet. In: Singapore International NDT Conference and Exposition 2011, Singapore International NDT Conference and Exposition, Singapore, November 2011
9. Kabiri, P., Makinejad, A.: Using PCA in acoustic emission condition monitoring to detect faults in an automobile engine. In: 29th European Conference on Acoustic Emission Testing (EWGAE2010), pp. 8–10 (2011)
10. Kumar, S.: Mobile sensor systems for field estimation and "hot spot" identification. Ph.D. thesis, Massachusetts Institute of Technology (2014)
11. Lane, N.D., Georgiev, P., Qendro, L.: Deepear: robust smartphone audio sensing in unconstrained acoustic environments using deep learning. In: Proceedings of the 2015 ACM International Joint Conference on Pervasive and Ubiquitous Computing, pp. 283–294. ACM (2015)
12. Liu, H., Motoda, H.: Feature selection for knowledge discovery and data mining, vol. 454. Springer Science & Business Media, Heidelberg (2012)
13. Logan, B., et al.: Mel frequency cepstral coefficients for music modeling. In: ISMIR (2000)
14. Merkisz, J., Bogus, P., Grzeszczyk, R.: Overview of engine misfire detection methods used in on board diagnostics. J. Kones Combust. Engines **8**(1–2), 326–341 (2001)
15. Merola, S.S., Vaglieco, B.M.: Knock investigation by flame and radical species detection in spark ignition engine for different fuels. Eng. Convers. Manag. **48**(11), 2897–2910 (2007)
16. Navea, R.F., Sybingco, E.: Design and implementation of an acoustic-based car engine fault diagnostic system in the android platform. In: International Research Conference in Higher Education. Polytechnic University of the Philippines (2013)
17. Regulation, section 1968.2 malfunction and diagnostic system requirements - 2004 and subsequent model year passenger cars
18. Schrank, D., Eisele, B., Lomax, T., Bak, J.: 2015 urban mobility scorecard. Texas A & M Transportation Institue & INRIX (2015). http://d2dtl5nnlpfr0r.cloudfront.net/tti.tamu.edu/documents/mobility-scorecard-2015.pdf

19. Service, B.U: Tech feature: Detecting misfires in OBD II engines. http://www.underhoodservice.com/tech-feature-detecting-misfires-in-obd-ii-engines/
20. Siegel, J.E.: Data Proxies, the Cognitive Layer, and Application Locality: Enablers of Cloud-Connected Vehicles and Next-Generation Internet of Things. Ph.D. thesis, Massachusetts Institute of Technology (2016)
21. Siegel, J.E., Bhattacharyya, R., Desphande, A., Sarma, S.E.: Smartphone-based vehicular tire pressure and condition monitoring. In: Proceedings of SAI Intellisys 2016 (2016)
22. Siegel, J.E., Bhattacharyya, R., Sarma, S., Deshpande, A.: Smartphone-based wheel imbalance detection. In: ASME 2015 Dynamic Systems and Control Conference. American Society of Mechanical Engineers (2015)
23. Sujono, A.: Utilization of microphone sensors and an active filter for the detection and identification of detonation (knock) in a petrol engine. Mod. Appl. Sci. 8(6), 112 (2014)
24. Tse, P.W., Tse, Y.L.: On-road mobile phone based automobile safety system with emphasis on engine health evaluation and expert advice. In: Technology Management for Emerging Technologies (PICMET), 2012 Proceedings of PICMET 2012, pp. 3232–3241. IEEE (2012)
25. United States Department of Transportation, Federal Highway Administration: traffic volume trends. https://www.fhwa.dot.gov/policyinformation/travel_monitoring/15dectvt/
26. Vulli, S., Dunne, J.F., Potenza, R., Richardson, D., King, P.: Time-frequency analysis of single-point engine-block vibration measurements for multiple excitation-event identification. J. Sound Vibr. 321(3), 1129–1143 (2009)
27. Zhang, Z., Saiki, N., Toda, H., Imamura, T., Miyake, T.: Detection of knocking by wavelet transform using ion current. In: ICICIC, pp. 1566–1569. IEEE (2009)

Nectar Track Contributions

From Plagiarism Detection to Bible Analysis: The Potential of Machine Learning for Grammar-Based Text Analysis

Michael Tschuggnall[(✉)] and Günther Specht

Department of Computer Science, University of Innsbruck, Innsbruck, Austria
{michael.tschuggnall,guenther.specht}@uibk.ac.at

Abstract. The amount of textual data available from digitalized sources such as free online libraries or social media posts has increased drastically in the last decade. In this paper, the main idea to analyze authors by their grammatical writing style is presented. In particular, tasks like authorship attribution, plagiarism detection or author profiling are tackled using the presented algorithm, revealing promising results. Thereby all of the presented approaches are ultimately solved by machine learning algorithms.

1 Introduction

One of the consequences of todays possibilities and ease to share information over the world wide web is the high availability of textual data, which is either created by social media users or made publicly available through large literary databases like Project Gutenberg[1]. Such data provides a huge source for scientific research in many different areas including text mining problems like web content mining or sentiment analysis [11], but also for social media text based recommender systems (e.g., [12]). A still very important field, which is discussed since the 19th century and which attempts to solve the problem to automatically detect (information about) the writer of a text is authorship attribution [6]. Typical metrics to build stylistic fingerprints include *lexical features* like character n-grams (e.g., [4]), word frequencies (e.g., [2]) or average word/sentence lengths (e.g., [13]), *syntactic features* like Part-of-Speech (POS) tag frequencies (e.g., [4]) or *structural features* like average paragraph lengths or indentation usages (e.g., [13]). A related problem emerges from the fact that the vast amount of available text collections makes it easier for a potential plagiarist to find fragments that can be copied. On the contrary it becomes steadily harder for detection systems to find misuses by just comparing text, and thus advanced algorithms have to be developed. This paper gives an overview of our recent grammar-based research in the broad field of author analysis, including authorship attribution, profiling, plagiarism detection and Bible analysis. All of those applications are based on a pure analysis of the grammar syntax of authors and processed by commonly used machine learning algorithms.

---

[1] https://www.gutenberg.org, visited April 2016.

© Springer International Publishing AG 2016
B. Berendt et al. (Eds.): ECML PKDD 2016, Part III, LNAI 9853, pp. 245–248, 2016.
DOI: 10.1007/978-3-319-46131-1_27

2   Grammar-Based Text Analysis

While constructing sentences, an author has to adhere to the syntactic rules
defined by a specific language. Nevertheless, the number of choices is large, which
leads to the assumption that writers intuitively reuse preferred patterns to build
their sentences. As a consequence, those patterns can be identified and utilized
as a style marker. All applications presented in this paper rely on the analysis
of sentences without considering the vocabulary used. Thereby, a parse tree (or
syntax tree) for each sentence is calculated, which consists of structured POS
tags and serves as the main processing unit to investigate the style of an author.
Figure 1 shows the parse trees of the Einstein quote *"Insanity: doing the same
thing over and over again and expecting different results"* (S_1) and a slightly
modified version (S_2). It can be seen that the trees differ significantly, although
the semantic meaning is the same. To quantify such differences of grammar trees,
the concept of pq-grams is used [1]. In a brief simplification pq-grams can be seen
as "n-grams for trees", as they represent structural parts of the tree. A pq-gram
consists of a stem p and a base q, whereby p defines how much nodes are included
vertically, and q defines the number of nodes to be considered horizontally. For
example, a valid pq-gram with $p = 2$ and $q = 3$ starting at the FRAG tag of the tree
for S_1 would be [FRAG-S-VP-CC-VP]. In order to obtain all possible pq-grams,
the base is shifted left and right additionally while marking non existing nodes
with *. Consequently, also the pq-grams [FRAG-S-*-*-VP], [FRAG-S-*-VP-CC],
[FRAG-S-CC-VP-*] and [FRAG-S-VP-*-*] are valid. Finally, the *pq-gram index*
contains all possible pq-grams of a grammar tree, starting at each node. Because
the presented approaches solely analyze the grammar, the leafs of the trees (i.e.,
the words) have been omitted. The main procedure is as follows:

1. Clean the document, split it into single sentences, calculate a parse tree for
   every sentence[2] and compute the corresponding pq-gram index.

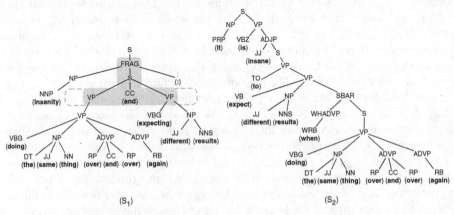

Fig. 1. Parse trees of sentences S_1 and S_2.

---

[2] Using the Stanford Parser [3].

2. Create a *profile* consisting of all[3] occurring pq-grams and transform the profile into a set of features.
3. Use the generated features as input for classifiers in order to, for example, assign authorships or predict the age of a writer.

A profile is calculated by normalizing the number of each occurring pq-gram and assigning it a rank by performing a sort in descending order. Table 1 shows an example using $p = q = 2$. Each profile is then transformed into a set of features which serve as input for

Table 1. Example of a pq-gram Profile.

pq-gram	Occurrence [%]	Rank
NP-NN-*-*	4.07	1
NP-DT-*-*	2.94	2
NP-NNS-*-*	2.90	3

machine learning algorithms, whereby each pq-gram results in two features: (a) the pq-gram with the occurrence rate, and (b) the pq-gram with its rank. As an example, the first line of Table 1 would be transformed into the two features: {'NP-NN-*-*': 4.07} and {'NP-NN-*-*--RANK': 1}. Depending on the document size, the number of distinct features utilized in the following applications ranges between 1,000 and 15,000, which have been processed by common classifiers like Naive Bayes or Support Vector Machines (LibSVM), included in the WEKA framework [5].

3 Approaches

The presented analysis has been applied to several problem types. At first it was used with authorship attribution [8], i.e., it was evaluated if the author of a document can be predicted by analyzing only the grammar syntax. Experiments on different datasets reveal promising accuracies between 75–100 %, which can be compared to other state-of-the-art approaches. Related to that, several approaches have been developed to reveal potential plagiarism [7]. Using machine-learned classifications of sliding windows, an accuracy (F-score) of up to 40 % (for "short" documents with less than 100 sentences even 54 %) could be gained, which is a very good value for so-called *intrinsic* plagiarism detectors. In addition it could be shown, that grammar-based machine learning algorithms can also be successfully used to predict meta-information like the gender or age of an author (accuracy~70 %, [9]), but also to attribute authors of Old Hebrew Bible passages [10] with a conformance rate of 80–100% compared to current literary criticism knowledge. Summarizing, grammar analysis in combination with machine learners provide a solid base for tackling the mentioned problems as well as general text analysis problems, as the pq-gram extraction is universally applicable to any written text.

---

[3] Depending on the approach, the total maximum number of pq-grams in a profile has been restricted, e.g., to the 200 most frequent pq-grams.

4   Conclusion

This paper gives an overview of the main idea to analyze authors by investigating the grammar style used to formulate sentences. The basic principle is to segment a text into sentences, calculate parse trees and to extract pq-grams, which represent the structure of the trees. Several approaches in different domains like authorship attribution or profiling reveal promising results by utilizing pq-gram features as input for common classifiers. Future work may focus on a fine-tuning of the configurations for the latter, as currently only the standard settings are used. Although it was shown that the grammar style is significant, it can additionally be assumed that the existing approaches can be enhanced by incorporating other commonly used features - in particular by features which include information about words and the vocabulary usage.

References

1. Augsten, N., Böhlen, M., Gamper, J.: The pq-gram distance between ordered labeled trees. ACM Trans. Database Syst. (TODS) **35**(1), 4 (2010)
2. Holmes, D.I.: The evolution of stylometry in humanities scholarship. Literary Linguist. Comput. **13**(3), 111–117 (1998)
3. Klein, D., Manning, C.D.: Accurate unlexicalized parsing. In: Proceedings of the 41st Annual Meeting on ACL, Sapporo, Japan, pp. 423–430 (2003)
4. Koppel, M., Schler, J., Argamon, S.: Computational methods in authorship attribution. J. Am. Soc. Inf. Sci. Technol. **60**(1), 9–26 (2009)
5. Hall, M., et al.: The WEKA data mining software: an update. ACM SIGKDD Explor. Newsl. **11**(1), 10–18 (2009)
6. Stamatatos, E.: A survey of modern authorship attribution methods. J. Am. Soc. Inf. Sci. Technol. **60**(3), 538–556 (2009)
7. Tschuggnall, M., Specht, G.: Using grammar-profiles to intrinsically expose plagiarism in text documents. In: Métais, E., Meziane, F., Saraee, M., Sugumaran, V., Vadera, S. (eds.) NLDB 2013. LNCS, vol. 7934, pp. 297–302. Springer, Heidelberg (2013)
8. Tschuggnall, M., Specht, G.: Enhancing authorship attribution by utilizing syntax tree profiles. In: Proceedings of the 14th Conference of the European Chapter of the ACL (EACL), Gothenburg, Sweden, pp. 195–199, April 2014
9. Tschuggnall, M., Specht, G.: On the potential of grammar features for automated author profiling. Adv. Intell. Syst. **8**(3&4), 255–265 (2015)
10. Tschuggnall, M., Specht, G., Riepl, C.: Algorithmisch unterstützte Literarkritik. Memorialband Richter, ATSAT 100, St. Ottilien (2016, to appear)
11. Vinodhini, G., Chandrasekaran, R.: Sentiment analysis, opinion mining: a survey. Int. J. **2**(6) (2012)
12. Zangerle, E., Gassler, W., Specht, G.: On the impact of text similarity functions on hashtag recommendations in microblogging environments. Soc. Netw. Anal. Min. **3**(4), 889–898 (2013)
13. Zheng, R., Li, J., Chen, H., Huang, Z.: A framework for authorship identification of online messages: writing-style features and classification techniques. J. Am. Soc. Inf. Sci. Technol. **57**(3), 378–393 (2006)

A KDD Process for Discrimination Discovery

Salvatore Ruggieri[✉] and Franco Turini

Dipartimento di Informatica, Università di Pisa,
Largo B. Pontecorvo 3, 56127 Pisa, Italy
{ruggieri,turini}@di.unipi.it

Abstract. The acceptance of analytical methods for discrimination discovery by practitioners and legal scholars can be only achieved if the data mining and machine learning communities will be able to provide case studies, methodological refinements, and the consolidation of a KDD process. We summarize here an approach along these directions.

1 The Way Ahead

Data mining and machine learning approaches to social discrimination discovery from historical decision records have recently gained momentum – see the surveys [1,6,8]. Most of the proposals are restricted to investigations of novel algorithms and models. In our opinion, the field still need major advancements towards: first, experimentation with real data; second, methodological refinements in compliance with legal rules and ethical principles; and third, the consolidation of a KDD process of discrimination discovery. Solving these issues is essential for the acceptance of discrimination discovery methods based on data mining and machine learning by practitioners and legal scholars. In the paper [7] we contributed in all those aspects by presenting: a case study on a real dataset about gender discrimination in scientific research proposals; an instantiation of the methodological approach of [4] based on the legal methodology of situation testing; a generalization of the case study to a KDD process in support of discrimination discovery. This is a summary of the last contribution.

2 Not only an Algorithm: An Analytical Process

Since personal data in decision records are highly dimensional, i.e., characterized by many multi-valued variables, a huge number of possible contexts may, or may not, be the theater for discrimination. In order to extract, select, and rank those that represent actual discriminatory behaviors, an anti-discrimination analyst should apply appropriate tools for pre-processing data, extracting prospective discrimination contexts, exploring in details the data related to the context, and validating them both statistically and from a legal perspective. Discrimination discovery consists then of an iterative and interactive process. Iterative because, at certain stages, the user should have the possibility of choosing different algorithms, parameters, and evaluation measures or to iteratively repeat some

© Springer International Publishing AG 2016
B. Berendt et al. (Eds.): ECML PKDD 2016, Part III, LNAI 9853, pp. 249–253, 2016.
DOI: 10.1007/978-3-319-46131-1_28

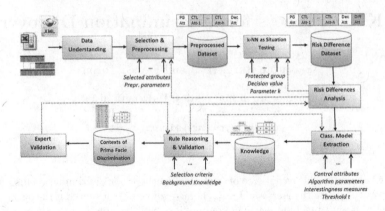

Fig. 1. The KDD process of situation testing for discrimination discovery.

steps to unveil meaningful discrimination patterns. Interactive because several stages need the support of a domain expert in making decisions or in analysing the results of a previous step. We propose in [7] to adopt the process reported in Fig. 1, which is specialized in the use of the situation testing for extracting contexts of possible discrimination. The process has been abstracted from the case study presented in the paper regarding gender discrimination in a dataset of scientific research proposals, and it consists of four major steps.

Data Understanding and Preparation. We assume a collection of data sources storing historical decisions records in any format, including relational, XML, text, spreadsheets or any combination of them. Standard data pre-processing techniques (selection, cleansing, transformation, outlier detection) can be adopted to reach a pre-processed dataset consisting of an *input relation* as the basis for the discrimination analysis. The grain of tuples in the relation is that of an individual (an applicant to a loan, to a position, to a benefit). Three groups of attributes are assumed to be part of the relation:

protected group attributes: one or more attributes that identify the membership of an individual to a protected group. Attributes such as sex, age, marital status, language, disability, and membership to political parties or unions are typically recorded in application forms, curricula, or registry databases. Attributes such as race, skin color, and religion may be not available, and must be collected, e.g., by surveying the involved people;

decision attribute: an attribute storing the decision for each individual. Decision values can be nominal, e.g., granting or denying a benefit, or continuous, e.g., the interest rate of a loan or the wage of a worker;

control attributes: one or more attributes on control factors that may be (legally) plausible reasons that may affect the actual decision. Examples include the financial capability to repay a loan, or the productivity of an applicant worker.

Risk Difference Analysis. Randomized experiments are the gold-standard for inferring causal influences in a process. However, randomized experiments are not possible or not cost-effective in discrimination analysis. An example of quasi-experimental approaches is situation testing [2], which uses pairs of testers who have been matched to be similar on all characteristics that may influence the outcome except race, gender, or other grounds of possible discrimination. In a legal setting, the tester pairs are then sent into one or more situations in which discrimination is suspected. In observational studies, [4] proposes to simulate the approach by contrasting the decisions of the tuple neighbors. For each tuple of the input relation denoting an individual of the protected group, the additional attribute *diff* is calculated as the risk difference between the decisions of its k nearest-neighbors of the protected group and the decisions for its k nearest-neighbors of the unprotected group (see Fig. 2). Risk difference is a measure of the degree of discrimination suffered by an individual. We call the output of the algorithm the *risk difference relation*. The value k is a parameter of the algorithm. A study of the distribution of *diff* for a few values of k is required. This means iterating the calculation of the *diff* attribute. Exploratory analysis of *diff* distributions may also be conducted to evaluate risk differences at the variation of: the protected group under consideration, e.g., discrimination against women or against youngsters; the compound effects of multiple discrimination grounds, e.g., discrimination against young women vs discrimination against women or youngsters in isolation; the presence of favoritism towards individuals of a dominant group, e.g., nepotism.

Discrimination Model Extraction. By fixing a threshold value t, an individual r of the protected group is then *labeled* as discriminated or not on the basis of the condition $diff(r) \geq t$. We introduce a new boolean attribute *disc* and set it to true for a tuple r meeting the condition above, and to false otherwise. A global description of who has been discriminated can now be extracted by resorting to a standard classification problem on the dataset of individuals of the protected group, where the class attribute is the newly introduced *disc* attribute. Accuracy of

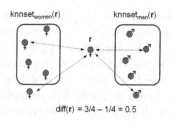

$$diff(r) = 3/4 - 1/4 = 0.5$$

Fig. 2. Example of risk difference $diff(r)$ for $k = 4$. Women are the protected group, $knnset_{women}(r)$ (resp., $knnset_{men}(r)$) is the set of female (resp., male) k-nearest neighbors of r. Red labels denote benefit denied, green labels denote benefit granted. (Color figure online)

the classifier is evaluated with objective interestingness measures, e.g., precision and recall over the *disc = true* class value. The choice of the value t should then be supported by laws or regulators. For instance, the *four-fifths rule* by the US states that a job selection rate lower than 80 % represents a *prima facie* evidence of adverse impact. Since the intended use of the extracted classifier is descriptive, classification models that are easily interpretable by (legal) experts and whose size is small should be preferred. In other words, one should trade

accuracy for simplicity. Classification rules and decision trees are natural choices
in this sense, since rules and tree paths can easily be interpreted and ranked.
The extracted classification models provide a global description of the *disc* class
values. They are stored in a knowledge base, for comparison purposes and for
the filtering of specific contexts of discrimination – as described next.

Rule Reasoning and Validation. The actual discovery of discriminatory situations and practices may reveal itself as an extremely difficult task. Due to time
and cost constraints, an anti-discrimination analyst needs to put under investigation a limited number of contexts of possible discrimination. In this sense,
only a small portion of the classification models can be analysed in detail, say
the top rules or the top paths of a decision tree [5]. We concentrate on rules
of the form: (cond_1) and ... and (cond_n) => disc=yes [prec] [rec] [diff],
where (cond_1) and ... and (cond_n) is obtained from a classification model.
Rules are ranked on the basis of one or more interestingness measures, including:
precision [prec], recall [rec], average value of *diff* [diff]. Statistical validation
is accounted for by relying on logistic regression, which is a well-known tool
in the legal and economic research communities. Earlier studies on discrimination discovery, instead, relied upon simple association or correlation measures.
Recently, the discrimination-aware data mining community has recognized the
importance of causal analysis [3,9].

3   Conclusion

The lesson learned by developing the case study in [7] is above all that discrimination discovery needs a structured process around an algorithmic approach,
and a solid compliance with legal rules and ethical principles. Not only this will
provide guidance to data scientists and decision makers, but it is the only way we
may hope to get acceptance of data mining and machine learning methods by the
users of such methods: legal communities, civil rights and digital rights societies,
regulation authorities, (inter)national agencies, and professional associations.

References

1. Barocas, S., Selbst, A.D.: Big data's disparate impact. California Law Rev. **104**
   (2016). SSRN: http://ssrn.com/abstract=2477899
2. Bendick, M.: Situation testing for employment discrimination in the United States
   of America. Horiz. Stratégiques **3**(5), 17–39 (2007)
3. Foster, S.R.: Causation in antidiscrimination law: beyond intent versus impact.
   Houston Law Rev. **41**(5), 1469–1548 (2004)
4. Luong, B.T., Ruggieri, S., Turini, F.: k-NN as an implementation of situation testing
   for discrimination discovery and prevention. In: Proceedings of the International
   Conference on Knowledge Discovery and Data Mining (KDD), pp. 502–510. ACM
   (2011)
5. Pedreschi, D., Ruggieri, S., Turini, F.: A study of top-k measures for discrimination
   discovery. In: Proceedings of ACM SAC 2012, pp. 126–131 (2012)

6. Romei, A., Ruggieri, S.: A multidisciplinary survey on discrimination analysis. Knowl. Eng. Rev. **29**(5), 582–638 (2014)
7. Romei, A., Ruggieri, S., Turini, F.: Discrimination discovery in scientific project evaluation: a case study. Expert Syst. Appl. **40**(10), 6064–6079 (2013)
8. Žliobaitye, I.: A survey on measuring indirect discrimination in machine learning. arXiv preprint arXiv:1511.00148v1 (2015)
9. Zhang, L., Wu, Y., Wu, X.: Situation testing-based discrimination discovery: a causal inference approach. In: Proceedings of IJCAI (2016, to appear)

Personality-Based User Modeling for Music Recommender Systems

Bruce Ferwerda[(✉)] and Markus Schedl

Department of Computational Perception, Johannes Kepler University,
Altenberger Street 69, 4040 Linz, Austria
{bruce.ferwerda,markus.schedl}@jku.at
http://cp.jku.at

Abstract. Applications are getting increasingly interconnected. Al-though the interconnectedness provide new ways to gather information about the user, not all user information is ready to be directly implemented in order to provide a personalized experience to the user. Therefore, a general model is needed to which users' behavior, preferences, and needs can be connected to. In this paper we present our works on a personality-based music recommender system in which we use users' personality traits as a general model. We identified relationships between users' personality and their behavior, preferences, and needs, and also investigated different ways to infer users' personality traits from user-generated data of social networking sites (i.e., Facebook, Twitter, and Instagram). Our work contributes to new ways to mine and infer personality-based user models, and show how these models can be implemented in a music recommender system to positively contribute to the user experience.

Keywords: Personalization · Music recommender systems

1 Introduction

An abundance of information about users is getting available with the increased interconnectedness of applications, which provide new ways to tackle problems that systems, such as recommender systems are facing (e.g., lacking behavioral data to infer preferences, such as with the "cold-start problem").[1] For example, the implementation of single sign-on (SSO) mechanisms[2] allow users to easily login and register to the application, but also let applications import user information from the connected application, which could be used for personalization.

Although with the interconnectedness of applications new information sources become available, not all the new information is directly applicable to

---

[1] The cold-start problem is most prevalent in recommender systems and occurs with new users of the application. It refers to that (almost) no information exists yet about the user to make inferences from.

[2] Buttons that allow users to register or login with accounts of other applications. For example, social networking services: "Login with your Facebook account."

© Springer International Publishing AG 2016
B. Berendt et al. (Eds.): ECML PKDD 2016, Part III, LNAI 9853, pp. 254–257, 2016.
DOI: 10.1007/978-3-319-46131-1_29

create personalized experiences with. Therefore, a general user model is needed to which users' behavior, preferences, and needs can be connected to in order to create personalized experiences for users. This allows the creation of only one user model that can be used across applications without the need of information that is directly related to a specific behavior, preference, or need of the user [1].

We model users based on their personality to make inferences about their behavior, preferences, and needs. Personality has shown to be a stable and enduring factor, which influences an individual's behavior, interest, and taste. As personality plays such a prominent role in shaping human preferences, one can expect similar patterns (i.e., behavior, interest, and taste) to emerge for people with similar personality traits, which makes it suitable for user modeling. In our works, we rely on the widely used five-factor model (FFM), which categorizes personality into five traits: openness to experience (O), conscientiousness (C), extraversion (E), agreeableness (A), and neuroticism (N) [10].

In the next sections we provide an overview of our works on user modeling, which comes in twofold: (1) understanding the relationship between personality traits of users and their behavior, preferences, and needs, and (2): implicit acquisition of users' personality traits from social media.

2   Understanding the User

In order to create personality-based recommender systems, the relationship with their behavior, preferences, and needs need to be identified first. We conducted several user studies on different aspects of the user experience in music recommender systems in order to identify relationships with users' personality.

Listening needs. In [4] we aimed to understand the music listening needs of users in order to provide better personalized recommendations. We investigated the relationship between personality traits and the preference for different kinds of music, and how these preferences change depending on users' emotional state. Our findings show that, in general, users like to listen to music in line with their emotional state. However, individual differences based on personality occur; especially in a negative emotional state (e.g., sadness). We found that when in a negative emotional state, those who scored high on openness to experience, extraversion, and agreeableness tend to cheer themselves up with happy music, while those who scored high on neuroticism tend to prefer to dwell a bit longer in this negative state by listening to sad music. This has important implications for playlist generation. By inferring users' emotional state (e.g., mining user-generated content), the next song can be better targeted toward their needs.

Meta information. In [14] we investigated the amount of meta information a user would want about the music pieces that is listened to. The results showed that the following personality traits tend to have a higher preference for more meta information: openness to experience, agreeableness, conscientiousness, and extraversion. This provides implications about the amount of meta information a system should present to the user without them experiencing information overload, which in turn, negatively affects the user experience of the user.

User interface. In [8] we simulated an online music streaming service to identify the relationship between personality traits and the way users browse for music. By exploring the most frequently used taxonomies to categorize music (i.e., by genre, activity, mood), we were able to identify distinct music browsing behavior based on users' personality, which could be used to create adaptive user interfaces. For example, findings indicate that those scoring high on openness to experience show a high preference for browsing for music by mood, while conscientious users show a preference for browsing by activity.

3   Acquisition of Users' Personality Traits

Besides identifying relationships between personality traits and users' behavior, preferences, and needs, we also looked into the implicit personality acquisition of users. We specifically focused on personality acquisition from social networking sites (SNSs: e.g., Facebook, Twitter, Instagram), as they are getting increasingly interconnected through SSO buttons. Besides accessing users' basic profile information, applications often ask for additional permissions to access other parts of the users profile [2]. By granting access, applications are able to unobtrusively infer users' personality traits. We report the RMSE on personality trait prediction (i.e., O, C, E, A, N) for each of our work below ($r \in [1,5]$).

Several works exist that show that it is possible to infer personality traits from user-generated data of SNSs (e.g., Facebook [11], and Twitter [9,12]). In [5,7] we add to the work on SNS analyses by inferring personality traits from users' Instagram picture features. We showed that personality traits are related to the way Instagram users modify their pictures with filters, and a reliable personality predictor can be created based on that ($RMSE$: $O = .68$, $C = .66$, $E = .90$, $A = .69$, $N = .95$). For example, open users tend to apply filters to their pictures in order to make them look more greenish. In [13] we tried to increase the prediction accuracy by fusing information from different SNSs (i.e., Instagram and Twitter). We show a significant improvement of the prediction accuracy when combining different sources ($RMSE$: $O = .51$, $C = .67$, $E = .71$, $A = .50$, $N = .73$).

One problem with the implicit acquisition of personality is that when users are not sharing information, the acquisition fails. We investigated this problem from two different directions: (1) understanding the underlying mechanisms of sharing information, (2) personality acquisition with limited user information.

In [3] we found that the lack of sharing and posting comes from the uncertainty of approval of the users viewing the posts. We were able to increasing sharing and posting by analyzing the user's social network and create proxy measures about how the shared or posted content would be received.

In [6] we looked at whether or not disclosing Facebook profile information reveals personality as well. By solely analyzing whether profile sections were disclosed or not (e.g., occupation, education), disregarding their actual content, we were able to create a personality predictor that is able to approximate the prediction accuracy of methods extensively analyzing content ($RMSE$: $O = .73$,

$C = .73$, $E = .99$, $A = .73$, $N = .83$). This provide opportunities to still being able to infer users' personality even when they are not disclosing information.

4 Conclusion

This paper gave an overview of our work on creating personalized experiences in music recommender systems. We revealed relationships between personality traits and different user behavior, needs, and preferences to improve the user experience, and showed how personality can be mined and inferred using the increased connectedness between applications and SNSs.

Acknowledgment. Supported by the Austrian Science Fund (FWF): P25655.

References

1. Cantador, I., Fernández-Tobías, I., Bellogín, A.: Relating personality types with user preferences in multiple entertainment domains. In: EMPIRE (2013)
2. Chia, P.H., Yamamoto, Y., Asokan, N.: Is this app. safe?: a large scale study on application permissions and risk signals. In: WWW. ACM (2012)
3. Ferwerda, B., Schedl, M., Tkalcic, M.: To post or not to post: the effects of persuasive cues and group targeting mechanisms on posting behavior. In: SocialCom (2014)
4. Ferwerda, B., Schedl, M., Tkalcic, M.: Personality & emotional states: Understanding users music listening needs. In: UMAP (2015)
5. Ferwerda, B., Schedl, M., Tkalcic, M.: Predicting personality traits with instagram pictures. In: EMPIRE (2015)
6. Ferwerda, B., Schedl, M., Tkalcic, M.: Personality traits and the relationship with (non-) disclosure behavior on facebook. In: WWW (2016)
7. Ferwerda, B., Schedl, M., Tkalcic, M.: Using instagram picture features to predict users' personality. In: MMM (2016)
8. Ferwerda, B., Yang, E., Schedl, M., Tkalcic, M.: Personality traits predict music taxonomy preferences. In: CHI Extended Abstracts (2015)
9. Golbeck, J., Robles, C., Edmondson, M., Turner, K.: Predicting personality from twitter. In: SocialCom (2011)
10. McCrae, R.R., John, O.P.: An introduction to the five-factor model and its applications. J. Pers. (1992)
11. Park, G., Schwartz, H.A., Eichstaedt, J.C., Kern, M.L., Kosinski, M., Stillwell, D.J., Ungar, L.H., Seligman, M.E.: Automatic personality assessment through social media language. J. Pers. Soc. Psychol. **108**(6), 934 (2015)
12. Quercia, D., Kosinski, M., Stillwell, D., Crowcroft, J.: Our twitter profiles, our selves: predicting personality with twitter. In: SocialCom (2011)
13. Skowron, M., Tkalčič, M., Ferwerda, B., Schedl, M.: Fusing social media cues: personality prediction from twitter and instagram. In: WWW (2016)
14. Tkalcic, M., Ferwerda, B., Hauger, D., Schedl, M.: Personality correlates for digital concert program notes. In: UMAP (2015)

Time and Again:
Time Series Mining via Recurrence Quantification Analysis

Stephan Spiegel[1]([⊠]) and Norbert Marwan[2]

[1] DAI Lab, Berlin Institute of Technology, Berlin, Germany
spiegel@dai-lab.de
[2] Potsdam Institute for Climate Impact Research, Potsdam, Germany
marwan@pik-potsdam.de

Abstract. Recurrence quantification analysis (RQA) was developed in order to quantify differently appearing recurrence plots (RPs) based on their small-scale structures, which generally indicate the number and duration of recurrences in a dynamical system. Although RQA measures are traditionally employed in analyzing complex systems and identifying transitions, recent work has shown that they can also be used for pairwise dissimilarity comparisons of time series. We explain why RQA is not only a modern method for nonlinear data analysis but also is a very promising technique for various time series mining tasks.

Keywords: Time series mining · Recurrence quantification analysis

1 Introduction and Background

A recurrence plot (RP) is an advanced technique of nonlinear data analysis [3]. Technically speaking, a recurrence plot R visualizes those times when the trajectory x of a dynamical system visits roughly the same phase space [3]: $R_{i,j} = \Theta(\epsilon - \|x_i - x_j\|)$, where ϵ is the similarity threshold, $\| \cdot \|$ a norm, $\Theta(\cdot)$ the unit step function, and $i, j = 1 \ldots N$ is the number of states. In addition, a cross recurrence plot (CRP) shows all those times at which a state $x_i \in \mathbb{R}^m$ in one dynamical system co-occurs $y_j \in \mathbb{R}^m$ in a second dynamical system [3]: $R_{i,j} = \Theta(\epsilon - \|x_i - y_j\|)$, where the dimension m of both systems must be the same, but the number of states can be different.

The recurrence quantification analysis (RQA) is a method of nonlinear data analysis which quantifies the number and duration of recurrences of a dynamical system presented by its state space trajectory [3]. RQA measures are derived from RP structures and can be employed to study the dynamics, transitions, or synchronization of complex systems [3,4]. The determinism measure (DET^μ), which is the fraction of recurrence points that form diagonal lines of minimum length μ, has e.g. been successfully applied to detect dynamical transitions [4].

2 Recent Trends and Advances

In time series mining, many algorithms are based on analogical reasoning or pairwise dissimilarity comparisons of (sub)sequences [13]. In general, the distance

© Springer International Publishing AG 2016
B. Berendt et al. (Eds.): ECML PKDD 2016, Part III, LNAI 9853, pp. 258–262, 2016.
DOI: 10.1007/978-3-319-46131-1_30

between time series needs to be carefully defined in order to reflect the underlying dissimilarity of the data, where the choice of distance measure usually depends on the invariance required by the domain [1].

Recent work [9–12] has introduced novel time series distance measures that use recurrence quantification analysis (RQA) techniques. The main idea [9] is to pairwise compare time series by (i) computing a cross recurrence plot (CRP) that reveals all times at which roughly the same states co-occur and, subsequently, (ii) quantifying the number and length of all diagonal line structures that indicate similar subsequences. Figure 1(a-b) shows a toy example, where a labeled time series is compared to two unlabeled data stream segments using CRPs as well as corresponding RQA measures.

It has been shown [9,11] that traditional RQA measures, such as the average diagonal line length and the determinism, can be used to compare time series that exhibit similar segments or subsequences at arbitrary positions. Time series with such an order invariance [9] can, for instance, be found in automotive engineering [11], where vehicular sensors observe driving behavior patterns in their natural occurring order and the recorded car drives are compared according to the co-occurrence of these patterns. Although the recurrence plot-based distance [11] was originally developed to determine characteristic driving profiles [12], this approach can be used to find representatives in arbitrary sets of single- or multi-dimensional time series of variable length [10].

In addition, it has been proposed to employ video compression algorithms for measuring the dissimilarity between un-thresholded recurrence plots and accordingly the time series that generated them [8]. This approach relies on the underlying assumption that video compression algorithms are able to detect similar structures in images or recurrence plots, which correspond to time series patterns. The result [8] show that the compression distance of recurrence plots works especially well for time series that represent shapes. A follow-up study [5] compared the performance of various MPEG video compression algorithms and furthermore introduced a compression distance for cross recurrence plots. Figure 1(c) contrasts two un-thresholded recurrence plots, which reveal structural dissimilarities between the examined time series.

Although recurrence plots have been adopted by the data mining community [2,5,8–12], their computation and quantification generally involve operations with quadratic time and space complexity. Hence, recent work [7,14] has introduced approximate RQA measures, which exhibit significantly lower complexity while maintaining high accuracy. Most important, these novel approximations [7,14] enable us to efficiently use recurrence quantification analysis for relatively long time series and fast time series streams. Figure 1(d) illustrates the fast computation of the approximate determinism ($aDET$) [7], which allows us, for example, to filter or identify time series segments with a certain behavior in an online fashion. The approximation of various RQA measures, such as laminarity and determinism, is explained at full length in a recent publication [14]

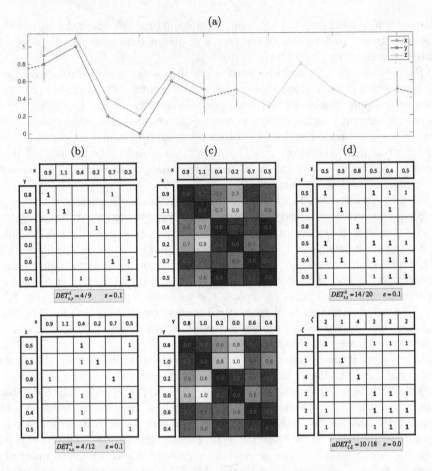

Fig. 1. Recurrence plot-based distances: (a) illustrates a time series mining scenario that assumes a labeled sequence x and a data stream with unlabeled segments y and z. In case (b) we compare time series x with segment y and z to assign labels. (b) shows two cross recurrence plots that indicate similar states ($\epsilon = 0.1$) for time series pairs (x,y) and (x,z), where recurrence points are represented by '1' entries and diagonal line structures are highlighted in bold font. According to the determinism, $DET_{x,y}^2 = 4/9 > 4/12 = DET_{x,z}^2$, the pair (x,y) is more similar than (x,z) [11], meaning x and y might be from the same class. (c) shows another way to determine the pairwise dissimilarity of time series. In this case (c) we create un-thresholded recurrence plots ($\epsilon = 0$), which facilitate pairwise comparisons by means of image processing and video compression algorithms [5,8]. The images in (c) resemble each other in structure since time series x and y have a similar shape. In case (d) we compute the approximate determinism to assess the 'complexity' of our sample data stream at time interval z and to filter/identify 'ir-/relevant' segments with a certain (nonlinear) behavior. (d) illustrates the recurrence plot of segment z and it's discretized version $\zeta = \lfloor \frac{z}{2\epsilon} \rfloor$. In our example (d) we achieve a fairly reasonable approximation of the determinism, $DET_{z,z}^2 = 14/20 \approx 10/18 = aDET_{\zeta,\zeta}^2$. Although the discretization step introduces some rounding errors, it allows us to approximate all traditional RQA measures in an efficient way without even creating and quantifying the RP [7,14].

3   Conclusion and Open Problems

Recurrence quantification analysis (RQA) is a method of nonlinear data analysis for the investigation of dynamical systems, which has its origin in theoretical physics [3,4]. Recently, RQA was adopted by the data mining community in order to: (i) define novel time series distance measures [5,8,11] and (ii) process massive data streams by means of approximate measures [7,14].

Although RQA has been successfully applied to data mining problems from engineering [12] and climatology [6,14], there exist open problems which prevent its widespread acceptance by the time series fraternity. The main problem with traditional RQA is that it excludes curved structures, which prevents us from comparing time series with local scaling or warping invariance [1]. This issue might be addressed by feeding un-thresholded RPs [5,8] into convolutional neural networks. In the case of the recently introduced approximate RQA [7,14], it is necessary to investigate time series representations and discretization techniques that enable us to bound the approximation error.

References

1. Batista, G., Keogh, E.J., Tataw, O.M., De Souza, V.M.A.: CID: an efficient complexity-invariant distance for time series. Data Min. Knowl. Disc. **28**, 634–669 (2014)
2. Gaebler, J., Spiegel, S., Albayrak, S.: MatArcs: an exploratory data analysis of recurring patterns in multivariate time series. In: Proceedings of ECML-PKDD (2012)
3. Marwan, N., Romano, M.C., Thiel, M., Kurths, J.: Recurrence plots for the analysis of complex systems. Phys. Rep. **438**, 237–329 (2007)
4. Marwan, N., Schinkel, S., Kurths, J.: Recurrence plots 25 years later - gaining confidence in dynamical transitions. Europhys. Lett. **101**, 20007 (2013)
5. Michael, T., Spiegel, S., Albayrak, S.: Time series classification using compressed recurrence plots. In: Proceedings of ECML-PKDD (2015)
6. Rawald, T., Sips, M., Marwan, N., Dransch, D.: Fast computation of recurrences in long time series. In: Marwan, N., Riley, M., Giuliani, A., Webber Jr., C.L. (eds.) Translational Recurrences. Springer Proceedings in Mathematics and Statistics, pp. 17–29. Springer, Switzerland (2014)
7. Schultz, D., Spiegel, S., Marwan, N., Albayrak, S.: Approximation of diagonal line based measures in recurrence quantification analysis. Phys. Lett. A **379**, 997–1011 (2015)
8. Silva, D.F., De Souza, V.M.A., Batista, G.: Time series classification using compression distance of recurrence plots. In: Proceedings of ICDM (2013)
9. Spiegel, S., Albayrak, S.: An order-invariant time series distance measure - position on recent developments in time series analysis. In: Proceedings of KDIR (2012)
10. Spiegel, S., Schultz, D., Albayrak, S.: BestTime: finding representatives in time series datasets. In: Calders, T., Esposito, F., Hüllermeier, E., Meo, R. (eds.) ECML PKDD 2014, Part III. LNCS, vol. 8726, pp. 477–480. Springer, Heidelberg (2014)
11. Spiegel, S., Jain, B.J., Albayrak, S.: A recurrence plot-based distance measures. In: Marwan, N., Riley, M., Giuliani, A., Webber Jr., C.L. (eds.) Translational Recurrences. Springer Proceedings in Mathematics and Statistics, pp. 1–15. Springer, Switzerland (2014)

12. Spiegel, S.: Discovery of driving behavior patterns. In: Hopfgartner, F. (ed.) Smart Information Services - Computational Intelligence for Real-Life Applications, pp. 315–343. Springer, Switzerland (2015)
13. Spiegel, S.: Time series distance measures: segmentation, classification and clustering of temporal data. Technische Universitaet Berlin (2015)
14. Spiegel, S., Schultz, D., Marwan, N.: Approximate recurrence quantification analysis in best code of practice. In: Webber Jr., C.L., Ioana, C., Marwan, N. (eds.) Recurrence Plots and Their Quantifications: Expanding Horizons. Springer Proceedings in Physics, pp. 113–136. Springer, Switzerland (2016)

Resource-Aware Steel Production
Through Data Mining

Hendrik Blom[✉] and Katharina Morik

TU Dortmund University, Dortmund, Germany
{hendrik.blom,katharina.morik}@tu-dortmund.de

Abstract. Today's steel industry is characterized by overcapacity and increasing competitive pressure. There is a need for continuously improving processes, with a focus on consistent enhancement of efficiency, improvement of quality and thereby better competitiveness. About 70 % of steel is produced using the BF-BOF (Blast Furnace - Blow Oxygen Furnace) route worldwide. The BOF is the first step of controlling the composition of the steel and has an impact on all further processing steps and the overall quality of the end product. Multiple sources of process-related variance and overall harsh conditions for sensors and automation systems in general lead to a process complexity that is not easy to model with thermodynamic or metallurgical approaches. In this paper we want to give an insight how to improve the output quality with machine learning based modeling and which constraints and requirements are necessary for an online application in real-time.

Keywords: Real time regression · Model predictive control · Prescriptive data analytics

1 Introduction

There are several ways to produce steel. A complete overview can be found in [6]. About 70 % of steel[1] is produced using the BF-BOF (Blast Furnace - Blow Oxygen Furnace) route [5]. The first step is to smelt ores to raw iron in a blast furnace. Coke is used as the primary energy source and as a reduction agent. The carbon will bind the oxygen of the iron oxides. At the end of the process liquid raw iron is produced and transported to the BOF. The produced liquid raw iron has a temperature of 1,200 °C and has a very high concentration of carbon and other unwanted substances. In the given use case [9], the BOF is charged with 150 tons of liquid raw iron and around 30 tons of scrap metal. The amount of unwanted contents (except carbon) will be bound in the slag by blowing pure oxygen on the mixture of liquid raw iron and scrap metal. The whole mixture is stirred by a bottom gas injection. During the process, the raw iron will be heated up to 1600 °C. The needed energy will be produced by the combustion of the contained carbon in the raw iron. After 20 to 30 min, the process will be stopped based on an analysis of the off-gas

---

[1] https://www.worldsteel.org.

© Springer International Publishing AG 2016
B. Berendt et al. (Eds.): ECML PKDD 2016, Part III, LNAI 9853, pp. 263–266, 2016.
DOI: 10.1007/978-3-319-46131-1_31

composition. The high temperature makes it very expensive and technically challenging [2] to measure the state of the BOF content during the process directly. Usually, there will only be a single measurement at the end of the process. In the given use-case, the quality of the output of the BOF process is described by the temperature, the carbon and phosphorus content of the raw steel and the iron content of the slag at the end of the process. Depending on the difference between the measured and the predefined target value, the process will be repeated until all quality indicators are within the specifications. With only a single measurement at the end of the process, only predictions of the quality indicators can be used to control the process. The prediction of a single quality indicator can be coined as a learning task. After one or multiple refinement steps, casting and rolling, the steel is delivered as coil, plate, sections or bars. The BOF process is the first step of controlling the composition of the steel. The quality of the output has an impact on all further processing steps and the overall quality of the end products. Thus, the quality requirements for the output are usually quite strict. It may happen that up to 20 % of the processes [2] have to be restarted at least once due to quality issues of the output. Hence, the improvement of the prediction is decisive to increase the efficiency and saving resources [7,10].

2   Process Control

There are multiple possibilities to control the outcome of the process directly. Corrective actions have the largest impact if they are executed as early as possible in the process. The most common approach is to precalculate the amount of blown oxygen and heating, cooling and slagging agents based on thermodynamic and metallurgical calculations [4]. The major challenges are presented by multiple sources of variance in the process. Wear and tear, weather, shift work, the unknown state and composition of the used input materials and the high volume of the BOF lead to conditions, that are hard to model with classical metallurgical approaches. Either these models are provided with numerous parameters and are therefore complex to handle or a too small number of parameters limits the reliability of the models. Nevertheless, the resulting predictions and the corrective actions of the operators deliver usually good results already. But even if the optimal metallurgical model would be used, the overall harsh conditions will lead to wear or failure of sensors and other automation equipment. If not handled properly, the reduced data quality and sensor reliability will reduce the quality of every prediction significantly.

3   The BOF Process from a Data Point of View

The data of the BOF process comprise of continuous and event-based data. These data streams are generated by two different data sources (Level 2 and 3 systems [3]). The data streams can be merged and partitioned in an sequence of BOF processes. The event-based data stream contains the results of the composition analysis and the results of the other external measurements, events like the addition of cooling or heating agent and meta-data about the state of the BOF itself. The continuous data stream contains all in-process measurements, like the off-gas composition, the oxygen and cooling water flow and multiple temperatures. A considerable proportion of the 100 raw features are not usable due to not sufficient positioning of the sensors. Until today data analysis only aimed at a better process understanding of the metallurgical experts. Even if learning algorithms were used to model the process no automatic extraction of features and application of learned models have been performed [11].

4 Offline Analysis and Online Application

The major improvement of successful predictions is tuning the features. For the first time, we have constructed multiple new features to describe the BOF process better and monitor the state more directly [7]. The promising results lead to an implementation and application of a prototype at the steel factory itself [9]. The online application of learned models should move beyond merely hand coding the model into a control program. In some factories there are up to 6 BOFs installed. Every BOF will be in a different physical state and the given input materials will be different for every factory. Consequently, every BOF requires a different set of models and different update policies. The manual management and update of the models would require great efforts. The wear and failure of equipment and sensors will lead to concept drifts [1] or a complete loss of raw data and all extracted features. Therefore, multiple models for the multiple sensor settings and an online monitoring and management of these models are needed. To the best of our knowledge, we are the first who developed an online model management module. We implemented a modular and scalable architecture, that is able to connect to multiple legacy systems, store all data efficiently and dynamically extract new features from these raw data, learn new models and apply these models in real-time [8].

5 From Predictions to Control Assistance

The predictions can be used by the operator to evaluate the potential outcome of multiple corrective actions directly. Moving beyond this manual operation on the basis of predictions, we improved the control assistance further. The improvement can be formulated as an multi-objective optimization problem [9]. The predictions are used as a surrogate function for the real value of the quality indicators. Similar to the metallurgical approach, the optimization algorithm uses the amount of oxygen and additions as variables. The costs of the used input materials can be used to calculate the costs of every potential corrective action and can be included into the optimization problem. The optimization problem is solved continuously (1 Hz) and the results can be used by the operator to adapt the amount of oxygen and additions as early as possible in the process.

6   Results and Conclusion

The prediction of the conditions at the BOF end-point have been improved over all development steps and have been constantly better than the classical approach. Nevertheless, the improvement of the prediction quality is only the first step for a successful control and monitoring of BOF processes. The different implementations have been executed successfully and reliable over multiple years. Only with a modular and scalable architecture and implementation it is possible to cope with the given harsh conditions and individual characteristics of every BOF in real-time.

Acknowledgement. This research was supported by SMS-Siemag and in part by the Deutsche Forschungsgemeinschaft (DFG) within the Collaborative Research Center SFB 876 Providing Information by Resource-Constrained Analysis, project B3.

References

1. Baena-García, M., del Campo-Ávila, J., Fidalgo, R., Bifet, A., Gavalda, R., Morales-Bueno, R.: Early drift detection method. In: Fourth International Workshop on Knowledge Discovery from Data Streams, vol. 6, pp. 77–86 (2006)
2. Chukwulebe, B.O., Robertson, K., Grattan, J.: The methods, aims and practices (map) for bof endpoint control. Iron Steel Technol. 4(11), 60–70 (2007)
3. International Electrotechnical Commission, et al.: Iec 62264–1 enterprise-control system integration-part 1: Models and terminology. IEC, Genf (2003)
4. Coudurier, L., Hopkins, D.W., Wilkomirsky, I.: Fundamentals of Metallurgical Processes: International Series on Materials Science and Technology, vol. 27. Elsevier (2013)
5. De Beer, J.: Future technologies for energy-efficient iron and steel making. In: Potential for Industrial Energy-Efficiency Improvement in the Long Term, pp. 93–166. Springer, Netherlands (2000)
6. Fruehan, R.J.: The Making, Shaping, and Treating of Steel: Ironmaking volume, vol. 2. AISE Steel Foundation (1999)
7. Morik, K., Blom, H., Odenthal, H.J., Uebber, N.: Resource-aware steel production through data mining. In: SustKDD Workshop at KDD (2012)
8. Schlüter, J., Odenthal, H.J., Uebber, N., Blom, H., Beckers, T., Morik, K., AG, S.S.: Reliable bof endpoint prediction by novel data-driven modeling. In: AISTech Conference Proceedings. AISTech (2014)
9. Schlüter, J., Odenthal, H.J., Uebber, N., Blom, H., Morik, K.: A novel data-driven prediction model for bof endpoint. In: Association for Iron & Steel Technology Conference, Pittsburgh, USA, vol. 6 (2013)
10. Wolff, B., Lorenz, E., Kramer, O.: Statistical learning for short-term photovoltaic power predictions. In: Lässig, J., Kersting, K., Morik, K. (eds.) Computational Sustainability. SCI, vol. 645, pp. 31–45. Springer, Heidelberg (2016). doi:10.1007/978-3-319-31858-5_3
11. Xu, L., Li, W., Zhang, M., Xu, S., Li, J.: A model of basic oxygen furnace (bof) endpoint prediction based on spectrum information of the furnace flame with support vector machine (svm). Optik-Int. J. Light Electron Optics **122**(7), 594–598 (2011)

Learning from Software Project Histories
Predictive Studies Based on Mining Software Repositories

Verena Honsel[✉], Steffen Herbold, and Jens Grabowski

Institute of Computer Science, University of Göttingen, Göttingen, Germany
{vhonsel,herbold,grabowski}@cs.uni-goettingen.de

Abstract. In software project planning project managers have to keep track of several things simultaneously including the estimation of the consequences of decisions about, e.g., the team constellation. The application of machine learning techniques to predict possible outcomes is a widespread research topic in software engineering. In this paper, we summarize our work in the field of learning from project history.

1 Introduction

The use of software repository data to investigate software evolution and for predictive studies is a wide-spread research topic, that spawned whole conferences like the MSR[1] in the software engineering community. The general approach depicted in Fig. 1 is similar for all application scenarios. First, the different data sources, i.e., repositories, are selected and the required data for the purpose of the investigation are combined. Researchers make use of the version control systems, issue tracking systems, mailing lists and similar systems as repositories. Second, a mental model of the software system is build which is filled with information from the repositories. These two steps can be summarized as data retrieval and modeling. Then, the usage of applicable tools for analysis accomplishes the mining process.

For almost a decade, our research group is interested in the application of theoretical methods to address problems from software repository mining.

- We applied a generalization of Probably Approximately Correct (PAC) learning to optimize metric sets [5].
- We worked on defect prediction in a cross-project context which leads to transfer learning problems [3,4].
- We developed an agent-based simulation model for software processes with automated parameter estimation [7,8].
- We created a model of the developer contribution behavior based on Hidden Markov Models (HMMs) [6].
- We implemented a smart data platform which can combine data collection and analysis through machine learning [9].

---

[1] http://msrconf.org/.

© Springer International Publishing AG 2016
B. Berendt et al. (Eds.): ECML PKDD 2016, Part III, LNAI 9853, pp. 267–270, 2016.
DOI: 10.1007/978-3-319-46131-1_32

Data Retrieval and Modeling   Data Analysis

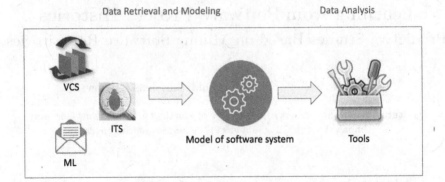

VCS

ITS

ML

Model of software system   Tools

Fig. 1. Mining software repositories (adopted from D'Ambros et al. [2]).

2   Optimization of Metric Sets with Thresholds

Our first contribution is an approach to optimize metric sets for classification using a threshold-based approach. Threshold-based classifications are an important tool for software engineering as they are easy to interpret by both developers and project managers and can be used to, e.g., define coding guidelines. In our work [5], we demonstrate that very few metrics are sufficient to apply threshold based approaches, if the metrics are selected carefully and the thresholds are optimized for the smaller metric set. To achieve this, we use a combination of a brute force search of the potential metric sets combined with a generalized PAC learning approach [1] to determine optimal thresholds.

3   Cross-Project Defect Prediction

Accurate defect prediction can be used to focus the effort of quality assurance and, thereby, ultimately reduce the costs of a project while still ensuring a high quality product. Cross-project defect prediction deals with the problem of using data from outside of the project scope where the prediction is applied to, i.e., across project context. Hence, cross-project defect prediction is a transfer learning challenge. Within our work, we proposed an approach for improving prediction models based on selecting a subset of the training data through relevancy filtering [3]. Using the filtered training data, standard classification models, e.g., Support Vector Machine (SVM), Naïve Bayes, and Logistic Regression were used to predict defects. Moreover, we provided the research community with a tool to benchmark prediction results [4].

4   Software Process Simulation

For the simulation of software processes we consider several facets of software processes over time and their impact on software quality. The general idea for building software process simulation models is to investigate repositories with the aim to find patterns which can describe evolutionary phenomena. For this, we applied statistical learning and machine learning, e.g., for the regression of growth trends. Our approach is agent-based, with the developers as active agents working on the software artifacts as passive agents. With our model, we simulated system growth, bugs lifespan, developer collaboration [7], and software dependencies [8].

5   Developer Contribution Behavior

Developers act in different roles in development projects, e.g., as core developer, maintainer, major developer or minor developer. We use of HMMs to describe involvement dynamics and the workload for the different developer types switching between different states (low, medium, and high) [6]. We take several actions of developers into account to model their workload: the monthly number of commits, bugfixes, bug comments, and mailing list posts. Figure 2 illustrates the learning process. We start with a sequence of monthly activity vectors as observations. We use the threshold learner described in Sect. 2 to classify the observations into low, medium, and high for each metric and with a majority vote for each observation. With the Baum-Welch and the Viterbi algorithm we calculate the transitions between the involvement states (e.g., low involvement to medium involvement) and the emissions for all states (i.e., the workloads). We build a HMMs for each developer of a project, as well as one general model for all developers. This way, we can describe the activity and workload of developers dynamically, which we will use to extend our simulation model to allow for changes in the project team during the simulation.

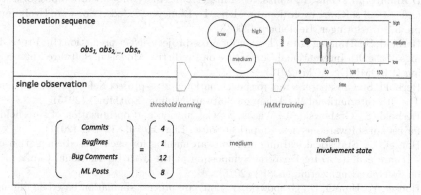

Fig. 2. Learning of developer's involvement state sequence.

6   Mining and Analysis Platform

A current problem in the state of practice of mining software repositories is the replicability and comparability of studies, which is a threat to the external validity of results. To address this, we created the prototype SmartSHARK [9]. SmartSHARK mines data from repositories automatically and provides users with the ability to analyze the data with Apache Spark, a big data framework. Through the MLlib of Apache Spark, we enable users to perform machine learning tasks on the collected data. Current examples on how to use the platform are, e.g., different models for defect prediction as well as a simple approach for effort prediction. SmartSHARK is available as a scalable Cloud platform that will provide a constantly growing amount of project data and, thereby, enable large scale experiments. By sharing Apache Spark jobs, the research become replicable and comparable.

7   Conclusion

Within our research, we show the manifold possibilities to apply machine learning techniques to problems from software engineering, ranging from PAC learning to determine thresholds over the transfer learning challenge cross-project defect prediction to simulation parameter estimation and modeling developers through HMMs, culminating in a smart data platform for software mining. We invite machine learning researchers to use their expertise to advance the state of the art of software engineering.

References

1. Brodag, T., Herbold, S., Waack, S.: A generalized model of pac learning and its applicability. RAIRO - Theor. Inf. Appl. **48**(2), 209–245 (2014)
2. D'Ambros, M., Gall, H., Lanza, M., Pinzger, M.: Analysing software repositories to understand software evolution. In: Mens, T., Demeyer, S. (eds.) Software Evolution, pp. 37–67. Springer, Heidelberg (2008)
3. Herbold, S.: Training data selection for cross-project defect prediction. In: Proceedings of the 9th International Conference on Predictive Models in Software Engineering (PROMISE). ACM (2013)
4. Herbold, S.: Crosspare: a tool for benchmarking cross-project defect predictions. In: The 4th International Workshop on Software Mining (SoftMine) (2015)
5. Herbold, S., Grabowski, J., Waack, S.: Calculation and optimization of thresholds for sets of software metrics. Empirical Softw. Eng. **16**(6), 812–841 (2011)
6. Honsel, V.: Statistical learning and software mining for agent based simulation of software evolution. In: Doctoral Symposium at the 37th International Conference on Software Engineering (ICSE) (2015)
7. Honsel, V., Honsel, D., Grabowski, J.: Software process simulation based on mining software repositories. In: The Third International Workshop on Software Mining (2014)
8. Honsel, V., Honsel, D., Herbold, S., Grabowski, J., Waack, S.: Mining software dependency networks for agent-based simulation of software evolution. In: The Fourth International Workshop on Software Mining (2015)
9. Trautsch, F., Herbold, S., Makedonski, P., Grabowski, J.: Adressing problems with external validity of repository mining studies through a smart data platform. In: 13th International Conference on Mining Software Repositories (MSR) (2016)

Practical Bayesian Inverse Reinforcement Learning for Robot Navigation

Billy Okal[1(✉)] and Kai O. Arras[1,2]

[1] Social Robotics Lab, University of Freiburg, Freiburg, Germany
okal@cs.uni-freiburg.de
[2] Bosch Corporate Research, Robert-Bosch GmbH, Renningen, Germany

Abstract. Inverse reinforcement learning (IRL) provides a concise framework for learning behaviors from human demonstrations; and is highly desired in practical and difficult to specify tasks such as normative robot navigation. However, most existing IRL algorithms are often ladened with practical challenges such as representation mismatch and poor scalability when deployed in real world tasks. Moreover, standard reinforcement learning (RL) representations often do not allow for incorporation of task constraints common for example in robot navigation. In this paper, we present an approach that tackles these challenges in a unified manner and delivers a learning setup that is both practical and scalable. We develop a graph-based spare representation for RL and a scalable IRL algorithm based on sampled trajectories. Experimental evaluation in simulation and from a real deployment in a busy airport demonstrate the strengths of the learning setup over existing approaches.

Keywords: Inverse reinforcement learning · Robot navigation · Representation

1 Introduction

The ability to learn behavior models from demonstrations is a powerful and much soxught after technique in many applications such as robot navigation, autonomous driving, robot manipulation among others. Concretely, a robot's decision making in such tasks is modeled using a Markov decision process (MDP). The reward function of the MDP is assumed to be the "representation of behavior". However, it is often difficult to manually design reward functions that encode desired behaviors; hence IRL formally introduced in [4] is commonly used to recover the reward function using human demonstrations of the task, as done in these examples [1,5,8]. This is because it is often much easier to demonstrate a task than rigorously specify all factors leading to desired behavior. Bayesian inverse reinforcement learning (BIRL) introduced in [7] is in particular suited for such task where a single reward function may not be sufficient, and expert demonstrations are often sub-optimal.

B. Berendt et al. (Eds.): ECML PKDD 2016, Part III, LNAI 9853, pp. 271–274, 2016.
DOI: 10.1007/978-3-319-46131-1_33

However, when using BIRL in practical tasks, we are faced with many challenges such as; very large, continuous and constrained state and action spaces, which make standard BIRL inference algorithms impractical. Constraints common in these tasks include the fact that often not all actions are executable on a robot. For example conventional cars cannot drive sideways, and not all robots can turn on the spot. Thus, naïve representation of such MDP using basic function approximation techniques such as grid discretization easily blow up in space and computational demands. Additionally, such discretization may also discard possible good policies by limiting possible actions available at a state, while still not accounting for the task constraints. We therefore develop a new graph based representation that significantly reduces the size of the state space and encodes task specific constraints directly into the action set of the MDP. Furthermore, standard BIRL inference algorithms such as policy walk (PW) of [7] based on Markov chain Monte Carlo (MCMC) or maximum a posteriori(MAP) approaches, often require iterating over all possible states and actions. This quickly becomes impractical when these spaces get very large as in our case. We thus develop a novel extension of the BIRL algorithm by defining a new likelihood function which does not require iterative over all states and actions, but instead uses samples of trajectories over possibly infinite state and action spaces.

2   Method

Our behavior learning setup consists of two stages; firstly, a flexible data-driven MDP representation called Controller graph (CG) detailed in Sect. 2.1, and secondly, reward learning step using sampled trajectory based BIRL.

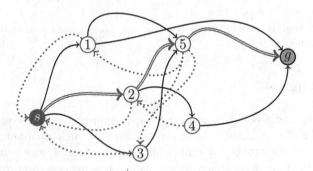

Fig. 1. Conceptual illustration of CG of a stochastic shortest path MDP with 7 states, s and g indicating start and goal states respectively. Policy is shown with double red line. Reverse edges are shown with dotted blue lines.

2.1   Flexible MDP Representation

We use CGs for efficiently representing very large, possibly continuous MDPs with action set already constrained to the target domain, by building upon [3,6].

A CG conceptually illustrated in Fig. 1, is a weighted labeled graph $\mathcal{G} = \langle \mathcal{V}, \mathcal{E}, \mathbf{W} \rangle$ with a vertex set $\mathcal{V} = \{v_i\}$, an edge set $\mathcal{E} = \{(v_i, v_j)_a\}$ and a transition matrix $\mathbf{W}$, such that $\mathcal{V} \subset \mathcal{S}$ and $\mathcal{E} = \{(v_i, v_j)_a \mid w_{i,j} > 0, \forall v_i, v_j \in \mathcal{V}, a \in \mathcal{A}\}$, where $\mathcal{S}$ and $\mathcal{A}$ are state and action spaces respectively of the underlying MDP.

Therefore, vertices are state samples summarised by a vector $\mathbf{x}_i$ and edges are short trajectories or "macro actions" $\mathbf{x}_{i:j}$ between vertices i and j, which we call *local controllers*. These local controllers can be any deterministic controller such as motion primitives [2]; hence directly encode task constraints. The transition weights $w_{i,j}$ can be estimated by simulating such local controllers a number of times and setting the weight as the log ratio of success in reaching the target vertex. The local controllers can also be interpreted as Markov options, in that once selected, a local controller completely defines the policy up to the next vertex. In practice, most robot control tasks already have fine tuned controllers that are almost deterministic.

To build a CG, an empty graph is initialized using samples from the expert demonstrations or alternatively random uniform samples from the state space. Additional vertex samples are then added iteratively by sampling around existing nodes; heuristically trading off exploration and exploitation. This trade off is guided by examining the variance of value of vertices around a local region, and whether or not a vertex is part of the iteration's best policy. In practice, this leads to very few states are shown in [5], where a $10\,\text{m}^2$ 2D area can be effectively represented using under 150 vertices. A grid discretization on the same area with 10cm resolution would already generate 10^4 states.

2.2 BIRL Using Sampled Trajectories

Building upon Ng and Russell [4] we develop an iterative BIRL algorithm that use trajectories randomly sampled from CGs to recover reward functions in very large (possibly infinite) spaces. We define a new likelihood function for BIRL that uses these sampled trajectories as shown in (1).

$$\Pr(\Xi \mid R) = \prod_{\xi^e \in \Xi} \left(\frac{\exp\left(\beta\zeta(\xi^e, R)\right)}{\exp\left(\beta\zeta(\xi^e, R)\right) + \sum_{i=1}^{k} \exp\left(\beta\zeta(\xi_i^g, R)\right)} \right) \quad (1)$$

where Ξ is the set of expert demonstrations, each being trajectory of state-action pairs. $\zeta(\xi, R) = \sum_{(s,a)\in\xi} Q^\pi(s, a)$, with policy π obtained using reward R. ξ_i^g is a trajectory sampled using a candidate policy at iteration i, while k is the current iteration. β is our confidence on the expert taking optimal actions when performing the demonstrations. Therefore, as the reward function improves, we are able to generate sample trajectories of increasing similarity to the expert. This new likelihood function is related to the original one of [7] when each trajectory is interpreted as a single action. The prior remains unchanged as given in [7]. The posterior is given by Bayes rule as $\Pr(R \mid \Xi) = 1/\eta \Pr(\Xi \mid R) \Pr(R)$, with $\eta = \int \Pr(\Xi \mid R) \Pr(R) \, dR$. To infer the reward posterior distribution, the same PW algorithm of [7] can be employed, or alternatively, MAP estimates also yield good results as we found out experimentally. Once the reward function is found, it can be used to generate costmaps for motion planning or directly embedded in planning algorithm objective functions. In our case, we additionally assume inline with [1,8] that the reward function is a linear combination of features of the state and action spaces; then infer the feature weights.

3   Experiments and Results

We conducted extensive experiments in simulation and on a real robot to demonstrate that the setup can indeed learn many complex navigation behaviors with practical constraints. We were able to learn five navigation behaviors useful for robot navigation in a busy airport scenario. These are: polite, sociable and rude navigation behaviors; and additionally, merging with flows and slipstream navigation. The behaviors were evaluated using objective and subjective metrics to assess potential trade offs in normativeness vs functionality. As shown in [5], we found that it is possible to have normative behavior with sacrificing functionality.

4   Conclusions

We have presented an approach that takes IRL algorithms developed in machine learning literature and develops compatible but practical extensions for application in real world robotics. This endeavor highlights the key challenges that need to be addressed to achieve more generalizable approaches. For the future, we are working on formal performance bounds for the new algorithm.

References

1. Henry, P., Vollmer, C., Ferris, B., Fox, D.: Learning to navigate through crowded environments. In: International Conference on Robotics and Automation (ICRA), Anchorage, Alaska (2010)
2. LaValle, S.M.: Planning Algorithms. Cambridge University Press, New York (2006)
3. Neumann, G., Pfeiffer, M., Maass, W.: Efficient continuous-time reinforcementlearning with adaptive state graphs. In: European Conference on Machine Learning (ECML), Warsaw, Poland (2007)
4. Ng, A.Y., Russell, S.J.: Algorithms for inverse reinforcement learning. In: International Conference on Machine Learning (ICML), Haifa, Israel (2000)
5. Okal, B., Arras, K.O.: Learning socially normative robot navigation behaviors using Bayesian inverse reinforcement learning. In: International Conference on Robotics and Automation (ICRA), Stockholm, Sweden (2016)
6. Okal, B., Gilbert, H., Arras, K.O.: Efficient inverse reinforcement learning using adaptive state-graphs. In: Learning from Demonstration: Inverse Optimal Control, Reinforcement Learning and Lifelong Learning Workshop at Robotics: Science and Systems (RSS), Rome, Italy (2015)
7. Ramachandran, D., Amir, E.: Bayesian inverse reinforcement learning. In: International Joint Conference on Artificial Intelligence (IJCAI), Hyderabad, India (2007)
8. Vasquez, D., Okal, B., Arras, K.O.: Inverse reinforcement learning algorithms and features for robot navigation in crowds: an experimental comparison. In: Int. Conf. on Intelligent Robots and Systems (IROS), Chicago, USA (2014)

Machine Learning Challenges
for Single Cell Data

Sofie Van Gassen[1,2(✉)], Tom Dhaene[2], and Yvan Saeys[1,3]

[1] VIB Inflammation Research Center, Ghent, Belgium
`sofie.vangassen@irc.vib-ugent.be`
[2] Department of Information Technology,
Ghent University, iMinds, Ghent, Belgium
[3] Department of Internal Medicine, Ghent University, Ghent, Belgium

Abstract. Recent technological advances in the fields of biology and medicine allow measuring single cells into unprecedented depth. This results in new types of high-throughput datasets that shed new lights on cell development, both in healthy as well as diseased tissues. However, studying these biological processes into greater detail crucially depends on novel computational techniques that efficiently mine single cell data sets. In this paper, we introduce machine learning techniques for single cell data analysis: we summarize the main developments in the field, and highlight a number of interesting new avenues that will likely stimulate the design of new types of machine learning algorithms.

Keywords: Bioinformatics · Single cell analysis · Machine learning

1 Introduction

Single-cell technologies have recently been shown to announce another level of complexity in genomics and medicine. Their high-throughput nature allows investigating millions of cells, allowing to better capture the dynamics of both single cells as well as cell populations. The most established method for high-throughput single cell data analysis is flow cytometry. Flow cytometry measures multiple parameters of cells that flow in a stream through a system of photonic detectors at a rate of 25,000 cells per second. This results in data sets containing millions of cells, and current instruments are able to measure up to 30 cell characteristics simultaneously. These characteristics are indicative of each cell's identity, for example allowing to quantify different types of immune cells in the blood. Recent advances in the field increase the number of characteristics, e.g. mass cytometry (up to 50 characteristics) and imaging flow cytometry (several hundreds of characteristics per cell). In parallel, recent developments from the field of genomics now allow performing transcriptomics experiments at the single cell level. This results in datasets where each cell is described by thousands of parameters, each of which corresponds to a certain gene's activity level.

© Springer International Publishing AG 2016
B. Berendt et al. (Eds.): ECML PKDD 2016, Part III, LNAI 9853, pp. 275–279, 2016.
DOI: 10.1007/978-3-319-46131-1_34

2   Novel Challenges for Machine Learning

From a machine learning perspective, single cell data analysis offers a number of challenges that require adaptation of existing or development of novel algorithms for efficient analysis. These fall broadly in two categories: algorithms for data exploration and algorithms for hypothesis generation. In the next sections, we briefly introduce data visualization and clustering for data exploration, predictive modeling and feature selection for hypothesis generation, and an example of a novel learning problem for single cell data: trajectory inference.

2.1   Structuring and Visualizing Millions of Cells

Each sample in a flow cytometry experiment consists of several thousands up to millions of cells, and to structure and visualize these data two main approaches can be used. Clustering based approaches first group the cells into cell types (clusters of similar cells) and subsequently visualize these grouped cells. This requires scalable and robust algorithms that can deal with millions of data points. An example is FlowSOM [1], shown in Fig. 1A, which clusters cells using a self-organizing map and visualizes the clusters in a minimal spanning tree structure to display information about the cell parameters (features). A second approach is to use dimensionality reduction techniques, which plot all the cells in a reduced (often two-dimensional) space. To keep running times acceptable, these techniques are often combined with subsampling schemes. As an example, Fig. 1B shows a visualization of t-stochastic neighbour embedding (tSNE [2]), which has been shown to perform well on single cell data, combined with color saturation plots to highlight the distribution of the cell characteristics.

To retrieve cell population structure from single cell data, a large range of both existing as well as novel clustering algorithms has been developed, benchmarked in [3]. Remaining challenges include the detection of very rare cell populations (less than 0.1 % of the total number of cells), dealing with heterogeneity (some cell type clusters are very compact while others may be very diffuse), mapping of clusters between different biological samples, and designing robust methods that can deal well with technical and biological variability between samples and noise.

2.2   Predictive Models and Feature Selection

A challenging aspect of single cell data is that every sample (e.g. a patient) is represented by a "bag" of cells, each with its own properties. This can be viewed as a particular type of multi-instance learning, requiring appropriate ways to aggregate the information in the bag of cells. This calls for novel combinations of unsupervised and supervised learning to first extract higher-level characteristics from the bag of cells and subsequently combine these with a predictive model. Benchmarks to compare predictive models have been set up in the FlowCAP challenges [3,4]. Combinations of unsupervised approaches with subsequent feature selection and model building, such as the FloReMi approach [5] have been shown to obtain superior performance in such contexts. Other techniques use the multi-instance data as-is as input for predictive models, e.g. using complex aggregates [6]. In the future, larger datasets for thousands of patients can be envisaged to become available, and each patient sample can contain millions of cells. This again will create a need for scalable and robust methods.

2.3 Unsupervised Trajectory Inference

Classical clustering algorithms that assign each cell to a single cell cluster are not ideal to model gradual processes, such as cellular development. Starting from a mixture of cells in different stages of a developmental process, unsupervised trajectory inference algorithms aim to automatically reconstruct the underlying developmental trajectory that cells are following (Fig. 1C-F). As an example

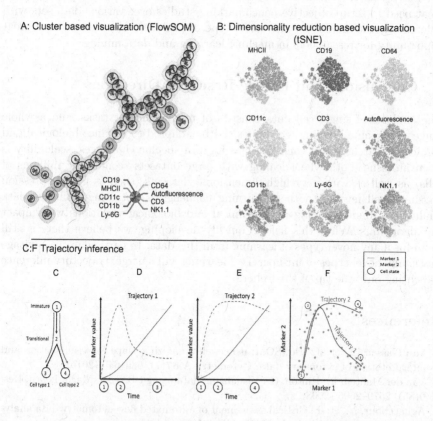

Fig. 1. A: FlowSOM visualization using star charts to denote cell characteristics in each cell type (circle). Background colors corresponds to a meta-level clustering that groups similar clusters. B: tSNE visualization with color saturation mapping to denote the presence of cell characteristics. Cells are clustered based on their characteristics. C-F: trajectory inference to model a branching path from a snapshot single cell data that contains cells in various phases of a developmental continuum. (Color figure online)

consider cells that start in an immature state 1 and subsequently go through an intermediate state 2 where they diverge into two distinct cell types (state 3 and 4, Fig. 1C). Subfigures D and E respectively denote how cell characteristics (marker 1 and 2) change when cells follow this developmental path. Finally, subfigure F shows a scatterplot of the data, showing the two trajectories, with a common path between state 1 and 2, after which they diverge to state 3 and 4. Automatically inferring these trajectories from a single data snapshot, and thus the underlying developmental structure, presents a new type of unsupervised learning with many challenges, as real single cell data may be described by thousands of parameters. Early approaches, such as the Wanderlust algorithm [7], use a K-nearest neighbor graph to model local similarities between cells, and already show promising results for linear (i.e. non-branching) trajectories. However, most realistic scenario's include branching, and sometimes even cyclicity, which still challenges the current state of the art. Furthermore, there is a great need to setup objective benchmarking studies over various data sets with different characteristics for this novel type of problem, which could be a great opportunity for researchers in machine learning and data mining.

3   Conclusions and Future Research Directions

The analysis of single cell data consists of many different tasks, and a whole range of machine learning techniques can be adapted to advance biological and medical research. While each task has its own specific challenges, scalability is often an issue, both when dealing with large datasets (containing millions of cells) as well as with very high-dimensional datasets (such as gene expression on single cell level). Another returning challenge is obtaining sufficient data quality and understanding how technical and biological variations will impact the algorithms. While this field is quickly developing, we believe there is still room for many novel types of learning from this data. In particular, we envisage that many novel types of unsupervised learning will emerge, trajectory inference presenting only the tip of the iceberg.

References

1. Van Gassen, S., et al.: FlowSOM: using self-organizing maps for visualization and interpretation of cytometry data. Cytometry A **87**(7), 636–645 (2015)
2. Van der Maaten, L., Hinton, G.: Visualizing data using t-SNE. J. Mach. Learn. Res. **9**(85), 2579–2605 (2008)
3. Aghaeepour, N., et al.: Critical assessment of automated flow cytometry data analysis techniques. Nat. Methods **10**(5), 445–445 (2013)
4. Aghaeepour, N., et al.: A benchmark for evaluation of algorithms for identification of cellular correlates of clinical outcomes. Cytometry A (2015). doi:10.1002/cyto.a.22732

5. Van Gassen, S., Vens, C., Dhaene, T., Lambrecht, B.N., Saeys, Y.: FloReMi: flow density survival regression using minimal feature redundancy. Cytometry A (2015). doi:10.1002/cyto.a.22734
6. Vens, C., Van Gassen, S., Dhaene, T., Saeys, Y.: Complex aggregates over clusters of elements. In: Davis, J., Ramon, J. (eds.) ILP 2014. LNCS(LNAI), vol. 9046, pp. 181–193. Springer, Heidelberg (2015)
7. Bendall, S.C., et al.: Single-cell trajectory detection uncovers progression and regulatory coordination in human B cell development. Cell 157(3), 714–725 (2014)

Multi-target Classification: Methodology and Practical Case Studies

Mark Last[✉]

Department of Information Systems Engineering, Ben-Gurion University of the Negev,
84105 Beer-Sheva, Israel
mlast@bgu.ac.il

Abstract. Most classification algorithms are aimed at predicting the value or values of a single target (class) attribute. However, some real-world classification tasks involve several targets that need to be predicted simultaneously. The Multi-objective Info-Fuzzy Network (M-IFN) algorithm builds an ordered (oblivious) decision-tree model for a multi-target classification task. After summarizing the principles and the properties of the M-IFN algorithm, this paper reviews three case studies of applying M-IFN to practical problems in industry and science.

Keywords: Multi-target classification · Multi-objective info-fuzzy networks · Information theory · Decision trees

1 Introduction

As indicated in [1, 2], the assumption that a learning task has only one objective is very restrictive. Data instances in many real-world datasets may be simultaneously assigned multiple class labels related to multiple tasks, which may be strongly related, completely unrelated, or just weakly related to each other. Examples include a student's grades in several courses, multiple diagnoses of a given patient, multiple outputs of a software system, multi-trait prediction from genomic data [3], etc. More examples of concurrent learning tasks are discussed in [4].

In [5], we have presented a unified framework for single-objective and multi-objective classification, called an *extended classification task*, which includes the following components:

- $R = (A_1,..., A_n)$ - a non-empty set of *n candidate input features* ($n \geq 1$), where A_i is an attribute i. The values of these attributes (features) can be used to predict the values of *class dimensions* (see below).
- $O = (C_1,..., C_m)$ - a non-empty set of *m class dimensions* ($m \geq 1$). This is a set of tasks (targets) to predict. The *extended classification task* is to build an accurate model (or models) for predicting the values of *all* class dimensions, based on the corresponding *dependency subset* (or subsets) $I \subseteq R$ of selected input features. A special case of this task is *multi-label classification*, which allows multiple labels of *the same dimension* to be assigned to a given instance.

© Springer International Publishing AG 2016
B. Berendt et al. (Eds.): ECML PKDD 2016, Part III, LNAI 9853, pp. 280–283, 2016.
DOI: 10.1007/978-3-319-46131-1_35

Section 2 of this paper outlines the methodology for inducing a multi-target model called Multi-objective Info-Fuzzy Network (M-IFN) and discusses its main characteristics. Section 3 reviews several case studies of applying M-IFN to practical problems in diverse branches of industry and science. Finally, in Sect. 4, we briefly discuss open challenges in multi-target classification.

2 Multi-objective Info-Fuzzy Networks

As indicated in [5], a multi-objective info-fuzzy network (M-IFN) is a multi-target extension of a single-objective info-fuzzy network (IFN). Similar to IFN, M-IFN has a single root node and an "oblivious read-once decision graph" structure, where all nodes of a given layer are labeled by the same feature and each feature is tested at most once along any path. It also has a target layer with a target node for each class label of every target. Every internal M-IFN node is shared among all targets, which makes it an extreme case of a Shared Binary Decision Diagram [6]. This implies that each terminal (leaf) node is connected to at least one target node associated with a value of every target.

Unlike CART [7], C4.5 [8], and EODG [9], the M-IFN construction algorithm has only the growing (top-down) phase, which iteratively chooses predictive features maximizing the decrease in the total conditional entropy of all targets. The top-down construction is pre-pruned by the Likelihood-Ratio Test. The details of M-IFN construction procedure are presented in [5].

In [10], we show the M-IFN algorithm to have the following important properties:

- The average conditional entropy of m targets in an n-input m-dimensional M-IFN model is not greater than the average conditional entropy over m single-target models S_i $(i = 1,..., m)$ based on the same n input features. This inequality is strengthened if the multi-target M-IFN model is built upon more features than the single-target models. Consequently, we may expect that the *average accuracy* of a multi-target M-IFN model in predicting the values of m targets will not be lower, or even will be higher, than the average accuracy of m single-target models using the same set of predictive features.
- If all class dimensions (targets) are either mutually independent or totally dependent on each other, the input features selected by the M-IFN algorithm will minimize the joint conditional entropy of all targets, i.e. will provide the most accurate classification model for all target classes. The case of mutual independence extends the scope of multitask (transfer) learning [1], where all "extra" tasks (targets) are assumed to be related to the main classification target.

3 Case Studies

Our first case study [11] refers to prediction of grape and wine quality in a multi-year dataset provided by Yarden - Golan Heights Winery in Israel. For each grape field in every season, the Winery keeps record of 27 quality parameters (target variables) along with 135 candidate input features. Thus predicting grape and wine quality is clearly a multi-target classification task. We have used M-IFN to identify the most significant predictive factors of grape and wine quality parameters. We have also shown that on average, single-target IFN models are significantly more accurate on this data than C4.5 decision-tree models whereas the M-IFN models are even more accurate than the single-target IFN models. This result agrees with the previously mentioned observation that the average accuracy of a single multi-target M-IFN model is not expected to be worse than the average accuracy of multiple single-target models using the same set of predictive features.

The second case study [12], partially supported by General Motors, deals with predicting the probability and the timing of vehicle failures based on an integrated database of sensor measurements and warranty claims. We have applied the IFN and M-IFN induction algorithms to a dataset of 46,418 records representing periodical battery sensory readings for 21,814 distinct vehicles of a high-end model. The prediction models have been evaluated by the area under ROC (Receiver Operating Characteristics) curves, also known as the Area under Curve, or AUC. Though the IFN and the M-IFN ROC curves for the target attribute *Battery Failure* are nearly identical, the multi-target approach has shown a clear advantage in terms of model comprehensibility as it reduced the total number of prediction rules by 33 %.

The third, more recent case study [13] is aimed at predicting the number and the maximum magnitude of seismic events in the next year based on the seismic events recorded in the same region during the previous years. The predictive features include six seismic indicators commonly used in earthquake prediction literature as well as 20 new features based on the moving annual averages of the number of earthquakes. We have evaluated eight classification algorithms on a catalog of 9,042 earthquake events, which took place between 01/01/1983 and 31/12/2010 in 33 seismic regions of Israel and its neighboring countries. The M-IFN algorithm has clearly shown the best result in terms of the Area under Curve (AUC) criterion, explained by its unique capability to take into account the relationship between two target variables: the total number of earthquakes and the maximum earthquake magnitude during the same year.

4   Conclusions

In this paper, we have presented the M-IFN (Multi-objective Info-Fuzzy Network) algorithm for inducing multi-target classification models. The algorithm's effectiveness and broad applicability have been demonstrated via case studies in three diverse fields: winemaking, predictive maintenance, and seismology. The multi-target classification domain is facing a number of exciting challenges such as semi-supervised learning from a subset of targets, handling delayed target values, and adapting deep learning algorithms for the multi-target classification task.

References

1. Caruana, R.: Multitask learning. Mach. Learn. **28**, 41–75 (1997)
2. Suzuki, E., Gotoh, M., Choki, Y.: Bloomy decision tree for multi-objective classification. In: PKDD 2001 (2001)

3. He, D., Kuhn, D., Parida, L.: Novel applications of multitask learning and multiple output regression to multiple genetic trait prediction. Bioinformatics **32**(12), i37–i43 (2016)
4. Caruana, R.: Multitask learning: a knowledge-based source of inductive bias. In: Proceedings of the 10th International Conference on Machine Learning, ML 1993, University of Massachusetts, Amherst (1993)
5. Last, M.: Multi-objective classification with info-fuzzy networks. In: Boulicaut, J.-F., Esposito, F., Giannotti, F., Pedreschi, D. (eds.) ECML 2004. LNCS (LNAI), vol. 3201, pp. 239–249. Springer, Heidelberg (2004)
6. Bryant, R.E.: Graph-based algorithms for boolean function manipulation. IEEE Trans. Comput. **C-35-8**, 677–691 (1986)
7. Breiman, L., Friedman, J., Olshen, R., Stone, P.: Classification and Regression Trees. Wadsworth, Belmont (1984)
8. Quinlan, J.R.: C4.5: Programs for Machine Learning. Morgan Kaufmann, San Francisco (1993)
9. Kohavi, R., Li, C.-H.: Oblivious decision trees, graphs, and top-down pruning. In: Proceedings of International Joint Conference on Artificial Intelligence (IJCAI) (1995)
10. Last, M., Friedman, M.: Black-box testing with info-fuzzy networks. In: Artificial Intelligence Methods in Software Testing, pp. 21–50. World Scientific, Singapore (2004)
11. Last, M., Elnekave, S., Naor, A., Shonfeld, V.: Predicting wine quality from agricultural data with single-objective and multi-objective data mining algorithms. In: Recent Advances on Mining of Enterprise Data: Algorithms and Applications, pp. 323–365. World Scientific, Singapore (2007)
12. Last, M., Sinaiski, A., Subramania, H.S.: Condition-based maintenance with multi-target classification models. New Gener. Comput. **29**(3), 245–260 (2011). Special Issue on Hybrid and Ensemble Methods in Machine Learning
13. Last, M., Rabinowitz, N., Leonard, G.: Predicting the maximum earthquake magnitude from seismic data in Israel and its neighboring countries. PLoS ONE **11**(1), e0146101 (2016)

Query Log Mining for Inferring
User Tasks and Needs

Rishabh Mehrotra[✉] and Emine Yilmaz

Department of Computer Science, University College London, London, UK
r.mehrotra@cs.ucl.ac.uk, emine.yilmaz@ucl.ac.uk

Abstract. Search behavior, and information seeking behavior more generally, is often motivated by tasks that prompt search processes that are often lengthy, iterative, and intermittent, and are characterized by distinct stages, shifting goals and multitasking. Current search systems do not provide adequate support for users tackling complex tasks due to which the cognitive burden of keeping track of such tasks is placed on the searcher. In this note, we summarize our recent efforts towards extracting search tasks from search logs. Based on recent advancements in Bayesian Nonparametrics and distributional semantics, we propose novel algorithms to extract task and subtasks from a query collection. The models discussed can inform the design of the next generation of task-based search systems that leverage user's task behavior for better support and personalization.

1 Introduction

Search behavior, and more generally, information-seeking behavior is often motivated by tasks that prompt search processes that are often lengthy, iterative, intermittent, and characterized by distinct stages, shifting goals and multitasking. Current search engines do not provide adequate support for tackling complex tasks (e.g. planning a trip, surveying a topic), due to which the cognitive burden of keeping track of such tasks and completing them is placed on the searcher. Ideally, a search engine should be able to decipher the underlying reason that led the user to submit a query (i.e., the actual task that caused the query to be issued), and be able to guide the user to achieve their task by incorporating this knowledge about the actual information need.

In this research, we hypothesize that developing a comprehensive understanding of user's tasks would help in providing better support and recommendations to users based on their contextual information and as a result, help users accomplish the task. As part of the proposed research, we consider the challenge of extracting tasks from a given collection of search log data and present task extraction techniques which rely on recent advancements in bayesian non parametrics and word embeddings. We evaluate the performance of such techniques using a number of techniques based on crowdsourced judgments as well as labelled ground truth data.

© Springer International Publishing AG 2016
B. Berendt et al. (Eds.): ECML PKDD 2016, Part III, LNAI 9853, pp. 284–288, 2016.
DOI: 10.1007/978-3-319-46131-1_36

2 Task Based Information Retrieval

Our efforts at developing task based retrieval systems have focussed around three major themes, (i) understanding searcher's behaviors, (ii) developing task extraction techniques and (iii) showing the benefits of task information via improved personalization. We next describe each of them in detail.

2.1 Understanding Searcher's Task Behavior

While a major share of prior work have considered search sessions as the focal unit of analysis for seeking behavioral insights [7–9], search tasks are emerging as a competing perspective in this space. In a recent work [1], we quantify multi-tasking behavior of web search users and show that over 50 % of search sessions have more than 2 tasks. Further, we provide a method to categorize users into focused, multi-taskers or supertaskers depending on their level of task-multiplicity and show that the search effort expended by these users varies across the groups. Additionally, in a follow up work [3] we relate user's multitasking propensities to tasks and topics. Specifically, we analyze user-disposition, topic and user-interest level heterogeneities that are prevalent in search task behavior. We find that not only do users have varying propensities to multi-task, they also search for distinct topics across single-task and multi-task sessions. The findings from our analysis provide useful insights about task-multiplicity in an online search environment and hold potential value for search engines that wish to personalize and support search experiences of users based on their task behavior.

2.2 Extracting Hierarchies

An important first step in developing task based systems is task extraction. In a recently published work [4], we considered the challenge of extracting hierarchies of search tasks and their associated subtasks from a given search log given just the log data without the need of any manual annotation of any sort. We present an efficient Bayesian nonparametric model for discovering task hierarchies and propose a tree based bayesian hierarchical task construction algorithm to discover this rich hierarchical structure embedded within search logs. Our model organises the queries into a nested hierarchy T of tasks/subtasks, with all queries in one node at the root and singleton queries at the leaves. We interpret a tree (T) as a mixture of partitions over those group of queries (Q). We define the probability of a group of such queries as:

$$p(Q|T) = \sum_{\phi} p(\phi(t))p(Q|\phi(t)) \tag{1}$$

where $p(\phi(T))$ is the mixing proportion of partition $\phi(T)$, and $p(Q|\phi(t))$ is the probability of the group of queries Q given a partitioning by $\phi(T)$. In general the number of partitions consistent with T can be exponentially large. To make computations tractable, we define the mixture model in such a way that $p(Q|\phi(t))$ can be computed using dynamic programming over T:

$$p(Q|T) = \pi_T f(Q) + (1 - \pi_t) \prod_{T_i \in ch(T)} p(leaves(T_i)|T_i) \qquad (2)$$

In the beginning, each query is regarded as a tree on its own. For each step, the algorithm selects two trees T_i and T_j and merges them into a new tree T_m. Unlike binary hierarchical clustering, we allow three possible merging operations: **(i) Join**: $T_m = \{T_i, T_j\}$, such that the tree T_m has two children now; **(ii) Absorb**: $T_m = \{children(T_i) \cup T_j\}$, i.e., the children of one tree gets absorbed into the other tree forming an absorbed tree with >2 children; and **(iii) Collapse**: $T_m = \{children(T_i) \cup children(T_j)\}$, all the children of both the sub-tree get combined together at the same level. Such a setting allows each task to be composed of an arbitrary number of sub-tasks without restricting tasks to contain only binary subtasks.

The tree is built in a bottom-up greedy agglomerative fashion, and the algorithm finishes when just one tree remains. At each iteration a pair of trees in the forest F is chosen to be merged by considering the pair and type of merger that yields the largest **Bayes factor improvement** over the current model. Further details of the work are available in our research paper [4].

2.3 Decomposing Complex Search Tasks

Quite often, search tasks (e.g. planing a trip) are complex and conceptually decompose into a set of sub-tasks (e.g. booking flights, finding places of interest etc.), each of which warrants the user to further issue multiple queries to solve. Given a collection of *on-task* queries (extracted using standard task extraction algorithm), we proposed a distance dependent Chinese Restaurant process model to extract these sub-tasks from a given collection of *on-task* queries.

In our sub-task extraction problem, each task is associated with a dd-CRP and its tables are embellished with IID draws from a base distribution over mixture component parameters. Let z_i denote the ith query assignment, the index of the query with whom the ith query is linked. Let d_{ij} denote the distance measurement between queries i and j, let D denote the set of all distance measurements between queries, and let f be a decay function. The distance dependent CRP independently draws the query assignments to sub-tasks conditioned on the distance measurements,

$$p(z_i = j|D, \alpha) \propto \begin{cases} f(d_{ij}) & \text{if } j \neq i \\ \alpha & \text{if } j = i \end{cases}$$

Here, d_{ij} is an externally specified distance between queries i and j, and α determines the probability that a customer links to themselves rather than another customer. Given a decay function f, distances between queries D, scaling parameter α, and an exchangeable Dirichlet distribution with parameter λ, N M-word queries are drawn as follows,

1. For $i \in [1, N]$, draw $z_i \sim dist - CRP(\alpha, f, D)$.
2. For $i \in [1, N]$,
   (a) If $z_i \notin R^*_{q_{1:N}}$, set the parameter for the ith query to $\theta_i = \theta_{q_i}$. Otherwise draw the parameter from the base distribution, $\theta_i \sim Dirichlet(\lambda)$.
   (b) Draw the ith query terms, $w_i \sim Mult(M, \theta_i)$.

Further details of the work are available in our research paper [2].

2.4 Task Based Personalization

In order to demonstrate the usefulness of a task based system, in recent work [5,6] we presented a novel approach to couple user's topical interest information with their search task information & their term usage behavior to learn a joint user representation technique. We demonstrated that coupling user's task information with their topical interests indeed helps us build better user models. We show through extensive experimentation that our task based method outperforms existing query term based and topical interest based user representation methods. By evaluating the quality of our approach on a variety of tasks for personalisation including collaborative query recommendation, cluster based recommendation and user cohort analysis, we demonstrate that the proposed methods result in better user profiles.

3 Conclusion

In this note, we offered insights about the shift in focus from sessions to tasks and presented a brief summary of our recent work aimed at extracting tasks from search logs. We believe that the task-based personalization and recommendation has the potential to shape the future of user interaction systems for the upcoming era of intelligent Web, and there is much to be done on this emerging topic. Some of the key problems to investigate in the future include using task based systems for improved recommendations and better predicting contextual needs of users for proactive recommendations.

References

1. Mehrotra, R., Bhattacharya, P., Yilmaz, E.: Characterizing users' multi-tasking behavior in web search. In: Proceedings of the ACM on Conference on Human Information Interaction and Retrieval (2016)
2. Mehrotra, R., Bhattacharya, P., Yilmaz, E.: Deconstructing complex search tasks: a bayesian nonparametric approach for extracting sub-tasks. In: Proceedings of NAACL-HLT, pp. 599–605 (2016)
3. Mehrotra, R., Bhattacharya, P., Yilmaz, E.: Sessions; tasks & topics - uncovering behavioral heterogeneities in online search behavior. In: Proceedings of the 39th International ACM SIGIR Conference on Research and Development in Information Retrieval. ACM (2016)

4. Mehrotra, R., Yilmaz, E.: Towards hierarchies of search tasks & subtasks. In: WWW (2015)
5. Mehrotra, R., Yilmaz, E.: Terms, topics & tasks: enhanced user modelling for better personalization. In: Proceedings of the International Conference on the Theory of Information Retrieval, pp. 131–140. ACM (2015)
6. Mehrotra, R., Yilmaz, E., Verma, M.: Task-based user modelling for personalization via probabilistic matrix factorization. In: RecSys Posters (2014)
7. Odijk, D., White, R.W., Hassan Awadallah, A., Dumais, S.T.: Struggling and success in web search. In: CIKM (2015)
8. White, R.W., Bennett, P.N., Dumais, S.T.: Predicting short-term interests using activity-based search context. In: CIKM (2010)
9. Xiang, B., Jiang, D., Pei, J., Sun, X., Chen, E., Li, H.: Context-aware ranking in web search. In: SIGIR (2010)

Data Mining Meets HCI: Data and Visual Analytics of Frequent Patterns

Carson K. Leung[1(✉)], Christopher L. Carmichael[1], Yaroslav Hayduk[1,2], Fan Jiang[1], Vadim V. Kononov[1], and Adam G.M. Pazdor[1]

[1] University of Manitoba, Winnipeg, MB, Canada
kleung@cs.umanitoba.ca
[2] Université de Neuchâtel, Neuchâtel, Switzerland

Abstract. As a popular data mining tasks, frequent pattern mining discovers implicit, previously unknown and potentially useful knowledge in the form of sets of frequently co-occurring items or events. Many existing data mining algorithms return to users with long textual lists of frequent patterns, which may not be easily comprehensible. As a picture is worth a thousand words, having a visual means for humans to interact with computers would be beneficial. This is when human-computer interaction (HCI) research meets data mining research. In particular, the popular HCI task of *data and result visualization* could help data miners to visualize the original data and to analyze the mined results (in the form of frequent patterns). In this paper, we present a few systems for data and visual analytics of frequent patterns, which integrate (i) data analytics and mining with (ii) data and result visualization.

1 Introduction and Related Works

Over the past two decades, many *frequent pattern mining* algorithms [1] have been developed for *data analytics* [3]. These algorithms usually produce long textual lists of frequent patterns, which may not be easily comprehensible. As a picture is worth a thousand words, a visual representation (i) matches the power of the human visual and cognitive system, and (ii) enables human to interact with computers effectively. This is when *human-computer interaction* (*HCI*) meets data mining. Specifically, HCI researches the design and usage of computer technology, with a focus on the interfaces between humans and computers. As a popular HCI tasks, *data and result visualization* could help data miners or data analysts to (i) visualize the original data and (ii) analyze the mined results (i.e., frequent patterns). This leads to *visual analytics* [2], which is the science of analytical reasoning supported by interactive visual interfaces.

Over the past two decades, several visualizers have been developed. Many of them (e.g., VisDB [6]) were designed for visualizing data only. Some were built for visualizing results of data mining tasks such as cluster analysis or anomaly detection. In the next section, we present and summarize some visualizers that have been developed for visual analytics of frequent patterns, which integrate (i) data analytics and mining with (ii) data and result visualization. Note that a

B. Berendt et al. (Eds.): ECML PKDD 2016, Part III, LNAI 9853, pp. 289–293, 2016.
DOI: 10.1007/978-3-319-46131-1_37

challenge of visualizing frequent patterns is the ability to show the patterns and their prefix-extension relationships (e.g., $\{a\}$ and $\{a, b\}$ are prefixes of $\{a, b, c\}$, whereas $\{a, b, c, d\}$ and $\{a, b, c, e\}$ are extensions of $\{a, b, c\}$). Another challenge is the ability to show the frequency of each pattern.

2   Frequent Pattern Visualizers

FIsViz [8] visualizes frequent k-itemsets (i.e., patterns consisting of k items) as polylines connecting k nodes in a two-dimensional space with (x, y)-coordinates, in which domain items are listed on the x-axis and frequency values are indicated by the y-axis. The x-locations of all nodes in the polyline indicate the domain items contained in a frequent pattern Z, and the y-location of the rightmost node of a polyline for Z indicates the frequency of Z. Hence, prefix-extension relationships can be observed by traversing along the polylines. See Fig. 1(a). In addition, to facilitate exploration of data and mining results, FIsViz also provides users with interactive detail-on-demand features. When the mouse hooves on a polyline connecting two nodes u and v, FIsViz shows a list of itemsets containing both u and v. Similarly, when the mouse hovers over a node, FIsViz shows a list of all patterns contained in all polylines starting or ending at this node.

As polylines in FIsViz can be bent and crossed over each other, it may not be easy to distinguish one polyline from another. To solve this problem, WiFIsViz [9] and FpVAT [7] were designed. As shown in Fig. 1(b), **WiFIsViz** uses two half-screens to visualize frequent patterns. Both half-screens are wiring-type diagrams (i.e., orthogonal graphs), which represent frequent patterns as horizontal lines connecting k nodes in a two-dimensional space (where the x-axis lists all the domain items). The left half-screen provides the frequency information by using the y-location of the horizontal line to indicate the frequency of the frequent pattern. The right half-screen lists all frequent patterns in the form of a trie.

FpVAT [7] also uses wiring-type diagrams to visualize frequent patterns. However, FpVAT shows all the frequent patterns and their frequencies on the same full-screen. See Fig. 1(c).

The above three visualizers show *all* frequent patterns. When handling very large datasets, the number of frequent patterns to be displayed can be huge due to pattern explosion. To improve this situation, **CloseViz** [5] extends WiFIsViz and FpVAT by providing users with explicit and easily-visible information among the *closed* patterns, which greatly reduces the number of displayed patterns without losing any frequency information. Note that a frequent pattern Z is *closed* if no superset of Z has the same frequency as Z. As shown in Fig. 1(d), CloseViz represents closed patterns as horizontal lines in a two-dimensional graph.

The above four visualizers show frequent patterns from a single database instance. However, there are situations in which users may be interested in differences between the results returned from two database instances. For example, a store manager may be interested in finding out the difference between popular sets of merchandise items sold in the summer and in the winter in order to detect the (temporal) changes in frequencies of the mined frequent patterns as

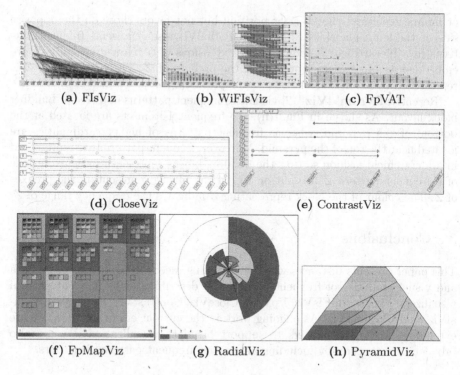

Fig. 1. Frequent pattern visualizers

well as their trends from one database instance to another. Similarly, a regional manager may want to find out the (spatial) difference between the popular sets of merchandise items sold in two different locations. To handle with these real-life situations, **ContrastViz** [4] extends WiFIsViz and FpVAT by helping users to visually contrast two collections of frequent patterns. As shown in Fig. 1(e), ContrastViz visualizes and analyzes all the frequent patterns, their frequencies, as well as changes in frequencies.

Instead of polylines or wiring-type diagrams (i.e., orthogonal graphs), Fp-MapViz [11], RadialViz [10] and PyramidViz [12] use alternative design with emphasis on showing the prefix-extension relationships among the frequent patterns. For example, inspired by the tree map representation of hierarchical information, **FpMapViz** represents frequent patterns as squares in a hierarchical fashion so that extensions of a frequent pattern Z are embedded within squares representing the prefixes of Z. The colour of the square representing Z indicates the frequency range of Z. See Fig. 1(f).

As shown in Fig. 1(g), **RadialViz** [10] also visualizes frequent patterns but in a radial layout, which leads to a benefit of being *orientation-free*. As such, the legibility of the represented frequent patterns is not be impacted by the orientation. Hence, RadialViz is ideal for the collaborative environment (cf. traditional two-dimensional rectangular space, which favors the viewer who visualizes data or mining results at the up-right position but not favors those on the opposite side or the left/right sides). Moreover, RadialViz also represents frequent patterns in a hierarchical fashion so that extensions of a frequent pattern Z are embedded within sectors representing the prefixes of Z. The frequency of Z is represented by the radius of the sector representing Z.

Recently, **PyramidViz** [12] visualizes frequent patterns in a tree or building block layout. As shown in Fig. 1(h), the frequent 1-itemsets are located at the bottom of the pyramid, whereas frequent patterns of higher cardinalities are located near the top of the pyramid. Moreover, frequent patterns are represented in a hierarchical fashion so that the building blocks representing the extensions of a frequent pattern Z are put on top of the blocks representing the prefixes of Z. The colour of the block representing Z indicates the frequency range of Z.

3   Conclusions

This paper presents instances when data mining meets HCI, with focus on data and visual analytics of frequent patterns by describing eight frequent pattern visualizers: FIsViz, WiFIsViz, FpVAT, CloseViz, ContrastViz, FpMapViz, RadialViz, and PyramidViz. As ongoing work in the current era of big data, we are extending existing visualizers to support big data visualization. We are also broadening our study by including alternative frequent pattern visualizers.

Acknowledgement. This project is partially supported by NSERC (Canada).

References

1. Agrawal, R., Srikant, R.: Fast algorithms for mining association rules in large databases. In: VLDB 1994, pp. 487–499 (1994)
2. Andrienko, N.V., Andrienko, G.L., Fuchs, G., Jankowski, P.: Visual analytics methodology for scalable and privacy-respectful discovery of place semantics from episodic mobility data. In: Bifet, A., May, M., Zadrozny, B., Gavalda, R., Pedreschi, D., Bonchi, F., Cardoso, J., Spiliopoulou, M. (eds.) ECML PKDD 2015 Part III. LNCS, vol. 9286, pp. 254–258. Springer, Heidelberg (2015)
3. Börner, M., Rhode, W., Ruhe, T., Morik, K.: Discovering neutrinos through data analytics. In: Bifet, A., May, M., Zadrozny, B., Gavalda, R., Pedreschi, D., Bonchi, F., Cardoso, J., Spiliopoulou, M. (eds.) ECML PKDD 2015. LNCS, vol. 9286, pp. 208–212. Springer, Heidelberg (2015)
4. Carmichael, C.L., Hayduk, Y., Leung, C.K.: Visually contrast two collections of frequent patterns. In: IEEE ICDM Workshops 2011, pp. 1128–1135 (2011)
5. Carmichael, C.L., Leung, C.K.: CloseViz: visualizing useful patterns. In: ACM SIGKDD Workshop on UP 2010, pp. 17–26 (2010)
6. Keim, D.A., Kriegel, H.-P.: Visualization techniques for mining large databases: a comparison. IEEE TKDE **8**(6), 923–938 (1996)
7. Leung, C.K., Carmichael, C.L.: FpVAT: a visual analytic tool for supporting frequent pattern mining. ACM SIGKDD Explor. **11**(2), 39–48 (2009)

8. Leung, C.K., Irani, P.P., Carmichael, C.L.: FIsViz: a frequent itemset visualizer. In: Washio, T., Suzuki, E., Ting, K.M., Inokuchi, A. (eds.) PAKDD 2008. LNCS (LNAI), vol. 5012, pp. 644–652. Springer, Heidelberg (2008)
9. Leung, C.K., Irani, P.P., Carmichael, C.L.: WiFIsViz: effective visualization of frequent itemsets. In: IEEE ICDM, pp. 875–880 (2008)
10. Leung, C.K., Jiang, F.: RadialViz: an orientation-free frequent pattern visualizer. In: Tan, P.-N., Chawla, S., Ho, C.K., Bailey, J. (eds.) PAKDD 2012, Part II. LNCS, vol. 7302, pp. 322–334. Springer, Heidelberg (2012)
11. Leung, C.K., Jiang, F., Irani, P.P.: FpMapViz: a space-filling visualization for frequent patterns. In: IEEE ICDM Workshops, pp. 804–811 (2011)
12. Leung, C.K., Kononov, V.V., Pazdor, A.G.M.: PyramidViz: visual analytics and big data visualization of frequent patterns. In: IEEE DASC/DataCom 2016 (2016)

Machine Learning for Crowdsourced Spatial Data

Musfira Jilani[1]([⊠]), Padraig Corcoran[2], and Michela Bertolotto[1]

[1] School of Computer Science, University College Dublin, Dublin, Ireland
musfira.jilani@ucdconnect.ie, michela.bertolotto@ucd.ie
[2] School of Computer Science, Cardiff University, Cardiff, UK
CorcoranP@cardiff.ac.uk

Abstract. Recent years have seen a significant increase in the number of applications requiring accurate and up-to-date spatial data. In this context crowdsourced maps such as OpenStreetMap (OSM) have the potential to provide a free and timely representation of our world. However, one factor that negatively influences the proliferation of these maps is the uncertainty about their data quality. This paper presents structured and unstructured machine learning methods to automatically assess and improve the semantic quality of streets in the OSM database.

Keywords: Probabilistic graphical modelling · Crowdsourced spatial data · Street networks · Semantics

1 Introduction

We live in an age where the demand for accurate and up-to-date spatial data has never been greater. However, obtaining and maintaining such spatial databases is a challenging and expensive task. In this context, crowdsourced maps such as OpenStreetMap (OSM)[1] can be a viable solution for obtaining a free and up-to-date representation of our world. While an extensive number of applications have been developed around OSM, concerns exist regarding the quality of OSM data. The predominant method for assessing OSM data quality is based on comparing the OSM data with some form of authoritative maps such as the Ordnance Survey UK [2], Google Maps, etc. However, we argue that this process of comparing a crowdsourced (heterogenous) database with authoritative maps is ineffective. Instead we propose the use of machine learning techniques for assessing and possibly improving the data quality of crowdsourced maps without referencing to external repositories.

Specifically, in this paper we focus on the semantic type quality of streets in the OSM where semantic type refers to the class of a street such as motorway, pedestrian, etc. We hypothesize that the semantic types of streets are a

---

[1] The OSM project was started in 2004 with a goal of creating a free and editable map of the entire world. www.openstreetmap.org.

© Springer International Publishing AG 2016
B. Berendt et al. (Eds.): ECML PKDD 2016, Part III, LNAI 9853, pp. 294–297, 2016.
DOI: 10.1007/978-3-319-46131-1_38

function of their geometrical and topological features and develop structured and unstructured machine learning models that can learn the semantic types of streets given such features. Interestingly, the structured learning models can also exploit the inherent spatial relationships within a street network.

2   Methodology

2.1   Data Representation

Appropriate data representation is a fundamental step toward useful knowledge discovery. Therefore, as a first step a novel multi-granular graph-based street network representation system is developed. All streets having same name and same semantic type correspond to a single node in a multi-granular graph. Such a representation makes the various features of a street explicit as opposed to implicit. More details of the multi-granular representation system can be found in [3].

2.2   Feature Extraction

Several topological and geometrical features of streets were extracted using the multi-granular street network representation obtained above. These include length, linearity, number of dead-ends, number of intersections, semantic types of adjacent streets (using a BoW model), node degree, and betweenness centrality.

2.3   Unstructured Learning

Next, we develop an unstructured (or classical) supervised machine learning model to learn the various semantics types of streets in the OSM database. The development of this model involves assessing the performance of the commonly used machine learning classifiers such as naive bayes, SVM, neural networks, and random forests in terms of their generalization performance on test data. More details on the implementation of the unstructured learning of the problem can be found in [4].

2.4   Structured Learning

A street network is a structured input as it consists of several streets, where not only the streets themselves contain information such as geometry, but also the way in which the streets are connected to each other is important. For such a structured input, we obtain a structured output of semantic types of streets over all the streets in the network. We exploit the Conditional Random Field (CRF) framework for performing structured prediction. The CRF framework allows us to leverage prior knowledge available to us in the form of crowdsourced semantics, the geometrical and topological features of individual streets, and the contextual (structural) relationships between various streets into a single unified model.

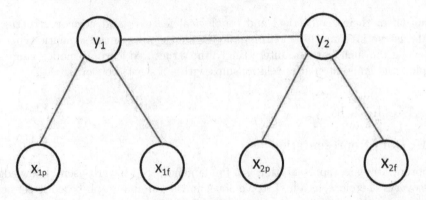

Fig. 1. Street network represented as a graphical model. x are the observed variables corresponding to the streets in the network and y are the labelling we want to infer. The green lines correspond to the unary potentials in the model and the red line to the pairwise potential.

Suppose we have a street network consisting of N streets $\mathbf{x} = \{x_1, x_2, \ldots, x_N\} \in \mathcal{X}$ and our goal is to predict the semantic type labellings $\mathbf{y} = \{y_1, y_2, \ldots, y_N\} \in \mathcal{Y}$ for these streets. Figure 1 shows our representation of such a street network as a graphical model where x_{ip} corresponds to the initial crowd sourced labels or priors and x_{if} corresponds to the geometric and topological features. Toward the goal of jointly learning the semantic type labelling $\mathbf{y}$, our model maximizes the conditional probability of $\mathbf{y}$ given $\mathbf{x}$ [6]:

$$\hat{\mathbf{y}} = \arg\max_{\mathbf{y} \in \mathcal{Y}} P(\mathbf{y}|\mathbf{x}; \mathbf{w}) \tag{1}$$

We use a max-margin approach for determining the model parameters w and a fusion moves approach for inferring the street labellings. More details on the structured learning of the problem can be found in [1,5].

3   Results and Discussion

We trained and tested our models on two non-overalapping regions from OSM London database. All 19 popular semantic types of streets used in OSM database for classifying a street were considered. An overall classification accuracy of 55.95 % was obtained using the unstructured learning model (random forest). This accuracy increased to 84.75 % when structured learning framework was used. Clearly, and naturally the structured learning framework outperforms the unstructured learning performance as it exploits the inherent structure in street networks. To the best of our knowledge, this is the first time that a structured learning framework has been used in the context of crowdsourced spatial data.

In this work, we considered all the 19 popular semantic types of streets used for classifying a street network. However, such a classification of street network is too fine-grained when compared with the commonly used and understood street network classifications where a street network is usually classified into 4–10 semantic types. In future we propose the development of a multi-layer conditional random field based model for simultaneously learning both the fine-grained (19) and coarse-grained (4–10) semantic types of streets. In addition, the models developed in this paper will also be extended to other map objects such as buildings, Points of Interests (PoIs), etc.

4   Dual Submissions

The work presented in this paper is a summary of the work already published at the following venues:

1. 23rd ACM SIGSPATIAL Conference, USA, 2015
2. 22nd ACM SIGSPATIAL Conference, USA, 2014
3. Intelligent Systems, Technologies, and Applications, Springer, 2016
4. UL-NUIG Research Day, 2016.
5. Related version submitted to the Indian Workshop on Machine Learning, 2016.

Acknowledgments. This work is supported by the Irish Research Council through the Embark Postgraduate Scholarship Scheme 2012.

References

1. Corcoran, P., Jilani, M., Mooney, P., Bertolotto, M.: Inferring semantics from geometry: the case of street networks. In: Proceedings of the 23rd ACM SIGSPATIAL International Conference on Advances in GIS. ACM (2015)
2. Haklay, M.: How good is volunteered geographical information? a comparative study of openstreetmap and ordnance survey datasets. Environ. Plann. **37**(4), 682–703 (2010)
3. Jilani, M., Corcoran, P., Bertolotto, M.: Multi-granular street network representation towards quality assessment of openstreetmap data. In: Proceedings of the Sixth ACM SIGSPATIAL International Workshop on Computational Transportation Science. ACM (2013)
4. Jilani, M., Corcoran, P., Bertolotto, M.: Automated highway tag assessment of openstreetmap road networks. In: Proceedings of the 22nd ACM SIGSPATIAL International Conference on Advances in GIS. ACM (2014)
5. Jilani, M., Corcoran, P., Bertolotto, M.: Probabilistic graphical modelling for semantic labelling of crowdsourced map data. In: Jilani, M., Corcoran, P., Bertolotto, M. (eds.) Intelligent Systems Technologies and Applications. AISC, vol. 385, pp. 213–224. Springer, Heidelberg (2016)
6. Koller, D., Friedman, N.: Probabilistic Graphical Models: Principles and Techniques. MIT press, Cambridge (2009)

Local Exceptionality Detection on Social Interaction Networks

Martin Atzmueller[(✉)]

Research Center for Information System Design (ITeG), University of Kassel,
Wilhelmshöher Allee 73, 34121 Kassel, Germany
atzmueller@cs.uni-kassel.de

Abstract. Local exceptionality detection on social interaction networks
includes the analysis of resources created by humans (e. g., social media)
as well as those generated by sensor devices in the context of (complex)
interactions. This paper provides a structured overview on a line of work
comprising a set of papers that focus on data-driven exploration and
modeling in the context of social network analysis, community detection
and pattern mining.

Keywords: Local exceptionality detection · Exceptional models ·
Subgroup discovery · Community detection · Social network analysis ·
Social interaction networks · Social media

1 Introduction

In ubiquitous and social environments, a variety of heterogenous multi-relational
data is generated, e. g., by sensors and social media. Then, a set of complex networks can be derived, in the form of social interaction networks [2], capturing
distinct facets of the interaction space [19]. In that context, *local exceptionality
detection* – based on *subgroup discovery* and *exceptional model mining* – provides flexible approaches for data exploration, assessment, and the detection of
unexpected and interesting phenomena.

Subgroup discovery [3,15,23] is an approach for discovering interesting subgroups – as an instance of *local pattern detection* [20]. The interestingness is
usually defined by a certain property of interest formalized by a quality function. In the simplest case, a binary target variable is considered, where the share
in a subgroup can be compared to the share in the dataset in order to detect
(exceptional) deviations. More complex target concepts consider sets of target
variables. In particular, *exceptional model mining* [3,12] focuses on more complex
quality functions. In the context of ubiquitous data and social media, interesting
target concepts are given, e. g., by densely connected graph structures (communities) [5], unexpected spatio-semantic distributions [8], or exceptional matches
between online-offline relations [13] for behavioral characterization.

© Springer International Publishing AG 2016
B. Berendt et al. (Eds.): ECML PKDD 2016, Part III, LNAI 9853, pp. 298–302, 2016.
DOI: 10.1007/978-3-319-46131-1_39

This paper focuses on formalizations and applications of subgroup discovery and exceptional model mining in the context of social interaction networks. We summarize recent work on community detection, behavior characterization and spatio-temporal analysis, and efficient implementation (comprising the papers [1,2,4–8,10,13]). In that way, we provide a compact and structured overview of recent scientific advances in this field, covering specific methods and their applications for analyzing social interactions.

2   Methods

Social interaction networks [2,17,18] focus on user-related social networks in social media capturing social relations inherent in social interactions, social activities and other social phenomena which act as proxies for social user-relatedness. Therefore, according to the categorization of Wassermann and Faust [22, p. 37 ff.] social interaction networks focus on *interaction* relations between *people* as the corresponding actors. This also includes interaction data from sensors and mobile devices, as long as the data is created by real users [1,2].

In such contexts, exploratory data analysis is an important approach, e. g., for getting first insights into the data. In particular, descriptive data mining aims to uncover certain patterns for characterization and description of the data and the captured relations. Typically, the goal of the methods is not only an actionable model, but also a human interpretable set of patterns [16].

Subgroup discovery and exceptional model mining are prominent methods for local exceptionality detection that can be configured and adapted to various analytical tasks. Local exceptionality detection especially supports the goal of explanation-aware data mining [9], due to its more interpretable results, e. g., for characterizing a set of data, for concept description, for providing regularities and associations between elements in general, and for detecting and characterizing unexpected situations, e. g., events or episodes. In the following, we summarize approaches and methods for local exceptionality detection on attributed graphs, for behavioral characterization, and spatio-temporal analysis. Furthermore, we address issues of scalability and large-scale data processing.

2.1   Description-Oriented Community Detection

Communities can intuitively be defined as subsets of nodes of a graph with a dense structure in the corresponding subgraph. However, for mining such communities usually only structural aspects are taken into account. Typically, no concise nor easily interpretable community description is provided.

In [5], we focus on description-oriented community detection using subgroup discovery. For providing both structurally valid and interpretable communities we utilize the graph structure as well as additional descriptive features of the graph's nodes. We aim at identifying communities according to standard community quality measures, while providing characteristic descriptions at the same time. We propose several optimistic estimates of standard community quality functions to be used for efficient pruning of the search space in an exhaustive branch-and-bound algorithm. We present examples of an evaluation using five real-world data sets, obtained from three different social media applications, showing runtime improvements of several orders of magnitude. The results also indicate significant semantic structures compared to the baselines. A further application of this method to the exploratory analysis of social media using geo-references in demonstrated in [2,6]. A scalable implementation of the described description-oriented community detection approach, i. e., the COMODO algorithm [5], is described in [7], which is also suited for large-scale data processing utilizing the Map/Reduce framework [11]. With that, we can apply the same method for in-memory datasets as well as for large-scale datasets supporting efficient processing.

2.2   Behavioral Characterization on Social Interaction Networks

Important structures that emerge in social interaction networks are given by subgroups. As outlined above, we can apply community detection in order to mine both the graph structure and descriptive features in order to obtain description-oriented communities. However, we can also analyze subgroups in a social interaction network from a compositional perspective, i. e., neglecting the graph structure. Then, we focus on the attributes of subsets of nodes or on derived parameters of these, e. g., corresponding to roles, centrality scores, etc. In addition, we can also consider sequential data, e. g., for characterization of exceptional link trails, i. e., sequential transitions, as presented in [4].

In [1], we discuss a number of exemplary analysis results of social behavior in mobile social networks, focusing on the characterization of links and roles. For that, we describe the configuration, adaptation and extension of the subgroup discovery methodology in that context. In addition, we can analyze multiplex networks by considering the match between different networks, and deviations between the networks, respectively. A description of characteristic (mis-)matches in a multiplex network, for example, is presented in [13] regarding relations between online and offline social interaction networks. Outlining these examples, we demonstrate that local exceptionality detection is a flexible approach for compositional analysis in social interaction networks.

2.3   Exceptional Model Mining for Spatio-Temporal Analysis

Exploratory analysis on ubiquitous data needs to handle different heterogenous and complex data types. In [2,8], we present an adaptation of subgroup discovery using exceptional model mining formalizations on ubiquitous social interaction networks. Then, we can detect locally exceptional patterns, e. g., corresponding to bursts or special events in a dynamic network. Furthermore, we propose subgroup discovery and assessment approaches for obtaining interesting descriptive patterns and provide a novel graph-based analysis approach for assessing the relations between the obtained subgroup set. This exploratory visualization approaches allows for the comparison of subgroups according to their relations to other subgroups and to include further parameters, e. g., geo-spatial distribution indicators. We present and discuss analysis results utilizing a real-world ubiquitous social media dataset.

3 Conclusions and Outlook

Subgroup discovery and exceptional model mining provide powerful and comprehensive methods for knowledge discovery and exploratory analyis in the context of local exceptionality detection. In this paper, we presented according approaches and methods, specifically targeting social interaction networks, and showed how to implement local exceptionality detection on both a methodological and practical level.

Interesting future directions for adapting and extending local exceptionality detection in social contexts include extended postprocessing and presentation options, e. g., [3]. In addition, extensions to predictive modeling, e. g., link prediction [2, 21] are interesting options to explore. Furthermore, extending the analysis of sequential data in online or offline social contexts, e. g., based on Markov chains as exceptional models [4, 10], or network dynamics [14] are further interesting options for future work.

References

1. Atzmueller, M.: Mining social media: key players, sentiments, and communities. WIREs Data Min. Knowl. Discovery (DMKD) **2**(5), 411–419 (2012)
2. Atzmueller, M.: Data mining on social interaction networks. JDMDH **29**, 1–21 (2014)
3. Atzmueller, M.: Subgroup discovery. WIREs DMKD **5**(1), 35–49 (2015)
4. Atzmueller, M.: Detecting community patterns capturing exceptional link trails. In: Proceedings IEEE/ACM ASONAM. IEEE Press, Boston, MA, USA (2016)
5. Atzmueller, M., Doerfel, S., Mitzlaff, F.: Description-oriented community detection using exhaustive subgroup discovery. Inf. Sci. **329**, 965–984 (2016)
6. Atzmueller, M., Lemmerich, F.: Exploratory pattern mining on social media using geo-references and social tagging information. IJWS **2**(1/2), 80–112 (2013)
7. Atzmueller, M., Mollenhauer, D., Schmidt, A.: Big Data analytics using local exceptionality detection. In: Enterprise Big Data Engineering, Analytics, and Management. IGI Global, Hershey, PA, USA (2016)
8. Atzmueller, M., Mueller, J., Becker, M.: Exploratory subgroup analytics on ubiquitous data. In: Atzmueller, M., Chin, A., Scholz, C., Trattner, C. (eds.) MUSE/MSM 2013, LNAI 8940. LNCS, vol. 8940, pp. 1–20. Springer, Heidelberg (2015)
9. Atzmueller, M., Roth-Berghofer, T.: The mining and analysis continuum of explaining uncovered. In: Proceedings 30th SGAI International Conference on Artificial Intelligence (2010)
10. Atzmueller, M., Schmidt, A., Kibanov, M.: DASHTrails: an approach for modeling and analysis of distribution-adapted sequential hypotheses and trails. In: Proceedings WWW 2016 (Companion). IW3C2/ACM (2016)
11. Dean, J., Ghemawat, S.: MapReduce: simplified data processing on large clusters. Commun. ACM **51**(1), 107–113 (2008)

12. Duivesteijn, W., Feelders, A.J., Knobbe, A.: Exceptional model mining. Data Min. Knowl. Discovery **30**(1), 47–98 (2016)
13. Kibanov, M., Atzmueller, M., Illig, J., Scholz, C., Barrat, A., Cattuto, C., Stumme, G.: Is web content a good proxy for real-life interaction? a case study considering online and offline interactions of computer scientists. In: Proceedings of the IEEE/ACM ASONAM. ACM (2015)
14. Kibanov, M., Atzmueller, M., Scholz, C., Stumme, G.: Temporal evolution of contacts and communities in networks of face-to-face human interactions. Sci. China Inf. Sci. **57**, 32103 (2014)
15. Klösgen, W.: Explora: a multipattern and multistrategy discovery assistant. In: Advances in Knowledge Discovery and Data Mining, pp. 249–271. AAAI Press (1996)
16. Mannila, H.: Theoretical frameworks for data mining. SIGKDD Explor. **1**(2), 30–32 (2000)
17. Mitzlaff, F., Atzmueller, M., Benz, D., Hotho, A., Stumme, G.: Community assessment using evidence networks. In: Atzmueller, M., Hotho, A., Strohmaier, M., Chin, A. (eds.) MUSE/MSM 2010. LNCS, vol. 6904, pp. 79–98. Springer, Heidelberg (2011)
18. Mitzlaff, F., Atzmueller, M., Benz, D., Hotho, A., Stumme, G.: User-Relatedness and Community Structure in Social Interaction Networks. CoRR/abs 1309.3888 (2013)
19. Mitzlaff, F., Atzmueller, M., Hotho, A., Stumme, G.: The social distributional hypothesis. J. Soc. Netw. Anal. Min. **4**(216), 1–14 (2014)
20. Morik, K.: Detecting interesting instances. In: Hand, D.J., Adams, N.M., Bolton, R.J. (eds.) Pattern Detection and Discovery. LNCS (LNAI), vol. 2447, pp. 13–23. Springer, Heidelberg (2002). doi:10.1007/3-540-45728-3_2
21. Scholz, C., Atzmueller, M., Barrat, A., Cattuto, C., Stumme, G.: New insights and methods for predicting face-to-face contacts. In: Proceedings ICWSM. AAAI, Palo Alto, CA, USA (2013)
22. Wasserman, S., Faust, K.: Social Network Analysis: Methods and Applications. No. 8 in Structural Analysis in the Social Sciences, 1st edn. Cambridge University Press, New York (1994)
23. Wrobel, S.: An algorithm for multi-relational discovery of subgroups. In: Komorowski, J., Zytkow, J. (eds.) PKDD 1997. LNCS, vol. 1263, pp. 78–87. Springer, Heidelberg (1997). doi:10.1007/3-540-63223-9_108

Author Index

Aberer, Karl II-79
Aggarwal, Charu C. I-697
Ahmed, Mohamed II-79
Akrour, Riad II-559
Alabdulmohsin, Ibrahim I-749
Allmendinger, Thomas III-193
Amer-Yahia, Sihem I-296
Andreoli, Rémi III-63
Andrienko, Gennady III-22, III-27, III-32
Andrienko, Natalia III-27, III-32
Aoga, John O.R. II-315
Arras, Kai O. III-271
Artikis, Alexander I-232
Assunção, Renato M. II-739
Atzmueller, Martin III-298
Aussem, Alex I-619
Awate, Suyash P. I-731

Bailey, James II-162
Bakker, Jorn III-67
Balamurugan, P. I-215
Bandyopadhyay, Bortik II-641
Banerjee, Arindam I-648
Bao, Hongyun I-443
Barddal, Jean Paul II-129
Becerra-Bonache, Leonor III-55
Bécha Kaâniche, Mohamed II-17
Bekmukhametov, Rustem III-193
Bellinger, Colin I-248
Benecke, Gunthard I-714
Bensafi, Moustafa III-17
Benyahia, Oualid III-41
Berahas, Albert S. I-1
Bertolotto, Michela III-294
Bifet, Albert II-129
Bilgic, Mustafa III-209
Blockeel, Hendrik III-55
Blom, Hendrik III-263
Bobkova, Yulia III-8
Bockermann, Christian III-22
Boley, Mario II-805
Boracchi, Giacomo III-145
Bosc, Guillaume III-17
Bothe, Sebastian II-805

Bouhoula, Adel II-17
Boulicaut, Jean-François III-17
Bourigault, Simon II-265
Boutsis, Ioannis III-22, III-177
Braun, Daniel A. II-475
Brown, Gavin II-442
Budziak, Guido III-27
Bui, Hung I-81
Busa-Fekete, Róbert II-511

Cachucho, Ricardo III-12
Calbimonte, Jean-Paul II-79
Cao, Bokai I-17, I-476
Cao, Xuezhi I-459
Carbonell, Jaime II-706
Carin, Lawrence I-98, I-777
Carmichael, Christopher L. III-289
Carrera, Diego III-145
Cataldi, Mario III-50
Chae, Minwoo I-509
Chakrabarti, Aniket II-641
Chan, Jeffrey II-162
Chang, Hong I-345
Chang, Xiaojun I-281
Chau, Duen Horng (Polo) II-623
Chawla, Sanjay I-49
Chen, Changyou I-98, I-777
Chen, Gang II-772
Chen, Huanhuan I-313
Chen, Ke II-347
Chen, Ling I-665
Chen, Rui II-112
Chen, Wenlin I-777
Chen, Xilin I-345
Cheng, Yun III-129
Chung, Hyoju I-509
Çilden, Erkin II-361
Cisse, Moustapha I-749
Claypool, Kajal III-112
Comte, Blandine I-572
Corcoran, Padraig III-294
Cordeiro, Robson II-623
Crestani, Fabio II-426
Cui, Ruifei II-377

Cule, Boris   I-361
Cunha, Tiago   II-393

Daly, Elizabeth   III-22
Das, Kamalika   III-209
David, Yahel   I-556
De Bie, Tijl   II-214, III-3
de Brito, Denise E.F.   II-739
de Carvalho, André C.P.L.F.   II-393
de Oliveira, Derick M.   II-739
Dembczyński, Krzysztof   II-511
Demir, Alper   II-361
Deng, Zhihong   III-36
Dhaene, Tom   III-275
Di Caro, Luigi   III-50
Diligenti, Michelangelo   II-33
Doan, Minh-Duc   III-193
Domeniconi, Carlotta   II-47
Dos Santos, Ludovic   II-606
Drummond, Christopher   I-248
Du, Changde   I-165
Du, Changying   I-148, I-165
Dutot, Pierre-Francois   I-296
Dutta, Sourav   II-195

Ehrenberg, Isaac   III-226
Enembreck, Fabrício   II-129

Faloutsos, Christos   I-264
Fang, Yuan   II-145
Fazayeli, Farideh   I-648
Feremans, Len   I-361
Ferwerda, Bruce   III-254
Flach, Peter   II-492
Fournier-Viger, Philippe   III-36
Fowkes, Jaroslav   II-410
Fragneto, Pasqualina   III-145
Frank, Eibe   II-179

Gal, Avigdor   III-22
Gallinari, Patrick   II-265, II-282, II-606
Galván, Maria   III-55
Gama, João   III-96
Gao, Lixin   II-722
Gasse, Maxime   I-619
Geilke, Michael   I-65
Genewein, Tim   II-475
Geras, Krzysztof J.   I-681
Ghavamzadeh, Mohammad   I-81

Ghoniem, Mohammad   III-59
Gionis, Aristides   II-674, II-690
Gisselbrecht, Thibault   II-282
Goethals, Bart   I-361
Gomariz, Antonio   III-36
Gori, Marco   II-33
Grabowski, Jens   III-267
Grau-Moya, Jordi   II-475
Greene, Derek   III-71
Grissa, Dhouha   I-572
Groot, Perry   II-377
Gualdron, Hugo   II-623
Gueniche, Ted   III-36
Gunopulos, Dimitrios   III-22, III-177
Guns, Tias   II-315
Guo, Tian   II-79
Guo, Xiawei   I-426
Guo, Yunhui   I-378

Haghir Chehreghani, Morteza   I-182
Haghir Chehreghani, Mostafa   I-182
Halft, Werner   III-22
Han, Bo   I-665
Hao, Hongwei   I-443
Hasan, Mohammad Al   I-394
Hayduk, Yaroslav   III-289
He, Jia   I-148
He, Lifang   I-476, I-697
He, Ling   III-79
He, Qing   I-148
Henao, Ricardo   I-98
Henelius, Andreas   I-329
Herbold, Steffen   III-267
Heskes, Tom   II-377
Hong, Junyuan   I-313
Honsel, Verena   III-267
Hooi, Bryan   I-264
Hope, Tom   II-299
Hou, Lei   II-331
Hüllermeier, Eyke   II-511, II-756

Ifrim, Georgiana   III-45
Ishutkina, Mariya   III-112

Jacquenet, François   III-55
Japkowicz, Nathalie   I-248
Jeong, Kuhwan   I-509
Jeudy, Baptiste   III-41
Jiang, Fan   III-289

Jilani, Musfira   III-294
Jin, Hongxia   II-112
Jin, Jianbin   II-331
Jin, Xin   I-165
Johri, Aditya   II-47
Jordan, Richard   III-112

Kahng, Minsuk   II-623
Kakimura, Naonori   I-132
Kalogeraki, Vana   III-22, III-177
Kalogridis, Georgios   II-492
Kamp, Michael   II-805
Kan, Andrey   II-162
Kang, Bo   II-214, III-3
Kang, Byungyup   I-509
Kang, U.   II-623
Karaev, Sanjar   II-576
Karimi, Hamed   I-795
Karlsson, Isak   I-329
Karwath, Andreas   I-65
Katakis, Ioannis   III-22, III-177
Kawarabayashi, Ken-ichi   I-132
Kaytoue, Mehdi   III-17
Kefi, Takoua   II-17
Keskar, Nitish Shirish   I-1
Khoa, Nguyen Lu Dang   I-49
Kim, Yongdai   I-509
Kinane, Dermot   III-22
Kloft, Marius   I-714
Knobbe, Arno   III-12
Kononov, Vadim V.   III-289
Kotłowski, Wojciech   II-511
Koushik, Nishanth N.   I-731
Koutsopoulos, Iordanis   I-588
Kramer, Stefan   I-65
Ksantini, Riadh   II-17
Kull, Meelis   II-492
Kumar, Sumeet   III-226
Kutzkov, Konstantin   II-79
Kveton, Branislav   I-81
Kwok, James T.   I-426

Lam, Hoang Thanh   III-36
Lamprier, Sylvain   II-265, II-282
Largeron, Christine   III-41
Last, Mark   III-280
Lauw, Hady W.   II-145
Le, Duc-Trong   II-145
Leathart, Tim   II-179
Leckie, Christopher   II-162

Leibfried, Felix   II-475
Leow, Alex D.   I-17
Leung, Carson K.   III-289
Li, Fengxiang   II-247
Li, Juanzi   II-331
Li, Xiao-Li   II-331
Li, Xiucheng   III-129
Li, Xue   I-281
Li, Yan   III-129
Li, Yang   I-313
Li, Yanhua   III-112
Li, Yitong   II-458
Li, Yucheng   I-165
Li, Yun   I-540
Liang, Kongming   I-345
Liebig, Thomas   III-22
Lijffijt, Jefrey   II-214, III-3
Lin, Jerry Chun-Wei   III-36
Lió, Pietro   II-789
Liu, Kaihua   III-12
Liu, Wei   II-162
Long, Guoping   I-148, I-165
Lu, Chun-Ta   I-17
Lu, Sanglu   I-524
Luo, Jiebo   III-79
Luo, Jun   III-112
Lynch, Stephen   III-22, III-177

Ma, Guixiang   I-476
Malinowski, Simon   I-632
Mandt, Stephan   I-714
Mannor, Shie   III-22
Mareček, Jakub   III-22
Marwan, Norbert   III-258
Matsushima, Shin   I-604
Matthews, Bryan   III-209
Mauder, Markus   III-8
Mayeur, Basile   II-559
Médoc, Nicolas   III-59
Méger, Nicolas   III-63
Mehrotra, Rishabh   III-284
Meira Jr., Wagner   II-739
Melnikov, Vitalik   II-756
Mendes-Moreira, João   I-410, III-96
Mengshoel, Ole J.   I-761
Michelioudakis, Evangelos   I-232
Miettinen, Pauli   II-576
Mock, Michael   II-805
Moeller, John   II-657
Moon, Seungwhan   II-706

Moreira-Matias, Luis   III-96
Morik, Katharina   III-22, III-263
Mukherjee, Subhabrata   II-195
Murilo Gomes, Heitor   II-129
Murphy, Lee   III-79
Muthukrishnan, S.   I-81

Nadif, Mohamed   III-59
Napoli, Amedeo   I-572
Natarajan, Sriraam   II-527
Navab, Nassir   III-193
Nguyen, Tuan   III-63
Nie, Feiping   I-281
Nielsen, David   III-209
Nieuwenhuijse, Alexander   III-67
Nijssen, Siegfried   III-12
Nogueira, Sarah   II-442
Ntoutsi, Eirini   III-8
Nutini, Julie   I-795

O'Brien, Brendan   III-22, III-177
Odashima, Shigeyuki   II-63
Odom, Phillip   II-527
Ohsaka, Naoto   I-132
Okal, Billy   III-271
Omidvar-Tehrani, Behrooz   I-296
Oza, Nikunj   III-209

Pal, Dipan K.   I-761
Paliouras, Georgios   I-232
Panagiotou, Nikolaos   III-22, III-177
Papapetrou, Panagiotis   I-329
Parkinson, Jon   II-347
Parthasarathy, Srinivasan   II-641
Pazdor, Adam G.M.   III-289
Pechenizkiy, Mykola   III-67
Pelechrinis, Konstantinos   III-161
Perry, Daniel J.   II-543
Pesaranghader, Ali   II-96
Pfahringer, Bernhard   II-129, II-179
Piatkowski, Nico   III-22
Pinto, Fábio   I-410
Piwowarski, Benjamin   II-606
Plantevit, Marc   III-17
Poghosyan, Gevorg   III-45
Polat, Faruk   II-361
Pölsterl, Sebastian   III-193
Pothier, Catherine   III-63
Pratanwanich, Naruemon   II-789

Prestwich, Steven   II-593
Pujos-Guillot, Estelle   I-572
Puolamäki, Kai   I-329, II-214, III-3

Qi, Zhenyu   I-443
Qian, Zhuzhong   I-524
Qureshi, M. Atif   III-45, III-71

Rafailidis, Dimitrios   II-426
Ragin, Ann B.   I-476
Rahman, Mahmudur   I-394
Rai, Piyush   I-777
Rajasegarar, Sutharshan   II-162
Ramamohanarao, Kotagiri   II-162
Rashidi, Lida   II-162
Ren, Kan   I-115
Revelle, Matt   II-47
Rigotti, Christophe   III-63
Rinzivillo, Salvatore   III-32
Rodrigues-Jr., Jose   II-623
Rossi, Beatrice   III-145
Rozenshtein, Polina   II-674, II-690
Ruggieri, Salvatore   III-249
Rundensteiner, Elke A.   III-112

Saeys, Yvan   III-275
Sandouk, Ubai   II-347
Sarma, Sanjay   III-226
Sawasaki, Naoyuki   II-63
Schaus, Pierre   II-315
Schedl, Markus   III-254
Scheffer, Tobias   I-714
Schifanella, Claudio   III-50
Schilling, Nicolas   I-33, I-199
Schmidt, Mark   I-795
Schmidt-Thieme, Lars   I-33, I-199
Schnitzler, Francois   III-22
Schoenauer, Marc   II-559
Scoca, Vincenzo   II-33
Sebag, Michèle   II-559
Shahaf, Dafna   II-299
Shan, Shiguang   I-345
Sharma, Manali   III-209
Shen, Yilin   II-112
Sheng, Quan Z.   I-281
Shevade, Shirish   I-215
Shi, Chuan   II-458
Shimkin, Nahum   I-556
Shimosaka, Masamichi   II-230

Shin, Kijung I-264
Siegel, Joshua III-226
Skarlatidis, Anastasios I-232
Soares, Carlos I-410, II-393
Soltani, Azadeh III-36
Song, Hao II-492
Souza, Roberto C.S.N.P. II-739
Specht, Günther III-245
Spentzouris, Panagiotis I-588
Spiegel, Stephan III-258
Srihari, Sargur N. II-772
Srijith, P.K. I-215
Srikumar, Vivek II-657
Stange, Hendrik III-22
Stegle, Oliver II-789
Stenneth, Leon II-247
Sun, Siqi I-81, II-1
Sutton, Charles I-681, II-410
Swaminathan, Sarathkrishna II-657

Tarim, S. Armagan II-593
Tavenard, Romain I-632
Theocharous, Georgios I-81
Tominaga, Shoji II-230
Toto, Ermal III-112
Trystram, Denis I-296
Tsang, Ivor W. I-665
Tschuggnall, Michael III-245
Tsubouchi, Kota II-230
Tsukiji, Takeshi II-230
Turini, Franco III-249

Ueki, Miwa II-63
Ukkonen, Antti I-329

Van Gassen, Sofie III-275
Vembu, Shankar I-493
Venkatasubramanian, Suresh II-657
Verago, Rudi III-22
Viktor, Herna L. II-96
Visentin, Andrea II-593
von Landesberger, Tatiana III-27

Waegeman, Willem II-511
Wang, Jun I-115
Wang, Sen I-281
Wang, Senzhang II-247

Wang, Sheng II-1
Wang, Wenlin I-777
Wang, Xin I-378
Wang, Yuchen I-115
Webb, Dustin II-657
Weber, Hendrik III-27
Wei, Xiaokai I-17
Weidlich, Matthias III-22
Weikum, Gerhard II-195
Whitaker, Ross T. II-543
Wistuba, Martin I-33, I-199
Wu, Bin II-458

Xiao, Han II-690
Xu, Bo I-443
Xu, Congfu I-378
Xu, Jiaming I-443
Xu, Jinbo II-1
Xu, Ran II-772

Yamaguchi, Yutaro I-132
Yan, Peng I-540
Yao, Lina I-281
Yilmaz, Emine III-284
Yin, Jiangtao II-722
Younus, Arjumand III-71
Yu, Jia Yuan III-22
Yu, Philip S. I-17, I-476, I-697, II-247
Yu, Yong I-115, I-459

Zacheilas, Nikos III-22, III-177
Zadorozhnyi, Oleksandr I-714
Zaïane, Osmar R. III-41
Zambon, Daniele III-145
Zhan, Qianyi I-697
Zhang, Bolei I-524
Zhang, Fan III-112
Zhang, Jiawei I-476, I-697
Zhang, Jie I-443
Zhang, Ke III-161
Zhang, Weinan I-115
Zhang, Xiangliang I-749
Zhang, Yizhe I-98
Zhao, Huidong II-458
Zheng, Suncong I-443
Zhuang, Fuzhen I-148, II-458
Zilles, Sandra I-493
Zygouras, Nikolas III-22, III-177

Printed in the United States
By Bookmasters